Advances in Material Research and Technology

Series Editor

Shadia Jamil Ikhmayies, Physics Department, Isra University, Amman, Jordan

This Series covers the advances and developments in a wide range of materials such as energy materials, optoelectronic materials, minerals, composites, alloys and compounds, polymers, green materials, semiconductors, polymers, glasses, nanomaterials, magnetic materials, superconducting materials, high temperature materials, environmental materials, Piezoelectric Materials, ceramics, and fibers.

Shadia Jamil Ikhmayies
Editor

Advances in Minerals Research

Springer

Editor
Shadia Jamil Ikhmayies
Department of Physics
School of Science
University of Jordan
Amman, Jordan

ISSN 2662-4761 ISSN 2662-477X (electronic)
Advances in Material Research and Technology
ISBN 978-3-031-49174-0 ISBN 978-3-031-49175-7 (eBook)
https://doi.org/10.1007/978-3-031-49175-7

This Springer imprint is published by the registered company Springer Nature Switzerland AG
The registered company address is: Gewerbestrasse 11, 6330 Cham, Switzerland

Paper in this product is recyclable.

Preface

Minerals are the backbone of all types of industry and energy technologies, providing them with the necessary materials. The book advances in minerals research is a rich reference for all those interested in minerals research and industry. Researchers will find in this book recent experimental results and in-depth discussions. Academics, graduate, and undergraduate students will find reviews and fundamental material. Beginners and research followers will find introductory material and basics to start their way. The book includes ten chapters described briefly in the following paragraphs.

Chapter One, "Introduction to Mineral Research" by Can Gungoren et al. is an introductory chapter that provides general information about minerals, mining, and mineral processing activities. This chapter is a detailed guide for undergraduate and graduate students, junior researchers, research fellows, mining engineers, and early-career metallurgists. This chapter, first, defines the concept of mineral, and explains the properties that a mineral should possess. Next, the methods of mining mineral deposits in the earth's crust are classified systematically. Finally, mineral processing techniques applied to ores for increasing their quality are explained in detail at an introductory level. Chapter two, "Use of Ultrasound in Physical and Chemical Mineral Processing Operations" by Can Gungoren et al. reviews the use of ultrasound waves in mineral comminution and classification. The physical and chemical effects of ultrasound in aqueous media were explained in detail. Chapter three, "Advanced Techniques on Fine and Coarse Particle Flotation" by Oktay Şahbaz et al. presents the development of new technologies to increase the efficiency of the flotation of fine and coarse particles. Besides the most appropriate techniques and industrial applicability of the fine and coarse particle flotation are explained in detail.

Chapter four, "Differentiation Between Natural Quartz-Based on Thermoluminescence Properties" by Shrouk Farouk et al., first introduces the basics of thermoluminescence phenomena (TL), types of luminescence, and the kinetic models describing the TL phenomena including glow curves and kinetic parameters. The different methods used in analyzing these kinetic models are highlighted, and the physical properties and applications of quartz are thoroughly discussed. Then the phenomenon of thermoluminescence and dosimetric parameters of quartz collected from different places are investigated and discussed. In chapter five, "Silicones and Their Applications" by Rukhmah Riaz et al. the authors review silicones, state how silicones are classified, and discuss their various practical applications.

Chapter six, "Overview of Some Production Routes for Hydroxyapatite and Its Applications" by Medhat Ibrahim et al. discusses the different methods for producing Hydroxyapatite (HA), its various properties, and its applications. The authors state that the preparation of HA with certain characteristics is very complicated due to the potential production of toxic intermediate phases that may occur during the synthesis process. But they state that further investigations and descriptions of HA structure and fictionalization can be achieved with molecular modeling. The different bone grafting materials and their defects, 3D bioprinting of scaffolds, definition, and types of biomaterials, and host responses to biomaterials are also pointed out in this chapter. Therefore, this chapter focuses on giving readers interested in biomaterials an overview of these valuable materials, especially, hydroxyapatite, and its various properties and applications in different fields, particularly biomedical ones.

In chapter seven "Biomineralization of $CaCO_3$: A Biomimetic Approach Using In Vitro Model Systems" by Debojit Paul and Gopal Das, the authors focus on the biomimetic synthesis of $CaCO_3$, and try to understand the factors that influence the formation, with an insight into the various stages leading to crystallization. But in chapter eight, "Mineralization of Low-Carbon Bio-carbonate, Bio-phosphate, and Bio-composite Cements" by Xiaoniu Yu and Qiyong Zhang the authors present an experimental study on the preparation of different types of low-carbon bio-cement with gelling properties. The authors show that morphology and size can be regulated by bacterial liquid, bacterial body, secretion, solvent, sandy soil, and so on. In this study, the authors prepared bio-carbonate, bio-phosphate, and bio-composite cements, and discussed their composition, microstructure, and cementation properties. Finally in chapter nine, "Advances in Mineral-Based Form Stable Phase Change Materials" by Xiaobin Gua and Kaijun Dong the authors review the effects of minerals on the preparation, properties, and application of mineral-based form-stable phase change materials FSPCMs. The associated analysis is presented, and research prospects are also provided. This review chapter provides a clear understanding of the preparation, properties, and application of various mineral-based FSPCMs.

Amman, Jordan Shadia Jamil Ikhmayies

Contents

Introduction to Mineral Research

Can Gungoren, Ilgin Kursun Unver, and Orhan Ozdemir

Abstract Several minerals are collected in certain parts of the earth's crust as a result of geological processes. Minerals, which are used in manufacturing and therefore have an economic value, form ores together with their surrounding minerals. In its widest meaning, the processes of exploration, finding, and processing of these ores are called mining. Mining and civilization have been in a close relationship since the early ages of humanity. The industrialization and welfare level of societies is strongly related to the efficient execution of mining activities. In this chapter, general information about minerals, mining, and mineral processing activities has been reviewed and detailed systematically to guide students, researchers, and engineers in the mining industry, who are at the beginning of their professional lives.

Keywords Minerals · Mining · Mineral processing · Comminution and classification · Beneficiation

1 Introduction

Since the first ages of history, humankind has needed some materials in the earth's crust to carry out activities such as shelter, nutrition, and protection. In the broadest sense, the whole process of extracting these materials from the earth's crust and making them suitable for a specific purpose can be called mining. Therefore, mining has been a very important field of occupation since the earliest times of civilization [1, 2]. It even gave a name to ages. About 450 thousand years ago, the first flint tools were made in the Stone Age. Copper was used around 9500 BC and tin was added to

C. Gungoren (✉) · I. Kursun Unver
Engineering Faculty, Mining Engineering Department, Istanbul University-Cerrahpasa, 34500 Buyukcekmece, Istanbul, Türkiye
e-mail: can@iuc.edu.tr

O. Ozdemir
Faculty of Mines, Department of Mineral Processing Engineering, Istanbul Technical University, Ayazağa Campus, Maslak, Sarıyer, Istanbul, Türkiye

S. J. Ikhmayies (ed.), *Advances in Minerals Research*, Advances in Material Research and Technology, https://doi.org/10.1007/978-3-031-49175-7_1

the copper, leading to the Bronze Age. After that with the use of iron, the Iron Age began in 2000 BC.

In mining, the location where the desired natural material is collected in the earth's crust within the geological processes must be found, and then this material must be extracted from the earth's crust. The material obtained is mostly not usable in its initial form and is subjected to various processes in line with the needs.

The methods used in performing all these mining activities can be explained within a certain systematic. The first document prepared for this purpose is the book of De Re Metallica written by Georgius AGRICOLA in 1556 and translated into English in 1922 [3]. Over time, with the developments in science and technology, mining activities has become practiced today.

This chapter has been prepared as the first chapter of this book to serve as a guide on mineral research for undergraduate and graduate students, junior researchers, research fellows, mining engineers, and metallurgists at the beginning of their professional life.

In this chapter, first, the concept of mineral is introduced and the properties that a mineral should possess are explained. Then, the mining methods of mineral deposits in the earth's crust are classified systematically. Finally, the mineral processing techniques applied to the ores for increasing their quality are explained in detail at an introductory level.

2 Minerals

A mineral is a naturally occurring solid, inorganic element, or compound having a uniform composition and a regularly repeating internal structure. The only mineral in liquid form at room temperature (22 °C) is mercury. Minerals form crystals under appropriate pressure and temperature. A crystal is a homogeneous solid body with a three-dimensional internal arrangement over a wide range.

The chemical compositions of minerals are shown with chemical formulas and there are seven characterized crystal systems namely, cubic, tetragonal, orthorhombic, hexagonal, trigonal, monoclinic, and triclinic. There are about 3500 defined minerals in literature.

Minerals are divided into classes according to the main anion and anionic groups. Today, mineral classification is made according to both chemical composition and internal structure. The minerals can be classified into eight groups:

1. Native elements
2. Sulfides
3. Oxides, hydroxides, and halides
4. Carbonates, nitrates, and borates
5. Sulfates and chromates
6. Tungstens and molybdates
7. Phosphates, arsenates, and vanadates

8. Silicates

In the earth's crust, minerals can be found alone or often together with other mineral types in the formation of rocks. Rocks are divided into three main groups: magmatic, metamorphic, and sedimentary. The rocks which comprise one or more valuable minerals that can be extracted profitably are called ores [4].

Ores, which consist of economically valuable minerals and relatively non-valuable (gangue) minerals, are usually classified according to the type of valuable minerals. In native ores, the metals are found in their elemental form. Meanwhile, in sulfurous ores, valuable elements are found bound to sulfurous groups, and similarly, in oxidized ores, they are bound to oxygen groups. There are also some ores containing more than one recoverable valuable mineral which are called complex ores.

Eight elements (O, Si, Al, Fe, Ca, K, Na, and Mg) make up more than 99% of the earth's crust. The amount of only three elements of industrial importance (Al, Fe, and Mg) is over 2%. Other metals are present below 0.1%. For example, Ti has a ratio of 0.57%, Cr 0.01%, and Cu 0.0055%. Commonly accepted metals such as Zn (0.0070%) and Pb (0.0013%) are less common than rare earth elements (0.023%) such as cerium and thorium [5].

If minerals containing important metals for the industry were evenly distributed on the earth's crust, they would be present in such a small amount per unit volume that their economic extraction would not be possible. Fortunately, the formation of minerals in nature is regulated by geological conditions, and a particular mineral can be found mainly associated with a rock type. Therefore, ore deposits are usually present in sufficient concentrations to allow the profitable recovery of minerals [5]. Mineral deposits existed in various types such as lode, vein, and placer dependent upon their geological formation process.

Mineral deposits can be collected under three generally accepted groups. The mineral deposits whose existence is only estimated in the light of geological structures and geophysical anomalies are called possible reserves. Meanwhile, the mineral deposits whose amount and grade are partially calculated by geological projection and long-range drilling is called probable reserves, and the mineral deposits whose amount and grade are determined with an accuracy of more than 95% by detailed studies are called proved reserves [6].

3 Mining

Mining can be defined as the extraction of economically valuable minerals or other geological materials from the surface of the earth. Mining is necessary for nations to have adequate and safe supplies of minerals and materials to meet their economic and defense needs.

Processes in a typical mining operation can be grouped under four main headings (Fig. 1).

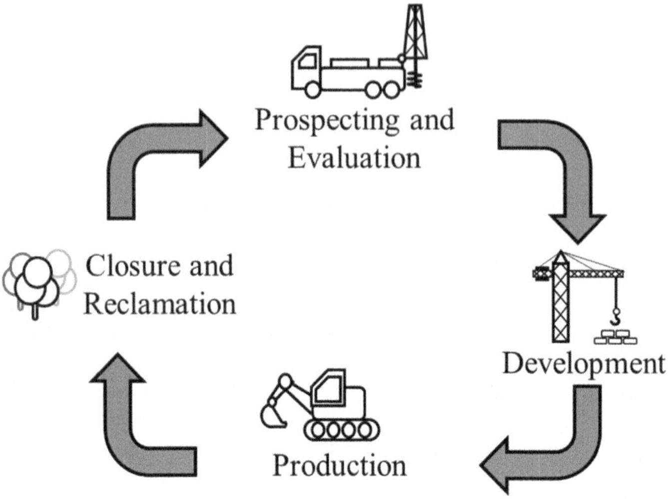

Fig. 1 Processes in a typical mining operation (Authors original drawing)

1. Exploration and Prospecting
2. Development
3. Production
4. Closure and Reclamation.

3.1 Prospecting and Evaluation

Prospecting involves the operations performed to locate a mineral deposit that is not visible from the earth's surface and can be divided into two groups: geological (direct) and geophysical (indirect) works.

Geological prospecting works begin with the examination of outcrops in the field and continue with the examination of aerial/satellite photographs and geological maps as well as the structural evaluation of the region. With these studies, geologists collect data about the location of the mineral deposit.

Geophysical works are aimed to determine the location of the mineral deposit by detecting gravitational, seismic, magnetic, electrical, electromagnetic, and radiometric anomalies by geophysical methods. These methods can be applied by an aircraft from the air or on the ground via proper devices.

In addition to these studies, geochemical and geobotanical studies can be carried out by doing quantitative elemental analysis of soil, stone, water, and plant samples in the field, to gather information about the deposits [3].

In the evaluation stage, reserve and grade calculations are made considering the geometric dimensions and depth of the ore bed found by exploration methods. In this phase, representative samples are obtained by chipping outcrops and drilling boreholes and subjected to chemical, and mineralogical analyses such as X-ray diffraction (XRD), X-ray fluorescence (XRF), inductively coupled plasma-mass spectrometry (ICP-MS), and atomic absorption spectroscopy (AAS).

3.2 Development

The development stage involves legal and economic considerations such as applying for permits for mineral rights, buying surface lands, and making financial arrangements in addition to the excavations in the bedrock such as driving wells and galleries to reach the ore deposit. This stage also includes the construction of the necessary facilities for the operation of the mine including access roads, office buildings and living spaces, power sources, mineral processing plants, and tailing ponds. These works can take years and may require large capital as ore usually cannot be produced and sold in this period.

3.3 Production

After sufficient development is done, the production stage is started in which the ore is extracted from the earth's crust with the appropriate mining method. There are many parameters such as the depth and type of the mineral deposit, topography, reserve, and grade values, and the strength of the bedrock and ore, which should be considered in the selection of the mining method. Considering these parameters, the most appropriate method is selected, considering the economic and safety factors. In general, a mineral deposit can be exploited in two ways: surface and underground mining [7, 8].

In surface mining, the overburden, which is the soil and/or rocks covering the deposit, is striped in steps to reach the ore deposit. Surface mining is the first preferred and the most common mining method used today because of its low cost and high productivity. However, since the amount of overburden to be excavated is increased with the depth of the ore, only the deposits that are relatively close to the earth's surface can be exploited by surface mining. Surface mines are usually operated as opencast or open-pit mines. While the deposit is excavated in a series of strips with the overburden from each successive strip moved into the preceding strip in opencast mining, in open-pit mining, a single large conical opening was excavated to produce the ore [4, 9].

The mineral deposits, which are not economical to be operated with surface mining since they are found at a deeper level, are recovered by underground mining methods. In underground mining, the ore is usually accessed by access shafts, spirals, or inclined tunnels. The most used underground mining techniques are explained in this chapter in the categories of unsupported, supported, and caving methods [10, 11].

3.3.1 Unsupported Methods

In these methods, no artificial pillars are used to assist in the support of the openings. However, a large amount of roof bolting and localized support are used. The unsupported methods of mining are used to extract mineral deposits that are roughly tabular and are generally associated with strong ore and surrounding rock.

The most common unsupported method is room-and-pillar mining, which is used primarily for flat-lying seams or bedded deposits like coal, trona, limestone, and salt. Support of the roof is provided by natural pillars of the mineral that are left standing in a systematic pattern.

Another widely used method is shrinkage stopping in which the ore is excavated in horizontal slices, starting from the bottom of the stope and advancing upward. A portion of the broken ore is allowed to accumulate in the stope to provide a working platform for the miners and is thereafter withdrawn from the stope through chutes.

As another unsupported method, sublevel stopping differs from shrinkage stopping by providing sublevels from which vertical slices are blasted. In this manner, the stope is mined horizontally from one end to the other [10, 12, 13].

3.3.2 Supported Methods

The supported mining methods are often used in mines with weak rock structures. Cut-and-fill stopping is the most common method and is used primarily in steeply dipping metal deposits. The cut-and-fill method is practiced both in the overhand (upward) and underhand (downward) directions. As each horizontal slice is taken, the voids are filled with a variety of fill types to support the walls. The fill can be rock waste, tailings, cemented tailings, or other suitable materials. Cut-and-fill mining is one of the most popular methods used for vein deposits.

Square-set stopping also involves backfilling mine voids. However, it relies mainly on timber sets to support the walls during mining. It is an underground mining method used in steeply dipping orebodies, under difficult wall rock conditions. Once the ore has been drilled and blasted, the stope is supported with large square timber sets that provide the platform for the next cycle [10, 12, 13].

3.3.3 Caving Methods

Caving methods are varied that involve caving the ore and/or the overlying rock. Longwall mining is a caving method particularly well adapted to horizontal seams, for instance, coal at some depth. In this method, a face of considerable length (wall) is maintained, and as the mining progresses, the overlying strata are caved, thus promoting the breakage of the coal itself.

Sublevel caving is employed for a dipping tabular or massive deposit. As mining progresses downward, each new level is caved into the mine openings with the ore materials being recovered while the rock remains behind.

Block caving is a large-scale or bulk mining method that is highly productive, low in cost, and used primarily on massive deposits that must be mined underground. It is mostly applicable to weak or moderately strong orebodies that readily break up when caved. Both block caving and longwall mining are widely used because of their high productivity [10, 12, 13].

3.4 *Closure and Reclamation*

The final stage in the operation of most mines is the closure of the mine and reclamation which is the process of closing a mine and re-contouring, revegetating, and restoring the water and land values. The best time to begin the reclamation process of a mine is before the first excavations are initiated [14, 15].

In planning for the reclamation of any given mine, many concerns must be addressed [16]. The first of these is the safety of the mine site, particularly if the area is open to the public. The removal of office buildings, processing facilities, transportation equipment, utilities, and other surface structures must generally be accomplished. Then, all mine shafts, adits, and other openings that may preset physical hazards are required to seal. Any existing high walls or other geologic structures may require mitigation to prevent geologic failures.

The second major issue to be addressed during the reclamation of a mine site is the restoration of the land surface with native plants, the water quality, and the waste disposal areas so that long-term water pollution, soil erosion, dust generation, or vegetation problems do not occur [3].

4 Particle Liberation

4.1 *Particle Comminution*

The ores produced from the mine (run-of-mine) contain invaluable (gangue) minerals as well as valuable minerals. Therefore, the grade of run-of-mine ore is generally low to be used in the industry directly. All of the processes carried out to separate valuable minerals from gangue minerals and thus to obtain a concentrate in the proper grade for the industry are called mineral processing. Ore dressing, ore beneficiation, mineral dressing, or milling, are some of the terms used with the same meaning [17].

In an ore particle, valuable and gangue minerals are usually bound to each other physically. Such particles are called locked or inter-locked particles and it is impossible to beneficiate an ore body consisting of these particles. For this reason, it is necessary to turn the ore into particles consisting of a single mineral. Such particles are called free or liberated particles and the process is called particle liberalization.

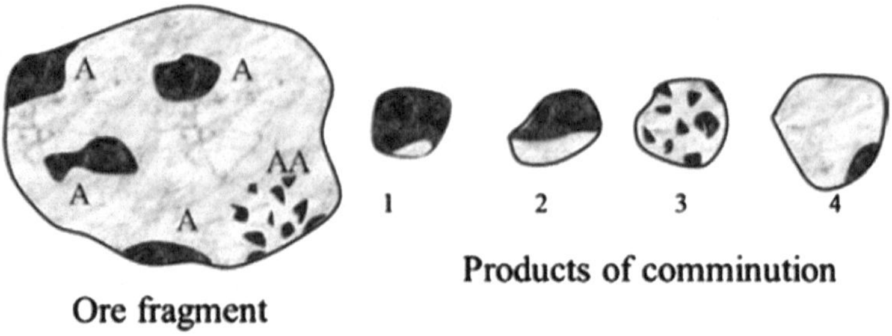

Fig. 2 Particles before and after comminution. Reprinted from Wills and Finch [5], with permission from Elsevier. Copyright (2016)

Particle liberalization is carried out by reducing the size of the particles by comminution techniques of crushing and grinding. Particles before and after comminution are shown in Fig. 2.

4.1.1 Crushing

Crushing is the first mechanical stage in the process of comminution. It is a dry process that results in particles of large sizes. Depending on the size of the ore excavated from the mine, it can be performed in up to three stages: primary, secondary, and tertiary. An ore particle of 1.5 m in size can be reduced to 10–20 cm in size by crushing. In comminution techniques, the ratio of feed size (F_{80}) to product size (P_{80}) is called reduction ratio and generally, an 80% passing size was taken as a reference [5].

Crushing is accomplished by compressing a particle between rigid surfaces or by striking it to a rigid surface. The devices in which the crushing is carried out are called crushers and there are various types of them. While primary crushing is usually carried out using jaw and gyratory crushers; cone, roll, and impact crushers are used for secondary and tertiary crushing.

4.1.2 Grinding

The final stage of the comminution process is grinding [18]. It can be carried out dry or wet for finer sizes (usually less than 25 mm). The machines by which the grinding process is carried out are called grinders or mills. There are different types of mills for laboratory applications that use rollers, discs, and pestles for the grinding of the ore. However, on the industrial scale, tumbling mills are preferred because of their higher efficiency. In tumbling mills, a mixture consisting of ore and grinding media are placed in a horizontal cylindrical mill body, and the mill is rotated at a certain speed [19]. Therefore, the ore is ground by the effect of impact, attrition, and

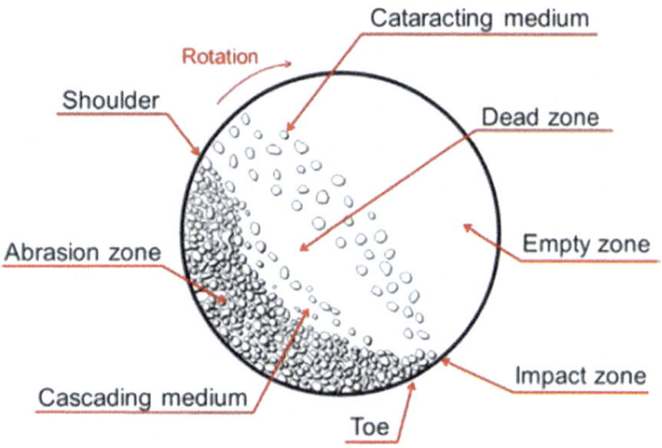

Fig. 3 The motion of charge in a tumbling mill. Reprinted from Wills and Finch [5], with permission from Elsevier. Copyright (2016)

abrasion. The motion of the charge (ore and grinding media) in a tumbling mill is seen in Fig. 3.

The grinding media generally consists of steel rods or balls. Ceramic balls or pebbles made of flint can be also used to reduce contamination. Moreover, in some cases, the coarse pieces of the ore are used to grind the smaller ore particles in autogenous mills. Sometimes the grinding in the autogenous mills should be supported with steel balls and this type of mill is called a semi-autogenous mill [20].

Crushers and mills can be operated in two systems: open or closed circuits. If the ore is fed to the crusher or mill without size classification and the comminuted product is sent directly to the next process, such circuits are called "open circuits" (Fig. 4a).

However, in some cases, the material is needed to be classified according to its size before or after the comminution process. If there is already a material below the desired size in the ore to be fed to the crusher or mill, the capacity of the device can be used more effectively by separating these particles before the comminution process (Fig. 4b).

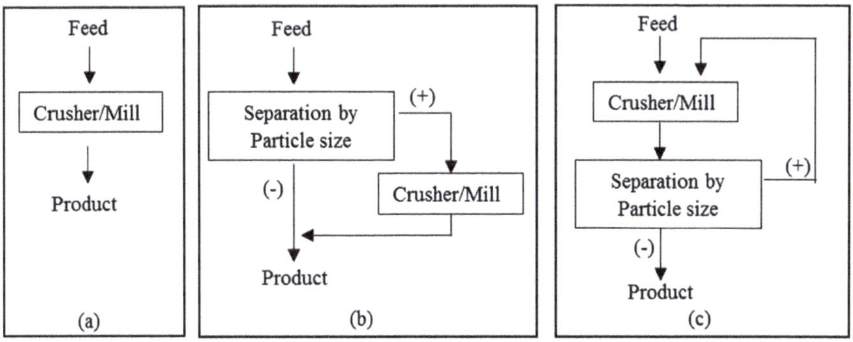

Fig. 4 **a** Open, **b** forward, and **c** reverse classification circuits. (+) and (–) represent the oversize and undersize, respectively (Authors original drawing)

Additionally, a comminution product may contain a material that is larger than the desired size. If the entire product is required to be under a certain size, the product must be classified according to size and the larger particles should be sent back to the comminution process. This method also prevents over-comminution, which has the potential to create a problem for subsequent beneficiation processes and saves energy. These circuits are called "closed circuits" (Fig. 4c).

4.2 Separation by Particle Size

Separation by particle size is the process of separating particles between certain sizes in a material therefore the material is divided into two or more size fractions. Separation by particle size is carried out by two techniques: sieving (screening) and size classification.

4.2.1 Sieving

Sieving is a process of size separation based on the ability of the particles to pass through openings of certain sizes or not (Fig. 5a). Sieve openings can be in various shapes. In sieve sizing, the diameter of circular openings, the edge length of square openings, and the short edge length of rectangular openings are used in cm, mm, or μm. As an alternative, the other definition used in sieves is mesh, which refers to the number of openings on a 1-inch-long sieve surface.

Sieves in the American Society for Testing and Materials (ASTM) and TYLER standards are generally used in the industry. Both sieve standards take the 74 μm (200 mesh) sieve as a reference, and sieve openings vary in proportion to $\sqrt{2}$ and $\sqrt[4]{2}$ times, respectively.

Industrial sieving can be applied from approximately 300–40 μm. However, the sieving efficiency decreases significantly with the decrease in particle size. It becomes important to use water in sieving (wet sieving) when working with fine particles to increase the sieving efficiency and reduce dust formation [5, 21, 22].

Apart from size separation, sieving is also used for solid–liquid separation, de-sliming, and de-dusting purposes. There are various sieve types used in industry including vibrating, inclined, circular motion, grizzly, banana, trommel, and gyratory sieves [5].

4.2.2 Size Classification

The efficiency of sieving is decreased significantly in the separation of fine particles (<200 μm). Therefore, these particles are generally separated using size classification techniques [23]. Size classification is the process of separating particles of different sizes into two or more products using different settling rates in a fluid medium.

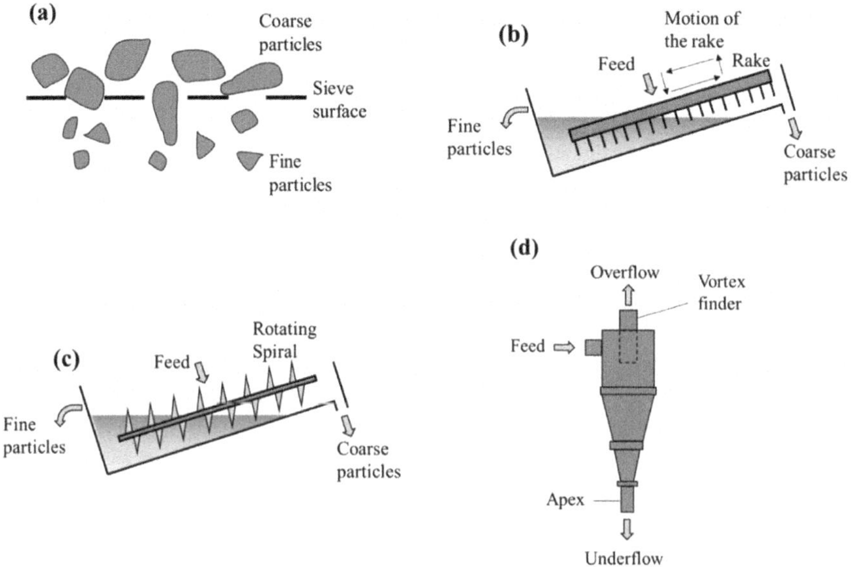

Fig. 5 Several size separation devices: **a** sieve **b** rake classifier **c** spiral classifier **d** hydrocyclone (Authors original drawing)

As known, in a vacuum environment, there is no resistance to the movement of the particles as it falls. Therefore, if a particle is subjected to a constant acceleration such as gravity, its speed will increase continuously regardless of its mass and density. However, in a viscous medium such as air or water, resistance is applied by the medium against the movement of the falling particle, and this resistance increases with the speed of the particle. When the accelerating force and the resistance (drag) force of the medium are equal to each other, the falling object reaches the terminal velocity and continues to fall at a constant speed [5].

Therefore, classification is a challenge between the accelerating force such as gravity and centrifuge, and the drag force that resists this force. The net force acting on a particle is different for fine and coarse particles. The larger one of the two particles with the same density has a higher terminal velocity. Therefore, it settles faster in a fluid.

The fluid medium in a classifier can be liquid (wet classification) or gas (dry classification). However, in mineral processing, wet classification was used more widely. It is possible to divide the classifiers into two: gravitational and centrifugal classifiers.

The most commonly used gravitational classifiers are rake and spiral classifiers. In rake classifiers, a rake positioned in an inclined tank makes an eccentric movement and pushes the collapsing coarse particles up for a short distance. Then, it goes up and returns to its initial position, and the cycle is repeated. Large particles are taken from the upper part and fine particles are taken from the lower part by overflow

(Fig. 5b). Similarly, in spiral classifiers, a continuously rotating spiral carries the settled material upwards (Fig. 5c).

A hydrocyclone, or shortly cyclone, is a device that uses centrifugal force to increase the sedimentation rate of the particles and separate them by size and is often used in mineral processing. The cyclone body consists of a cylindrical upper part that the pulp is fed tangentially and a conical lower part. It has an outlet opening both at the top and at the bottom (Fig. 5d). In the cyclone body, an overflow (vortex) is generated in the low-pressure region near the center, and an underflow (apex) is formed in the outer high-pressure region. A particle entering the cyclone body with tangential feeding is directed downwards by an obstacle called a vortex finder, thus preventing short-circuit by being caught by the overflow. Centrifugal force acts outward and drag force acts inward on this particle in a circular motion. In a cyclone, rapidly settling particles move towards the wall of the cyclone and exit the apex by being caught in the underflow. Slowly settling particles, on the other hand, move to the low-pressure region with the effect of drag force and are taken from the overflow [24].

5 Beneficiation Methods

After the minerals in an ore are liberated, they can be separated from each other by various beneficiation methods by utilizing the difference in a specific mineral property. The beneficiation methods used can be grouped as physical, chemical, and physicochemical methods according to the utilized mineral property.

5.1 *Physical Beneficiation*

5.1.1 Gravity Concentration

The beneficiation carried out by separating the mineral particles from each other by moving in different ways due to the specific weight difference in a fluid medium is called gravity concentration. In this method, the size and shape of the particles and the density and viscosity of the fluid medium play a significant role.

The concentration criterion ($\Delta\rho$) given in Eq. 1 can be used to make a rough estimate of the difficulty of gravity separation.

$$\Delta\rho = \frac{\rho_h - \rho_f}{\rho_l - \rho_f} \tag{1}$$

where ρ_h, ρ_l, and ρ_f are the densities of the heavy and light minerals, and fluid medium, respectively [5]. The difficulty level of separation related to the concentration criterion is seen in Table 1.

Table 1 The difficulty level of separation related to the concentration criterion. Redrawn after Wills and Finch [5], with permission from Elsevier. Copyright (2016)

Concentration criterion ($\Delta\rho$)	Separation difficulty level
2.50	Relatively easy
2.50–1.75	Possible
1.75–1.50	Difficult
1.50–1.25	Very difficult
<1.25	Not possible

There are various devices used for the separation of minerals by gravity. The simplest mechanical gravity separators are jigs that use the vertical movement of the separating medium to separate particles. Jigs consist of a reservoir containing the separating medium, usually water. The top section of this reservoir is divided into two chambers. Generally, there is a piston moving up and down in one chamber. The other chamber contains heavy and light particles located on a sieve. When the piston moves down, it exerts a pushing force (pulsation stroke) on the fluid. As the piston moves up, a suction stroke is applied to the fluid. In the pulsation stroke, the particles move upwards in the liquid. In this phase, heavy particles move more and light particles move less. In the ensuing suction stroke, the force from the suction movement is added to the settling speed of the particles. With the constant repetition of this movement, heavy and fine particles settle on the bottom layer, and heavy and coarse particles are located above them. Meanwhile, light and fine particles settle on these particles. Finally, light and coarse particles are located on the top of this particle bed. In this way, layering occurs according to the specific gravity difference between the particles.

Another widely used gravity separator is a shaking table. Shaking tables have rectangular or similar shapes such as parallel sides or trapezoids. The table is positioned properly so that its short axis makes an adjustable angle with the ground. There is a water distributor on the upper edge of the table and the water coming from this distributor flows on the inclined table surface with the force of gravity. Therefore, the table surface is covered with a water film throughout the separation process. The table is moved back and forth in the direction of its long axis by a motion mechanism. The backward movement of this movement is faster and its amplitude is adjustable.

The components of a shaking table are seen in Fig. 6a as well as the product belts. Heavy and light particles to be separated are fed to the table surface at a certain speed from the feed box located in one of the corners of the feed side of the table. While the force of gravity acting on the particles tries to move them toward the tailing side with the help of the water flow, the force created due to the movement of the table tries to move them to the concentrate end. The particles move under the resultant effect of these two forces. The lightest and coarsest particles travel the shortest distance, while the heaviest and finest particles travel the longest distance to the concentrate end. The surface of the table is covered with narrow and long obstacles parallel to each other, such as riffles to increase the efficiency of the separation process. As a result, particles of different specific gravity and sizes are collected on the table in certain belts and are collected separately from the table surface.

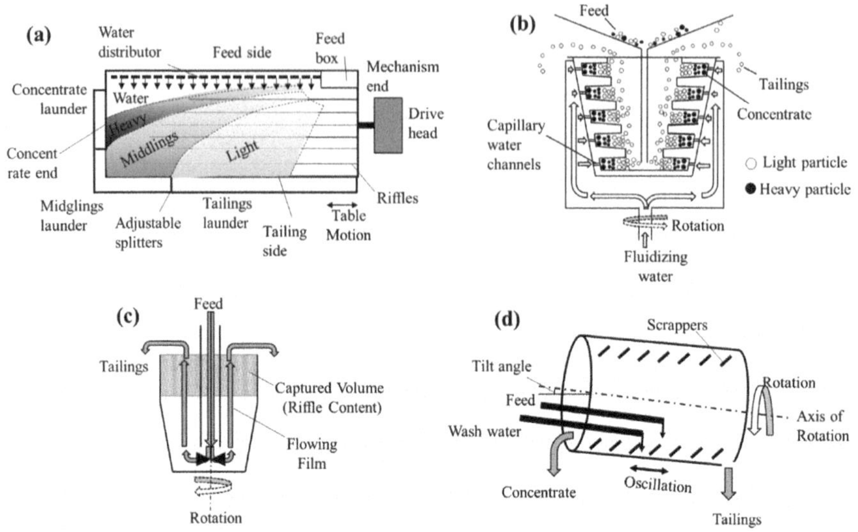

Fig. 6 Several gravity concentrators: **a** vibrating table **b** Knelson concentrator **c** Falcon concentrator **d** multi gravity separator (MGS) (Authors original drawing)

As the particle size decreases, the effect of the gravity force on the separation also decreases. For this reason, centrifugal separators have been developed to increase the force of gravity acting on the particle. The simplest centrifugal device is a fixed spiral called Humphrey's spiral. In this device, the particles are separated from each other by collecting in different parts of the spiral surface under the resultant effect of gravity and centrifugal force.

Another centrifugal device is the Kelsey jig, which is based on the working principles of a standard jig but operates at higher G-forces [25, 26]. With the Kelsey jig, forces of up to 80–100 G (gravity force) can be achieved in the recovery of fine (0.5–0.038 mm) particles [27, 28].

As a more complex separator, the Knelson concentrator (KC) consists of a roughly conical-shaped chamber with horizontal riffles inside and a pressurized water jacket around it. In the KC, the material is fed from the center of the conical chamber. The particles move upwards from the wall of the conical chamber at different velocities under the influence of centrifugal force. Meanwhile, water is supplied from the capillary openings located in the cavities of the riffles. Therefore, a fluidized particle bed is formed in the riffle cavities. Light particles continue their upward movement and are collected from the top of the chamber by overflow. On the other hand, heavy particles remain in the cavities of the riffles and are recovered from the bottom of the chamber after the device is stopped [29–31]. Figure 6b shows how beneficiation is done in a KC.

A Falcon concentrator (FC) is another gravity separator having similar working principles to the KC. The FC rotates slightly faster than the KC and the resulting G-force can be up to 300 G (gravity force). For this reason, it can be used to separate finer particles [28, 32]. Contrary to KC, FC does not have riffles for bed fluidization for the retaining of heavy minerals in the conical section. Instead, a thin, fluid slurry film forms along the wall of the conical section, and the particles are separated into layers due to hydraulic differences in the FC (Fig. 6c). Heavy particles moving along the inner surface of the chamber are discharged through small ports along the chamber wall. The light particles accumulating above the layer are discharged from the top of the chamber [33].

There is another device, which combines the centrifugal force and the oscillating motion of the shaking table named multi gravity separator (MGS). In MGS, an inclined drum rotates and makes an oscillating motion along its long axis. The heavy particles are transported upwards in the drum with the help of scrapers and recovered from the top of the drum. Meanwhile, light particles are discharged from the lower end of the drum with the help of washing water and the effect of gravity (Fig. 6d).

5.1.2 Magnetic Separation

Magnetic susceptibility of an atom is mainly due to the spin motion of the unpaired electrons, especially in the outer orbitals. However, the fact that an element has a magnetic susceptibility does not mean that a solid particle consisting of that element will be strongly magnetic.

Magnetic mineral particles can be classified as diamagnetic, paramagnetic, and ferromagnetic from weak to strong magnetic susceptibility. Diamagnetism is a weak magnetic property, which is in the opposite direction to the applied magnetic field. Meanwhile, paramagnetic minerals have randomly arranged magnetic dipoles. These dipoles are aligned partially or completely in the direction of the magnetic field depending on their strength. Therefore, paramagnetic minerals show a small but positive susceptibility. On the other hand, in some solids, magnetic moments are self-directed parallel to each other. These substances are called ferromagnetic minerals and their magnetic susceptibility is very high and positive.

Mineral particles of different magnetic susceptibility act differently in the influence of a magnetic force and auxiliary forces such as gravity, centrifugation, and friction. Therefore, they can be separated by magnetic separators [34, 35]. Magnetic separators can be classified as dry and wet according to the separation media and as low and high field intensity according to the applied magnetic field strength. While magnetic separators with low field intensity ($\lesssim$ 0.3 T) are sufficient to separate ferromagnetic minerals, high field strengths (>2 T) are needed to separate the paramagnetic ones [5]. Apart from these classifications, the magnetic separators can be named according to their structural features such as drum, band, and disc.

In drum and roll magnetic separators, permanent magnets are generally used to create magnetic fields. These magnets are arranged in the North (N) and South (S) order. Strongly magnetic minerals adhere to the drum or roll as they move. Meanwhile, weakly magnetic minerals are less affected by the attraction of magnets and therefore they move in a different trajectory. Non-magnetic minerals, on the

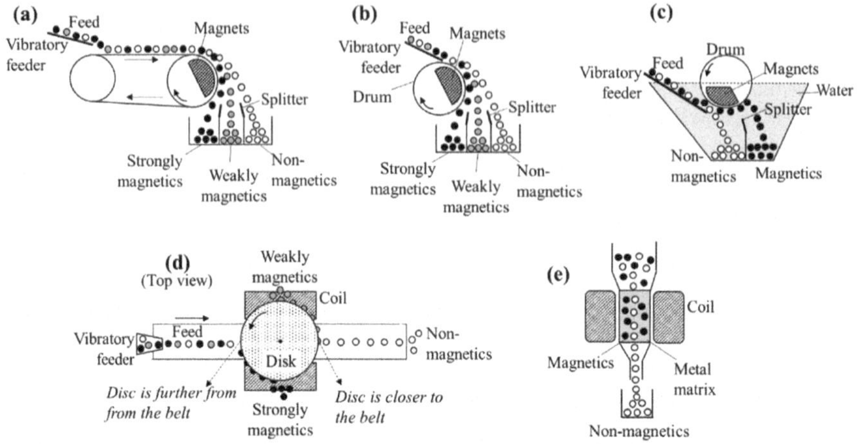

Fig. 7 Schematics of: **a** dry-roll **b** dry-drum **c** wet-drum **d** dry-disk **e** wet-metal matrix magnetic separators (Authors original drawing)

other hand, are not affected by the magnetic field and move under the influence of centrifugal and gravity forces (Fig. 7a–c).

The disc magnetic separator consists of an inclined rotating disk on a belt and a pair of coils that provide magnetic properties to the disc when current is passed through them. Since the disc is inclined, the distance between the disc and the belt differs as seen in Fig. 7d. The particles fed to the belt first pass through the point where the disc is further from the belt and the strongly magnetic particles are attracted by the disc in this region where the magnetic field is weaker. Among the particles that keep going on the belt, weakly magnetic particles are attracted from the region of the disc closer to the band, where the magnetic field strength is higher. As the disc rotates, it removes the adhered particles from the belt and the particles are separated from the disc with the help of scrapers. Meanwhile, the non-magnetic particles continue their movement on the belt without being affected by the magnetic field.

An example of a wet high-field strength separator is the metal matrix separator (Fig. 7e). By passing an electric current through the coils on both sides, the metal matrix becomes magnetized. Non-magnetic particles fed as pulp from the top of the matrix continue their falling and discharge from the bottom of the device. Meanwhile, the magnetic minerals adhere to the metal matrix. After the product collection chamber is changed, the electric current is cut off and the washing water is given from the top of the matrix. Thus, magnetic particles are also recovered.

5.1.3 Electrostatic Separation

The separation process that utilizes the electrical conductivity difference between the minerals is called electrostatic separation. In this method, electrical charge is imparted to mineral particles by two methods: conduction and ion bombardment. In these circumstances, particles are charged differently according to their conductivity properties. When these particles interact with other grounded or charged materials, the electrical force acting on them will be different, as well as the path they will follow under the influence of auxiliary forces such as gravity and centrifugation.

The electrostatic separators generally can be categorized as charging by induction and ion bombardment (corona discharge) according to the working principles of the separators in charging particles.

Electrostatic separators working with the principle of charging by induction consist of a grounded drum and an electrode (Fig. 8a). Particles with different electrical conductivity properties are fed from the upper part of the drum. By rotating the drum clockwise, all particles enter the electric field created by an electrostatic electrode. While the grounded part of the drum is neutral, the part of the drum close to the electrode is counter-charged to the charge of the electrode. All particles entering the electric field become polarized. Since electrons can move freely within the conductive particles, electrons can transfer between the particle and the drum. Thus, the oppositely charged particles are pushed by the drum and pulled by the electrode. This movement of the particles is called lifting.

On the other hand, the polarized insulator particles remain polarized because they cannot exchange electrons with the drum. Therefore, they adhere to the drum (pinning) and start to rotate. Finally, they are separated from the drum surface by a separator (brush). Meanwhile, the semiconductor particles follow a path between these two behaviors.

In the separators working with the principle of ion (electron) bombardment, a corona electrode is used instead of an electrostatic electrode (Fig. 8b). Due to the high voltage applied, the air between the corona electrode and the drum is ionized and the electrons begin to move toward the drum. Therefore, the particles entering the area between the drum and the electrode are exposed to electron bombardment and become negatively charged. The conductive particles give their electrons to the

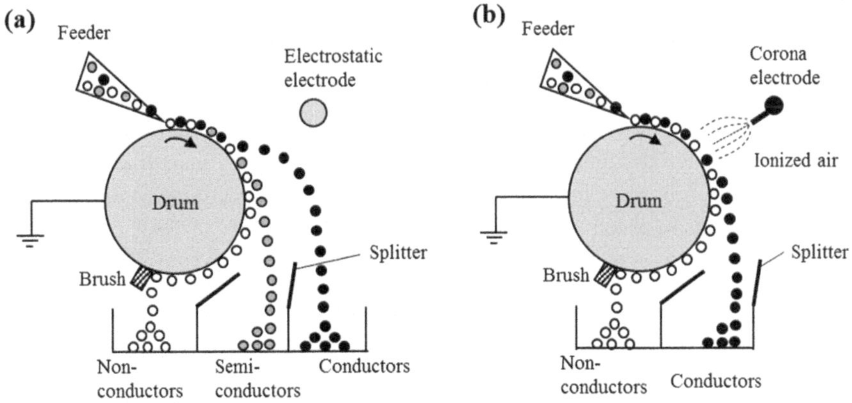

Fig. 8 Electrostatic separators work with the principle of **a** charging by induction. **b** ion bombardment (corona discharge) (Authors original drawing)

ground through the drum and fall under the influence of gravity. Since the non-conductor particles cannot manage this, they remain negatively charged, adhere to the drum, and rotate [36, 37].

There are also devices in which these two types of electrodes are used together to strengthen the separation [38].

5.2 Chemical Beneficiation

Chemical beneficiation is the process of obtaining pure compounds or metals from ores or pre-concentrates using chemical methods. Pyro-metallurgy, electrometallurgy, and hydrometallurgy (leaching) are chemical beneficiation processes. While pyro-metallurgical processes utilize the reactions occurring at high temperatures, in electrometallurgy, reactions are carried out in aqueous solutions using electrical energy. In hydrometallurgy, on the other hand, valuable minerals are selectively taken into solution using a proper solvent, and then pure metals or compounds are obtained by various purification and recovery techniques. Different leaching methods can be used in hydrometallurgy depending on the characteristics of the feed material (chemical composition, grade, size, etc.) [39, 40].

For instance, in in-situ leaching, the solution is given to the ore bed, and the pregnant solution is collected by drainage. In heap leaching, the crushed or ground ore is piled on the impermeable ground near the mine sites and the solvent is given with a sprinkler system. The pregnant solution is collected from the bottom of the heap. Meanwhile, in vat leaching, the ore is placed in a tank with a perforated tray at the bottom. In this method, the solvent can be fed from the top or bottom and the pregnant solution is discharged in a batch system. In agitation leaching, the ore (0.5 mm) is mixed mechanically or pneumatically under an appropriate pressure until it dissolves in autoclaves at a 40–70% solids ratio. The product is recovered using a batch system [41].

A series of processes are applied to recover the precious metal or metals taken into solution by the aforementioned methods. First, the pregnant solution is separated from its insoluble part by processes such as precipitation, thickening, filtration, washing, and clarification. The concentration of precious metals in the clarified solution is increased by methods such as ion exchange and solvent extraction. Then, precious metals are recovered from the loaded solution by precipitation. Finally, the empty solution is adjusted and fed back into the circuit.

5.3 Physico-chemical Beneficiation

As the most widely used beneficiation technique, in flotation, the particles are separated due to the difference between the wettability of their surfaces [42, 43]. The surfaces of minerals with few exceptions such as graphite, talc, sulfur, and molybdenite are naturally wettable due to the polar groups on their surfaces that can bond with the water molecules. While the minerals whose surfaces are wetted easily are called hydrophilic, the minerals with a limited affinity for wetting are called hydrophobic [5, 44]. The degree of hydrophobicity of minerals is usually determined by measuring the contact angle, which is the angle between the solid–liquid and air–liquid interfaces when the bilateral tensions between the solid, liquid, and air interfaces are equal.

In general, minerals do not have surface properties to attach to a bubble if they are hydrophilic (zero or close to zero contact angle, which means that the mineral surface is completely wet). However, bubble-particle attachment time measurements, which is the minimum time required for a particle to adhere to the bubble, is another important method used for the same purpose as contact angle measurements [45–47].

The beneficiation of minerals with the flotation method depends on making mineral surfaces hydrophobic selectively, the pH of the environment, and the appropriate size of air bubbles. As a result, chemicals added to the flotation medium to perform the flotation process are called flotation reagents that are collected in three main groups: Collectors, frothers, and control reagents which are activators, depressants, pH regulators, etc.

In flotation, minerals are made hydrophobic selectively. The reagents used to make the minerals attach to the bubble are called "collectors". Mineral particles in the liquid medium are separated from each other using air bubbles. Therefore, it is necessary to create an appropriate size and number of air bubbles in flotation [48]. The reagents that reduce the surface tension of the water and produce stable bubbles are called "frothers". Also, the reagents that facilitate the adsorption of the collectors to the desired minerals are called "activators", and the reagents that make it difficult are called "depressants". There are also some reagents called "dispersants", which prevent the mineral particles in the pulp from being covered with slime. And, some chemicals used to make the environment acidic or basic in flotation are called "pH regulators".

Since the air is strongly hydrophobic, the hydrophobic minerals adhere to the air bubbles by van der Waals forces. Bubble-hydrophobic particle clusters whose total density is less than the density of the suspension rise above the liquid phase and form a froth layer. Froth laden with hydrophobic minerals is recovered by stripping, while hydrophilic minerals remain in the flotation suspension.

Many ores can be recovered by flotation. Since the ores may contain minerals with different properties, the appropriate flotation method and the reagents differ. Minerals can be classified according to their flotation properties as follows [17]:

- Non-metals and solids with significant natural hydrophobicity
- Native metals and sulfides
- Oxidized minerals of non-ferrous metals
- Oxidizes, hydroxides, and silicates
- Sparingly soluble salts
- Soluble salts

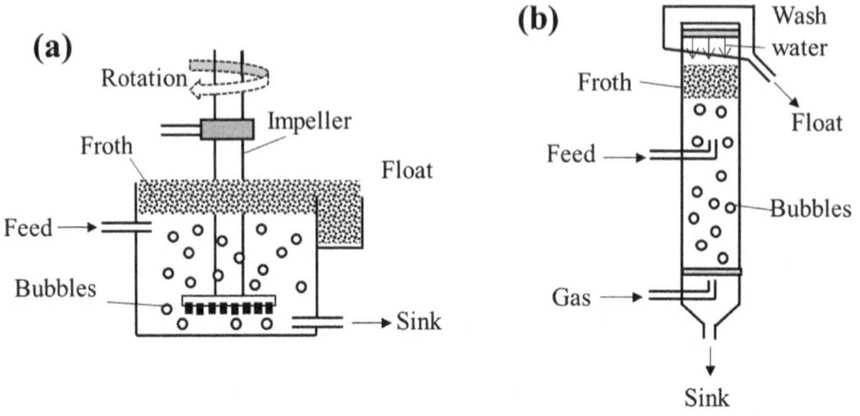

Fig. 9 **a** Mechanical and **b** pneumatic (column-type) flotation devices (Authors original drawing)

Flotation machines can be classified as mechanical, and pneumatic according to their working principles on the stirring of the pulp.

Mechanical flotation machines are the most widely used machines. They contain an impeller that stirs the suspension and disperses the air into bubbles. In these types of machines, the air is usually drawn into the cell by the rotation of the impeller naturally (self-aerated) or it can be supplied externally (supercharged) (Fig. 9a).

On the other hand, pneumatic flotation machines do not use a mechanical impeller. Instead, the pulp is suspended with the help of pressurized air. The column cell is the most popular type of pneumatic flotation device. It is more effective in separating fine particles and has lower investment and operating costs. It also takes up less space and is more prone to automatic control. A flotation column consists of two parts, the collection (recovery or pulp) zone, and the froth (washing or cleaning) zone. The collection zone is below the feed point, and the particles are suspended between the descending water phase and the rising bubble cluster created by a spray system. Hydrophobic particles attach to the bubbles and rise to the froth zone, while hydrophilic particles are removed from the bottom of the column. The particles that are transferred to the froth phase unselectively with the entrainment process are washed back to the collection zone with the washing water. Thus, the decrease in the grade of the concentrate is prevented (Fig. 9b) [49]. There are also hybrid flotation machines, although they are not widely used in the industry today [50–52].

As mentioned, a flotation system consists of two separate regions an intense mixing region where bubble-particle collision and attachment occur, and a quiescent region where the bubble-particle aggregates separate from the slurry. While the two regions exist together in a single container in conventional flotation machines, pneumatic machines use different chambers for these two processes. Jameson cell is a good example of these kinds of cells [5].

6 Conclusions

Minerals and their products extracted and processed by mining activities are used in all areas of our modern life. Since mining also provides the raw materials necessary for the operation of other industries, it is an indispensable industry for humanity. With the increasing demand for consumption, the high-grade mineral deposits are being depleted gradually and the necessity for the utilization of more complex ores increasing. These difficulties in the processing of these complex ores can be overcome with environmentally friendly technologies with higher separation efficiency. Therefore, technological advancements are crucial to maintaining a steady supply of essential raw materials.

These advancements can be made only by well-trained engineers and researchers. For this reason, our colleagues at the beginning of their professional lives should follow the literature closely and contribute to the industry by designing new techniques, methods, and devices with an innovative perspective. Therefore, this chapter is a useful resource for students, engineers, and researchers who want to improve their knowledge about minerals, mining, and mineral processing operations.

References

1. M. Tausova, K. Culkova, L. Domaracka, C. Drebenstedt, M.S. Muchova, J. Kosco, A. Behunova, M. Drevkova, B. Bencoova, Acta Montanistica Slovaca **22**(4), 359 (2017)
2. K. McQueen, Earth Sci. Hist. **31**(2), 316 (2012)
3. H.L. Hartman, J.M. Mutmansky, *Introductory Mining Engineering*, 2nd edn. (Wiley India Pvt. Limited, 2002), pp. 1–24
4. L.J. Thomas, *An Introduction to Mining* (Hicks Smith & Sons, Sydney, 1973), pp.1–30
5. B.A. Wills, J.A. Finch, *Wills' Mineral Processing Technology*, 8th edn. (Elsevier, 2016), pp. 1–27
6. T.M. McCoy, H.C.W. Parks, R.F. Tabor, Highly efficient recovery of graphene oxide by froth flotation using a common surfactant. Carbon **135**, 164–170 (2018). https://doi.org/10.1016/j.carbon.2018.04.032
7. F.S. Namin, A. Ghadi, F. Saki, A literature review of Multi Criteria Decision-Making (MCDM) towards mining method selection (MMS). Resour. Policy **77**, 1–13 (2022). https://doi.org/10.1016/j.resourpol.2022.102676
8. V. Mutambo, S. Kangwa, M. Fisonga, S.K. Shukla, Mining method selection for extracting moderately deep ore body using analytical hierarchy process at mindola sub-vertical shaft, Zambia. Cogent Eng. **9**(1) (2022). https://doi.org/10.1080/23311916.2022.2062877
9. A. Selyukov, S. Zhironkin, M. Cehlar, Improvement of continuous lateral surface mining method for coal extraction from closed mines' pillars. Acta Montanistica Slovaca **25**, 542–562 (2021). https://doi.org/10.46544/AMS.v25i4.09
10. A. Nieto, Selection process for underground soft-rock mining, in *SME Mining Engineering Handbook* (2011), pp. 377–384
11. S. Mijalkovski, K. Zeqiri, Z. Despodov, V. Adjiski, Underground mining method selection according to Nicholas methodology. Природни ресурси и технологии **16**(1), 5–11 (2022). https://doi.org/10.46763/NRT22161005m
12. R.L. Bullock, Comparison of underground mining methods, in *SME Mining Engineering Handbook* (SME, 2011), pp. 385–405

13. H. Harmin, Underground mining methods and applications, in *Underground Mining Methods: Engineering Fundamentals and International Case Studies* (SME, 2001), pp. 3–14
14. M.A. Pashkevich, in *Restoration and Reclamation of Mining Influenced Soils*, ed. by J. Bech, C. Bini, M.A. Pashkevich (Academic Press, 2017), pp. 1–32
15. A. Khaidir, Testing of bituminous coal and its impurity layers as in situ ameliorant of physico-chemical soil reclamation properties of coal mine. Agric. Sci. Dig. **42**(5), 621–625 (2022). https://doi.org/10.18805/ag.DF-449
16. G. Li, Z. Hu, P. Li, D. Yuan, Z. Feng, W. Wang, Y. Fu, Innovation for sustainable mining: integrated planning of underground coal mining and mine reclamation. J. Clean. Prod. **351** (2022). https://doi.org/10.1016/j.jclepro.2022.131522
17. J. Drzymala, *Mineral Processing, Foundations of Theory and Practice of Minerallurgy*, English edn. (Wroclaw University of Technology, Wroclaw, 2007)
18. S.C. Bouffard, Benefits of process control systems in mineral processing grinding circuits. Miner. Eng. **79**, 139–142 (2015). https://doi.org/10.1016/j.mineng.2015.06.006
19. A. Azizi, R. Rooki, N. Mollayi, Modeling and prediction of wear rate of grinding media in mineral processing industry using multiple kernel support vector machine. SN Appl. Sci. **2**(9) (2020). https://doi.org/10.1007/s42452-020-03212-0
20. J.A. Herbst, Y.C. Lo, B. Flintoff, Size reduction and liberation, in *Principles of Mineral Processing*, ed. by M.C. Fuerstenau, K.N. Han (SME, Littleton, Colorado, 2003), pp. 61–118
21. W. Wang, M. Pan, C. Duan, H. Jiang, Y. Zhao, H. Lu, Dry deep screening of spodumene and its mineral processing technology. Miner. Eng. **179** (2022). https://doi.org/10.1016/j.mineng.2022.107445
22. C. Yu, X. Wang, S. Gong, K. Pang, G. Zhao, Q. Zhou, D. Lin, N. Xu, Stability analysis of the screening process of a vibrating flip-flow screen. Miner. Eng. **163** (2021). https://doi.org/10.1016/j.mineng.2021.106794
23. S. Palaniandy, M. Yahyaei, M. Powell, Assessment of hydrocyclone operation in gravity induced stirred mill circuits. Miner. Eng. **108**, 83–92 (2017). https://doi.org/10.1016/j.mineng.2017.01.002
24. T. Zhang, J. Li, A. Wei, J. Huang, S. Li, J. Huan, F. Wang, H. Wang, Ultra-clean separation of micro-particles in lubricant oil based on short-flow control of mini-hydrocyclone. Sep. Purif. Technol. 304 (2023). https://doi.org/10.1016/j.seppur.2022.122370
25. P. Tucker, Technical note-modelling the Kelsey centrifugal jig. Miner. Eng. **8**(3), 333–336 (1995). https://doi.org/10.1016/0892-6875(94)00130-5
26. P. Yerriswamy, A.K. Majumder, J.P. Barnwal, B. Govindarajan, T.C. Rao, Study on Kelsey jig treating Indian coal fines. Miner. Process. Extr. Metall. **112**(3), 206–210 (2013). https://doi.org/10.1179/037195503225003654
27. R.K. Singh, A. Das, Analysis of separation response of Kelsey centrifugal jig in processing fine coal. Fuel Process. Technol. **115**, 71–78 (2013). https://doi.org/10.1016/j.fuproc.2013.04.005
28. O. Canieren, C. Karaguzel, A. Aydin, New technology gravity separators in mineral processing. Paper presented at the international conference on engineering and natural sciences, Sarajevo, 24–28 May 2016
29. Q. Chen, H. Yang, L. Tong, Z. Liu, G. Chen, J. Wang, Analysis of the operating mechanism of a Knelson concentrator. Miner. Eng. **158** (2020). https://doi.org/10.1016/j.mineng.2020.106547
30. L. Basnayaka, B. Albijanic, N. Subasinghe, Performance evaluation of processing clay-containing ore in Knelson concentrator. Miner. Eng. **152** (2020). https://doi.org/10.1016/j.mineng.2020.106372
31. Q. Chen, H. Yang, L. Tong, H. Niu, F. Zhang, G. Chen, Research and application of a Knelson concentrator: a review. Miner. Eng. **152** (2020). https://doi.org/10.1016/j.mineng.2020.106339
32. M.R. Andavarapu, A. Vidyadhar, R. Prasad, Beneficiation of difficult-to-wash Indian low volatile coking coal fines by Falcon concentrator. Int. J. Coal Prep. Util. **42**(12), 3618–3636 (2021). https://doi.org/10.1080/19392699.2021.1984902
33. A.K. Majumder, J.P. Barnwal, Modeling of enhanced gravity concentrators–present status. Miner. Process. Extr. Metall. Rev. **27**(1), 61–86 (2006). https://doi.org/10.1080/08827500500339307

34. P. Dai, J. Yang, Z. Wei, J. Zeng, Z. Xue, L. Chen, Magnetic properties of chalcopyrite and arsenopyrite for high-gradient magnetic separation with crystal-field theory. Miner. Eng. **189** (2022). https://doi.org/10.1016/j.mineng.2022.107893
35. J.S. Guiral-Vega, L. Pérez-Barnuevo, J. Bouchard, A. Ure, É. Poulin, C. Du Breuil, Particle-based characterization and classification to evaluate the behavior of iron ores in drum-type wet low-intensity magnetic separation. Miner. Eng. **186** (2022). https://doi.org/10.1016/j.mineng.2022.107755
36. L.O. Filippov, I.V. Filippova, G. Crumiere, R. Sousa, M. Machado Leite, A. Botelho de Sousa, C. Korbel, S.K. Tripathy, Separation of lepidolite from hard-rock pegmatite ore via dry processing and flotation. Miner. Eng. **187** (2022). https://doi.org/10.1016/j.mineng.2022.107768
37. D. Franke, T. Suponik, P.M. Nuckowski, K. Gołombek, K. Hyra, Recovery of metals from printed circuit boards by means of electrostatic separation. Manag. Syst. Prod. Eng. **28**(4), 213–219 (2020). https://doi.org/10.2478/mspe-2020-0031
38. A. Catinean, L. Dascalescu, M. Lungu, L.M. Dumitran, A. Samuila, Improving the recovery of copper from electric cable waste derived from automotive industry by corona-electrostatic separation. Part. Sci. Technol. **39**(4), 449–456 (2020). https://doi.org/10.1080/02726351.2020.1756545
39. A. Chagnes, Advances in hydrometallurgy. Metals **9**(2), 211 (2019). https://doi.org/10.3390/met9020211
40. D. Qi, in *Hydrometallurgy of Rare Earths*, ed. by D. Qi (Elsevier, 2018), pp. 671–741
41. A. Gupta, D. Yan, *Mineral Processing Design and Operations*, 2nd edn. (Elsevier, 2016), pp. 421–469
42. D. Wang, Q. Liu, Hydrodynamics of froth flotation and its effects on fine and ultrafine mineral particle flotation: a literature review. Miner. Eng. **173** (2021). https://doi.org/10.1016/j.mineng.2021.107220
43. D. Tao, Recent advances in fundamentals and applications of nanobubble enhanced froth flotation: a review. Miner. Eng. **183** (2022). https://doi.org/10.1016/j.mineng.2022.107554
44. J.S. Laskowski, *Coal Flotation and Fine Coal Utilization*, 1st edn. (Elsevier, Amsterdam, 2001), pp. 31–88
45. C. Gungoren, Y. Baktarhan, I. Demir, S.G. Ozkan, Enhancement of galena−potassium ethyl xanthate flotation system by low power ultrasound. Trans. Nonferrous Metals Soc. China **30**(4), 1102–1110 (2020). https://doi.org/10.1016/S1003-6326(20)65281-5
46. C. Gungoren, O. Guven, M. Cinar, O. Ozdemir, An investigation of the effect of clay type on coal flotation along with DLVO theoretical analyses. Int. J. Coal Prep. Util. 1–13 (2019). https://doi.org/10.1080/19392699.2019.1603146
47. C. Gungoren, O. Ozdemir, X. Wang, S.G. Ozkan, J.D. Miller, Effect of ultrasound on bubble-particle interaction in quartz-amine flotation system. Ultrason. Sonochem. **52**, 446–454 (2019). https://doi.org/10.1016/j.ultsonch.2018.12.023
48. C. Gungoren, E. Islek, Y. Baktarhan, I. Kurşun Unver, O. Ozdemir, A novel technique to investigate the bubble coalescence in the presence of surfactant (MIBC) and electrolytes (NaCl and $CaCl_2$). Physicochem. Probl. Miner. Process. **54**(4), 1215–1222 (2018). https://doi.org/10.5277/ppmp18158
49. A. Gupta, D. Yan, *Mineral Processing Design and Operations*, 2nd edn. (Elsevier, 2016), pp. 689–741
50. M. Azhin, K. Popli, A. Afacan, Q. Liu, V. Prasad, A dynamic framework for a three phase hybrid flotation column. Miner. Eng. **170** (2021). https://doi.org/10.1016/j.mineng.2021.107028
51. M. Azhin, K. Popli, V. Prasad, Modelling and boundary optimal control design of hybrid column flotation. Can. J. Chem. Eng. **99**(S1) (2021). https://doi.org/10.1002/cjce.24010
52. P.K. Tsave, M. Kostoglou, T.D. Karapantsios, N.K. Lazaridis, A hybrid device for enhancing flotation of fine particles by combining micro-bubbles with conventional bubbles. Minerals **11**(6) (2021). https://doi.org/10.3390/min11060561

Use of Ultrasound in Physical and Chemical Mineral Processing Operations

Can Gungoren, Safak Gokhan Ozkan, and Orhan Ozdemir

Abstract Ultrasound, which is a sound wave above the human perception limit, creates some extraordinary phenomena in liquids including acoustic streaming, and cavitation related to its various frequencies. While acoustic streaming causes the effective cleaning of solid surfaces and accelerates chemical reactions, cavitation, which is the formation, growth, and collapsing of bubbles in a liquid, may create liquid jets and short-lived hot spots that may reach extremely high local pressures (1000 atm) and temperatures (5000 °C). Because of these unique features, ultrasound is used in various industries including mineral processing to enhance separation recoveries and obtain better-quality products at acceptable and bearable costs. The current study also specifically focuses on reviewing the use of ultrasound on mineral comminution as well as its applications in physical and chemical mineral processing operations. Various previous studies about the use of ultrasound in the literature indicated that ultrasound could increase the comminution rate of the particles related to their mineral properties. In physical techniques, ultrasound provides surface cleaning and slime removal as well as causing aggregation of very finely sized particles. Ultrasound also enhances the efficiency of physical processes such as ore washing and solid/liquid separation. Furthermore, in the case of chemical processes including leaching and electrolysis, ultrasound also increases the process efficiencies at lower retention times and chemical dosages by increasing the surface area of the particles and creating more reactive surfaces. For all of these above-mentioned applications, the conditions for the ultrasonic applications such as device, frequency, time, power, and intensity levels are prominent as well as the properties of the medium including purity, viscosity, composition, and temperature. As most of these studies on the ultrasonic application

C. Gungoren
Engineering Faculty, Mining Engineering Department, Istanbul University-Cerrahpasa, 34500 Buyukcekmece, Istanbul, Türkiye

S. G. Ozkan
Department of Robotics and Intelligent Systems, The Institute of the Graduate Studies in Science and Engineering, Turkish-German University, Beykoz, Istanbul 34820, Türkiye

O. Ozdemir (✉)
Faculty of Mines, Department of Mineral Processing Engineering, Istanbul Technical University, Ayazağa Campus, Maslak,Sarıyer, Istanbul, Türkiye
e-mail: orhanozdemir@iuc.edu.tr

S. J. Ikhmayies (ed.), *Advances in Minerals Research*, Advances in Material Research and Technology, https://doi.org/10.1007/978-3-031-49175-7_2

in mineral processing are carried out at the laboratory or pilot scale, the scale-up studies need to be performed to utilize ultrasound in commercial applications in future studies.

Keywords Ultrasound · Mineral processing · Cavitation · Comminution and classification, · Physical beneficiation · Leaching.

1 Introduction

In mineral processing, commercially valuable minerals are separated selectively from their ores with physical, chemical, or physicochemical techniques. Rapid changes in technical and economic conditions in the mineral processing industry necessitate the investigation of various treatment techniques to increase the process efficiency and product quality as well as to minimize the operational costs and negative effects of the tailings generated. One of the methods used for these purposes in aqueous media is the application of ultrasound (US), a sound wave above human perception frequency, at an appropriate stage of the mineral processing operations (Fig. 1).

As known from the literature, sound waves are produced when mechanical energy propagates through the medium in the form of compression and rarefaction movements. When sound waves move parallel to the direction of motion of the particles, longitudinal (compressional, primary, P) waves are produced. Meanwhile, other types of waves, such as shear (transverse, secondary, S) waves can be generated when particles oscillate perpendicular to the direction of the wave propagation. The third type of wave is surface (Rayleigh) waves, which propagate across the surface of the medium [1, 2].

As sound is a kind of momentum energy that is transmitted through pressure fluctuations in the medium, its frequency, intensity, amplitude, wavelength, and propagation speed are important parameters for its consequences [3]. Soundwaves are

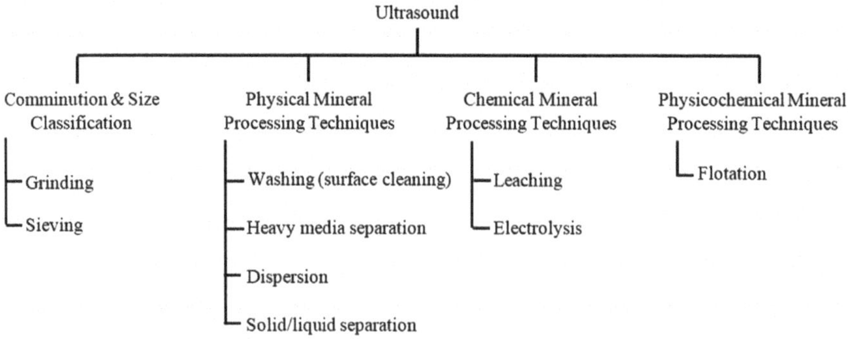

Fig. 1 Application of ultrasound in mineral processing operations (Authors original drawing)

classified as infrasound (<16 Hz), audible (16 Hz–20 kHz), and ultrasound (20 kHz–1 GHz) [4] related to their propagation frequencies, and ultrasound can be investigated in three frequency ranges, namely, low (20–100 kHz), high (2–10 MHz), and diagnostic (300–1000 kHz).

When ultrasound is applied to an aqueous medium, various effects can be observed according to the frequency and power of the ultrasonic wave. Ultrasound can accelerate chemical reactions and create bubbles at atmospheric pressures below the boiling temperatures of liquids. The latter process is called cavitation and it is responsible for most of the beneficial influences of ultrasound in the industry. Generally, cavitation can be classified into four types according to its generation technique: acoustic, hydrodynamic, optic, and particle. However, only acoustic and hydrodynamic cavitation have the potential of being used commercially due to the ease of operation and the possibility of the generation of the required intensities of cavitational conditions [5, 6]. As shown in Fig. 2, cavitation in an aqueous solution generates three regions, where different reactions occur [7].

Additionally, the frequency of ultrasound affects its application field. Ultrasound at a higher frequency (>1 MHz) is usually used for diagnostic purposes to avoid affecting the propagation medium with a low intensity (<1 W/cm^2) [8]. On the other hand, high power ultrasound (>0.1 W/cm^2) at 16–100 kHz frequency (most commonly 20–40 kHz) is used for industrial purposes to improve a wide range of processes as seen in Fig. 3 [8, 9].

In literature, there are various studies on the use of ultrasound in physicochemical processes, especially on the flotation of coal [10–18], shale [19], and various minerals including graphite [20–22], feldspar [23], quartz [24–26], phosphate [27,

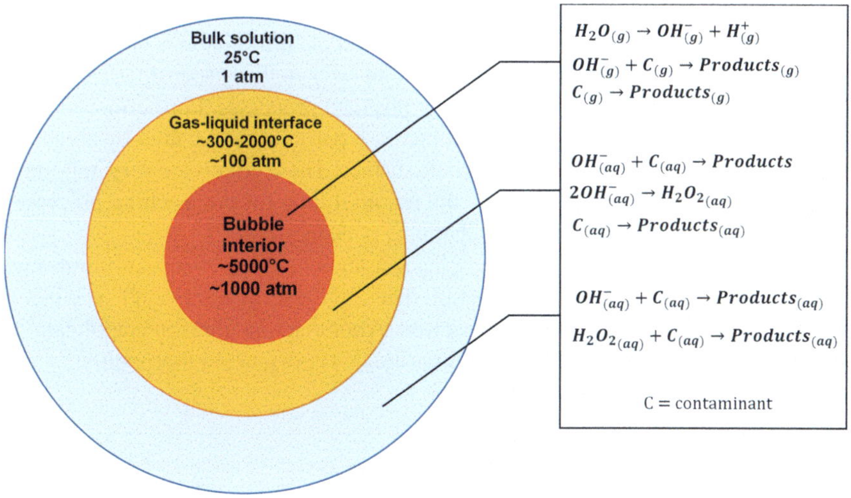

Fig. 2 Acoustic cavitation in water. Reprinted from de Andrade et al. [7], with permission from Elsevier. Copyright (2021)

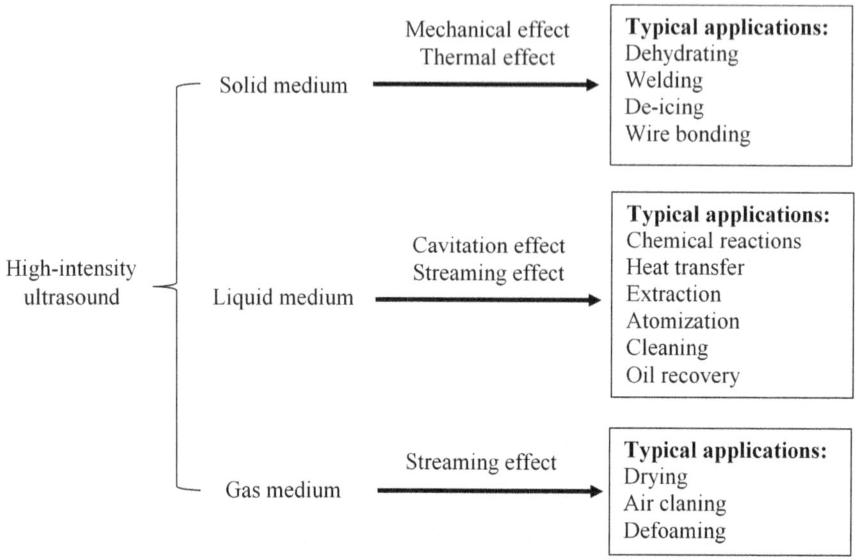

Fig. 3 Applications of power ultrasound in different media. Reprinted from Yao et al. [9], with permission from Elsevier. Copyright (2020)

28], potash [29], ilmenite [30, 31], zinc [32], copper [33], pyrite, chalcopyrite, and galena [26, 34–36].

Meanwhile, the review studies on the use of ultrasound on particle comminution, physical, and chemical beneficiation techniques are quite limited. However, ultrasound can create or widen fractures in the particles, which eases the comminution of particles. In addition, it can facilitate the removal of slime coatings on the particle surfaces; change the average particle and droplet sizes in the suspensions and emulsions, respectively, which are beneficial for physical processes. Furthermore, it can improve the transportation of chemicals, increase particle surface area by abrasion, and disrupt the diffusion layer on the particle surface. These effects are very important to increase the efficiency and decrease the required time for the leaching processes of the minerals at a reduced leaching chemical concentration.

Therefore, in this chapter, first, the physical and chemical effects of ultrasound in aqueous media are explained in detail. Then, the use of ultrasound in mineral comminution and classification was reviewed. After that, the ultrasonic applications in physical and chemical beneficiation processes were discussed thoroughly.

2 Effects of Ultrasound in an Aqueous Medium

Ultrasound is applied with two common devices in practice; ultrasonic baths and ultrasonic probes (horns) (Fig. 4). In ultrasonic baths, an ultrasonic transducer is attached to the outer surface of a liquid container. Meanwhile, in ultrasonic probes, an ultrasonic wave is radiated from a tip (immersed in a liquid) that is much smaller than the acoustic wavelength [37]. Both devices have some advantages and disadvantages that need to be considered for experimental studies [38].

Ultrasonic baths have low purchase and operational costs. They are more silent than probes. They usually do not require any additional equipment and are easy to use. Because of these advantages, they are more preferred in the studies. However, they have several disadvantages that affect the reproducibility of the experiments. Since they need additional water to transmit ultrasonic waves to the sample and the sample should be in another container in most cases, there are high energy losses in addition to the lack of a uniform distribution of the ultrasound. Furthermore, although they have a time adjustment knob in most cases, the power or amplitude cannot be controlled by ultrasonic baths.

Ultrasonic probes, on the other hand, focus the energy to a point, providing more effective cavitation. The power and amplitude can be adjusted. However, they are pricey and different kinds of probe extensions may be required according to the used sample container. They create more noise and operators may need ear protectors in prolonged applications. Most importantly, strong cavitation abrade the tips of the probes in time although they are made of titanium alloys (Fig. 5). Therefore, the tips have to be replaced periodically to obtain uniform energy distribution in all experiments.

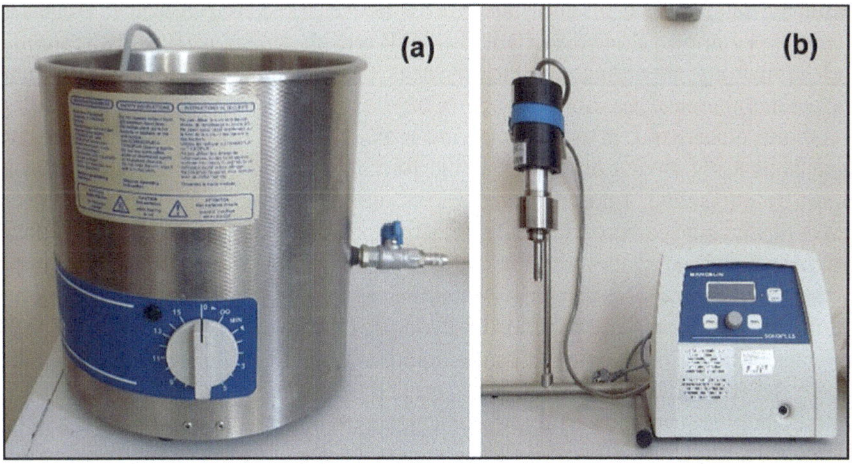

Fig. 4 **a** An ultrasonic bath (Bandelin, RK106, Germany) and **b** an ultrasonic probe (Bandelin, HD3200, Germany)

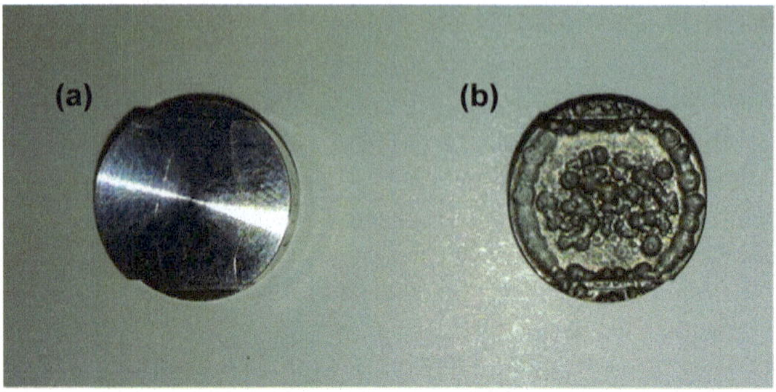

Fig. 5 The replaceable titanium probe tip (Bandelin, TT 13) **a** before and **b** after the abrasion caused by ultrasonic cavitation [39]

According to our experience with the use of ultrasound in various applications of mineral processes, ultrasonic baths are more useful in the processes in which less energy is adequate such as sieving (opening the clogged sieves) and surface cleaning (removal of slime coatings from the surface of the particles). On the other hand, probes are required for higher-energy applications such as comminution, changing the roughness and shape of the particles, and altering the adsorption pattern of chemicals.

Since ultrasound is a sound wave, it needs a material medium to propagate. While it does not show a remarkable effect when propagated in solid and gaseous media, it creates various extreme conditions in liquids through two major phenomena: acoustic streaming and acoustic cavitation.

Acoustic streaming is a term for unidirectional streams generated in the liquid media in the presence of ultrasonic waves above 100 kHz frequency. Acoustic streaming comprises three important effects (Rayleigh streaming, Eckert streaming, and Schlichting streaming) and it enhances the cleaning of solid surfaces and accelerates chemical reactions [40].

On the other hand, at frequencies lower than 100 kHz, the dominant mechanism is acoustic cavitation, which is the formation and collapse of bubbles in a liquid by ultrasound. Like all sound waves, ultrasound propagates in a series of waves that include compression and rarefaction phases. A sufficiently strong rarefaction phase overcomes the forces of attraction between the molecules of the liquid medium and exceeds the breaking strength of the liquid [41]. The magnitude of the required negative pressure depends on the type, purity, and viscosity of the liquid. The tensile strength in completely pure liquids is quite high. For instance, approximately 1000 atm of negative pressure is needed to create cavitation bubbles in pure water. However, even powerful ultrasonic generators can generate about 50 atm of negative pressure [42]. Thus, it is difficult to create cavitation in pure liquids. For the same reason, cavitation cannot occur at very high viscosities. In some applications, high ambient pressures are used deliberately to prevent cavitation [40].

The size of the freshly formed cavitation bubbles is around 100 μm [42]. These bubbles grow over several wave cycles until they reach an equilibrium size that matches the bubble resonance frequency by absorbing gas or vapor from the medium. Cavity growth is related to the intensity of the sound. If the cavity growth is too fast in the rarefaction phase of the wave, the bubble will not have enough time to shrink in the subsequent compression phase. Therefore, at high intensities, the bubble grows rapidly and collapses which is called transient cavitation.

On the contrary, at low intensities, cavity growth is slower (rectified diffusion). Under these conditions, the cavity oscillates around one dimension during the compression and rarefaction phases. The amount of gas or steam that passes into and out of the bubble along these oscillations is related to the surface area of the bubble, which is slightly larger during rarefaction than in the compression phase. Therefore, the growth of the bubble during rarefaction phases is greater than its shrinkage in the course of compression, and hence, the bubble grows through wave cycles. This type of cavitation is called permanent cavitation. The growing bubble eventually reaches a critical (resonant) size at which the ultrasound can effectively adsorb energy from the emission. This resonant size of a bubble depends on the properties of the liquid and the frequency of the sound wave [40, 42–44].

The gases and vapors inside a collapsing cavitation bubble are trapped. The compression of gas generates a large amount of heat. Heat generation during compression and collapsing of cavitation bubbles occurs faster than heat conduction. Thus, the temperature of the immediate vicinity of the bubble increases [45]. This causes the formation of regions called short-lived localized hot spots. Although the temperature of these regions is very high, their dimensions are very small compared to the total volume of the liquid, and the resulting heat is dissipated in a very short time without making any measurable change in the ambient conditions. Suslick [42] calculated that the heating-cooling variations are more than 1 billion °C/s, while Luque-Garcia and Luque de Castro [38] reported that it is 10 billion °C/s. It is impossible to detect the temperatures that occur and disappear in such a short time with a physical thermometer. Raylight [46] predicted that very high local temperature and pressure values could be reached in his mathematical model studies on bubble collapses in liquids [47]. Because of the difficulty of measuring the temperatures formed and dissipated during cavitation by physical methods, Suslick [42] used an alternative method in which the theoretical temperatures reached can be calculated by experimental methods. He thought that the temperature at which a particular chemical reaction takes place depends on the negative inverse logarithm of the reaction rate, therefore the temperature could be calculated by observing the reaction rate at which a known chemical reaction takes place. This technique, in which the temperature is calculated based on known chemical reactions, is called comparative-rate chemical thermometry. With this method, by measuring the speed of several different reactions under ultrasonic effect, they calculated the temperature formed by the collapse of the cavitation bubble and calculated that the temperatures formed during cavitation reached approximately 5000 °C and the pressures 1000 atm [48].

As known, chemistry is the interaction of energy and matter, in addition to this, chemical reactions need energy to go on. In other words, the properties of the energy source determine the functioning of the chemical reactions [47]. The application of ultrasound in chemical reactions and processes is known as sonochemistry. Ultrasound is quite different from conventional energy sources such as heat, light, or ionizing radiation in terms of time per molecule, pressure, and energy. Ultrasound provides a unique mechanism for chemical reactions with the high local temperature, pressure, and heating/cooling rates generated by cavitation [47].

There are two types of sonochemical reactions. The process is called homogeneous sonochemistry when the ultrasound-applied medium has a homogeneous structure. If there are volatile molecules in this kind of system, they are filled in bubbles. The high temperature and pressure produced during cavitation break the chemical bonds. Various radicals are formed in the liquid depending on the properties of the medium. The generated short-lived chemicals return to the liquid bulk phase at room temperature and react with other chemicals [6]. Furthermore, the temperature of the ultrasound-applied liquid medium increases in time with inverse proportion to its volume. The rise in the temperature can increase the solubility of several solids in the system. Since the pressure distribution around a collapsing bubble is uniform in homogeneous systems, the bubble remains spherical during the collapse.

On the other hand, when ultrasound is applied to a heterogeneous system, the process is called heterogeneous sonochemistry. The presence of an anisotropy driver such as a gravitational field, nearby boundaries, other bubbles, and stationary potential flow near a collapsing bubble disrupts the pressure distribution formed by the ultrasonic wave. Therefore, the bubble collapses asymmetrically and a liquid jet, also called a micro-jet, is formed at a speed of about 400–500 km/h [38, 44]. The formation of liquid jets under various anisotropy drivers is shown in Fig. 6. In the presence of solid surfaces as in Fig. 6a, the liquid jet moves asymmetrically from the collapsing bubble toward the solid surface [49]. The liquid jets that hit the solid surface can remove the slime coatings from the particle surface, cause abrasion, and change the particle shape by rasping sharp edges and corners. Furthermore, they can create or widen the fractures on the particle surface, and in some cases, they can break the particle into smaller pieces. Therefore, the average size of the particles decreased and therefore their surface area increased. Since the effect of cavitation is proportional to the amount of energy stored in the cavitation bubble, more energy is released during the collapse of larger bubbles in the presence of lower frequencies [44].

Furthermore, ultrasonic treatment of two immiscible liquids (such as oil and water) may result in emulsification. The pressure and rarefaction phases of ultrasound waves break the cohesive forces that hold large droplets together. Therefore, the droplets are divided into smaller ones and therefore one liquid is dispersed in another as microscopic droplets. As a result, the chemical reaction rate of the liquids increases due to their increased surface area [42].

In addition to these features, while a collapsing cavitation bubble creates a hot spot in various liquids, the molecules may rise to high energy levels and emit visible light as they return to their ground state [42]. This phenomenon was first discovered by Frenzel and Schultes [50] and called sonoluminescence.

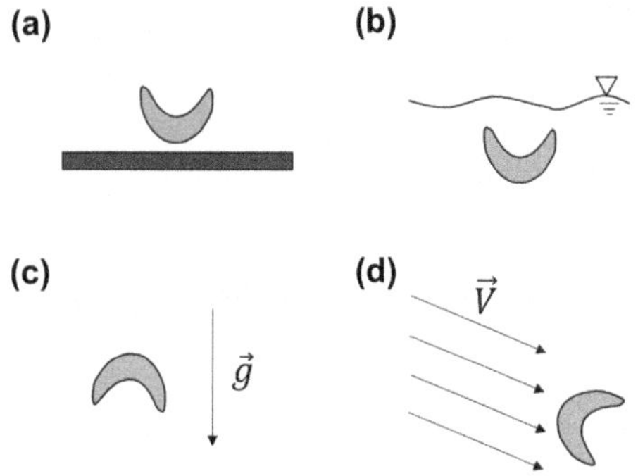

Fig. 6 Bubble collapse and jetting inside an anisotropic pressure field results from different causes: in the presence of **a** a rigid surface, **b** a free surface, **c** a gravity field, and **d** a shear flow. Reprinted from Orthaber et al. [49], with permission from Elsevier. Copyright (2020)

3 Effect of Ultrasound on Particle Comminution

Comminution is the mechanical breakdown of solid particles into smaller ones and ultrasound has been reported as a promising supporting technique for improving the comminution process [51–53]. The shock waves and liquid jets created by cavitation cause the abrasion of the particle surfaces, and the formation of fractures in the friable ores and coals. The widening and deepening of these fractures in the later stages result in the comminution of the particle. Ultrasonic treatment can reduce grinding energy by 85% for compressed coal samples [54]. As seen from the results of Sahinoglu and Uslu [55] in Fig. 7, the use of ultrasound (at 750 W, 20 kHz) reduced the size of coal particles significantly.

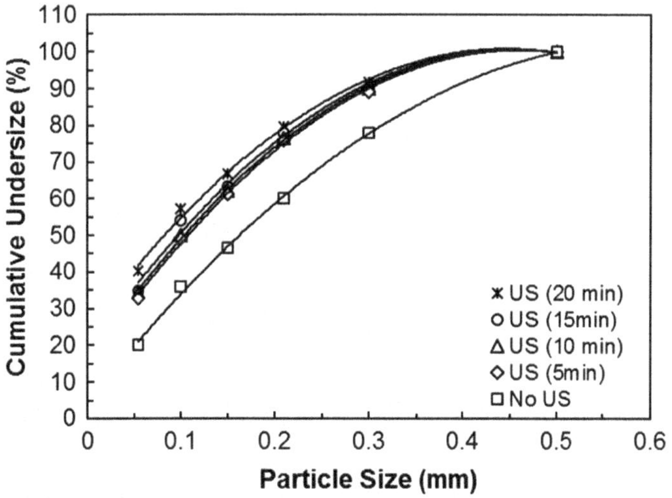

Fig. 7 Effect of ultrasonic treatment on the particle size distribution of coal at 113.6 W/cm^2 ultrasonic intensity and 10% solids ratio. Reprinted from Sahinoglu and Uslu [55], with permission from Elsevier. Copyright (2015)

Similarly, coal, iron, and manganese ore samples were ground at various ultrasonic intensities and treatment times by Singh et al. [54]. The results showed that d_{80} of particle diameters reduced about 6, 5.95, and 35.63% for the coal, iron, and manganese ore samples, respectively, during the grinding.

Li et al. [56] investigated the ultrasound-assisted wet grinding of $LiFePO_4$ (LFP), which is a promising cathode material for lithium-ion batteries. In the result of their study, 180 nm-sized LFP particles were produced from 35 μm-sized (d_{50}) particles with the help of ultrasound (350 W h), which is impossible with the conventional grinding process because of particle agglomeration at lower sizes. Ultrasound increased the grinding media/particle contact by deagglomerating these clusters and hence enhanced the grinding efficiency.

The frequency of ultrasound is also an influencing factor for the particle size distribution and the final particle size of the product. At low frequencies, where cavitation is more dominant, the particle size distribution becomes smaller. In this aspect, the effect of ultrasound frequency was investigated by Ambedkar et al. [57] on coal particle size (Fig. 8). As seen in Fig. 8, the low-frequency ultrasound (20 kHz ultrasonic probe and 25 kHz ultrasonic bath) causes a considerable decrease in the average particle size in a short time.

In the early stages of ultrasonic application, the particles have sharp corners, sharp edges, and micro-cracks. Sharp corners and edges allow cavitation bubbles to form. Micro-cracks, on the other hand, create weak spots and facilitate particle comminution. In time, these sharp corners and micro-cracks decrease, which causes a slowdown in comminution. Gibson et al. [58] indicated that larger particles (250–90 μm) were more prone to comminution than finer ones (63 × 38 μm).

In another study, Kang et al. [59] performed the sieve analysis before and after the ultrasonic treatment to determine its effect on the particle size of the coal sample.

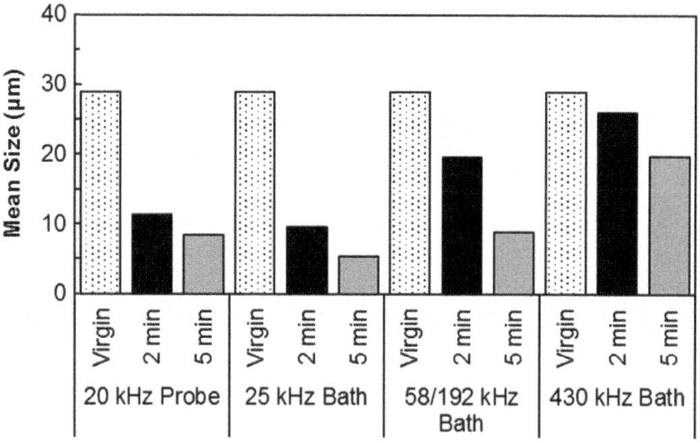

Fig. 8 Effect of ultrasound frequency on coal particle size at various time intervals. Reprinted from Ambedkar et al. [57], with permission from Elsevier. Copyright (2011)

It is seen from Table 1 that the content of coarser sizes decreased after the ultrasonic application.

The power and time of the ultrasonic application also play an important role in particle comminution as well as ambient temperature. As the number of cavitation bubbles increases at higher ultrasonic power levels, comminution enhances. Likewise, the longer the application time, the greater the comminution up to a certain time. Meanwhile, since temperature changes the viscosity, density, and vapor pressure of the liquid, it has a great effect on cavitation. Raman and Abbas [60] investigated particle comminution under ultrasonic effect (24 kHz) at different ultrasonic power (150–350 W) and temperature (10–50 °C) levels using aluminum oxide particles with 150 μm average particle size. As a result of their study, they reported that there was more size reduction at higher ultrasonic power levels and the optimum temperature for size reduction was between 25 and 37 °C.

Another important factor for the comminution by ultrasound is the viscosity of the medium. As the viscosity of the medium increases, the coarse particles move away from the bubble surface and hence escape from the shock waves. As a result, the increase in the medium viscosity reduces the comminution. In some cases, dispersion of the suspensions can be almost prevented completely by using highly viscous components [61].

In addition to comminution processes, ultrasound has also found use in size classification. Ultrasonic baths have been used for a long time to clean laboratory sieves. Moreover, manufacturers of the pilot and industrial-scale sieves also produce ultrasonic generator-mounted sieves. With the support of ultrasonic energy, sieve clogging can be prevented and continuity in sieving can be ensured, thus sieve capacities increase. Ultrasonic cleaning eliminates sieve wear that occurs with mechanical pore opening systems and therefore extends the sieve life. With ultrasound-assisted sieving systems, sieving can be carried out in smaller sizes compared to conventional vibrating sieves.

Table 1 Results for the particle size analysis of the coal sample before and after the ultrasonic treatment (Reprinted from Kang et al. [59])

Size (mm)	Before ultrasound application	After ultrasound application	Change rate (%)
+0.500	5.28	2.78	−47.35
0.500 × 0.250	15.26	9.68	−36.57
0.250 × 0.125	27.32	22.24	−18.59
0.125 × 0.075	16.09	18.52	15.10
0.075 × 0.045	11.01	13.03	18.35
−0.045	25.04	33.75	34.78
Total	100	100	

4 Effect of Ultrasound on Physical Mineral Processing Techniques

Ultrasound is used as an auxiliary treatment technique in mineral processing operations based on physical property differences such as washing, heavy media separation, dispersion, and solid/liquid separation (settlement, filtration, aggregation, flocculation, and precipitation).

In the removal of clay particles from bauxite ore, Ahmad et al. [62] used an ultrasonic washing process at 24 ± 1 kHz frequency and 23 W power level. After 5 min washing time, it was found that 70% mass and 77.5% alumina recovered at 36 μm cut-size, whereas 69% of total silica moved towards the tailings streams.

Similarly, ultrasound was also used in the beneficiation of boron minerals that are generally beneficiated by mechanical dispersion, abrasion, attrition, scrubbing, and washing processes. Sonmez et al. [63] investigated the usability of ultrasonic cleaning in the recovery of the colemanite tailings of these beneficiation processes (−3 mm, 24.17% B_2O_3). In the experiments, the parameters of processing time, pulp density, feeding rate, and particle size were investigated. In the studies, an ultrasonic bath with a power of 500 W and a frequency of 25–40 kHz, with 12 generators was used. Since there was no significant difference between the studies performed at 25 and 40 kHz frequencies, the experiments were carried out at 40 kHz frequency. The results of the ultrasonic experiments were compared with the results of the conventional mechanical abrasion and washing experiments. As a result, it was seen that ultrasonic applications gave better results compared to conventional experiments. By mechanical abrasion and washing, a concentrate with a grade of 33.01% B_2O_3 was obtained with the recovery of 84.93% from the feed with a grade of 24.17% B_2O_3. In the experiments with ultrasound, the concentrate grade reached 41.69% B_2O_3 and a recovery of 97.01% was obtained.

Another boron-containing mineral, tincal, was also beneficiated using ultrasound by Alp et al. [64]. In this study, an ultrasonic bath with a frequency of 35 kHz and an ultrasonic intensity of 1.24 W/cm^2 was used as an ultrasound source. In the experiments, the samples were subjected to ultrasonic treatment for the determined time in the ultrasonic bath, then washed, decanted, and the fine material was removed. The remaining ore was filtered and dried, then separated according to the particle size and the B_2O_3 content of each fraction was determined. As a result of this study, after 15 min of ultrasonic application, 35.29% B_2O_3 concentrate was obtained with a recovery of 95.69% from the feed (24.37% B_2O_3), and it was revealed that ultrasound could be used for clay removal from tincal ores.

Kyllonen et al. [65] examined the use of ultrasound prior to heavy media separation. Samples containing bullet fragments obtained from an old military shooting range were used in the experiments. Metal removal processes were performed with a 22 kHz frequency and 100–500 W power ultrasonic vibrating tray. As a result, it was observed that ultrasound facilitated the obtaining of clean products in both floating and sinking fractions in heavy media separation via removing the fine metal particles from the rocks.

Moreover, ultrasound is also used in the dispersion and aggregation processes. Fridman [61] investigated the interaction of the cavitation bubble with barite-calcite particles and investigated the acoustic dispersion and aggregation phenomena in solid-liquid and liquid-liquid media. As a result of their experiments, it was observed that fine barite-calcite particles (≤2 μm) were not attracted to the vibrating bubbles but they remained at a distance from the surface of the bubble and oscillated around a fixed point. The shock waves generated by the collapsing bubble first disrupt the coarse particles lying on the bubble surface. If the particles are finer than 2 μm in size, they are accumulated at a certain distance from the bubble surface; this accumulation causes the particle concentration to increase at that point and hence aggregation. Therefore, in the vicinity of a cavitation bubble two competing processes of dispersion and aggregation take place simultaneously.

It was observed from the experiments with barite-calcite and water suspension that 15–20 μm-sized of the particles comminuted to 2–4 μm after 40–50 min of ultrasonic application. Longer application time showed no significant effect on the particle size. When the same process was applied to a barite-calcite sample with a particle size of 0.5 μm, the dispersion was not observed. On the contrary, the particles aggregated into larger particles with a size of 2–4 μm. In other words, the final particle size is the same in both applications as seen in Fig. 9a. Furthermore, it was also observed that the increase in the ultrasonic power accelerated the particle comminution by augmentation in the number of cavitation bubbles produced. As seen in Fig. 9b, after a sudden decrease in the particle size at the early stages of ultrasonic application, the final size remains the same [61].

A similar phenomenon was observed in liquid-liquid systems. After a relatively short ultrasonic application time of 180 s, the oleic acid-water emulsion consisting of droplets of 5–8 μm and 15–20 μm size turned into a mono-dispersive state consisting of 2–4 μm-sized droplets [61].

Orthaber et al. [49] also investigated liquid/liquid systems in the presence of cavitation bubbles produced by ultrasound. Their results indicated that the liquid

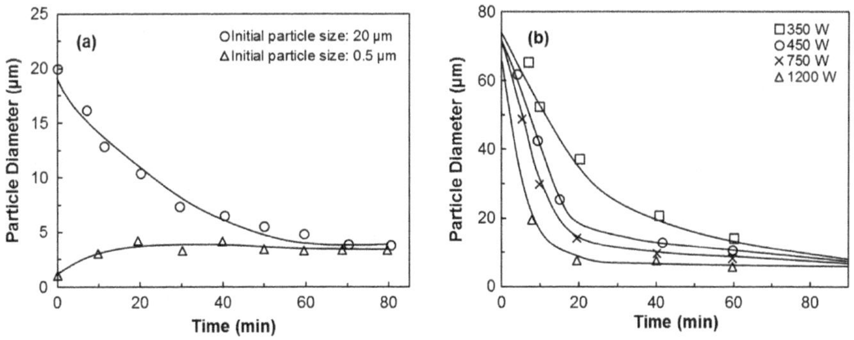

Fig. 9 **a** Dispersion and aggregation of barite-calcite suspensions with different initial particle sizes and **b** the effect of ultrasonic power on dispersion rate and final particle size of barite-calcite suspension. Reprinted from Fridman [61], with permission from Elsevier. Copyright (1972)

jets resulting from the collapse of cavitation bubbles always occurred toward the water/oil interface if it grew in the lighter liquid, and correspondingly away from the interface if it was initiated inside the denser liquid. Furthermore, they also stated that efficient emulsification could only be achieved when the jets towards the interface, hence the jets must be created inside the lighter (oil) phase. This can only be achieved by an ultrasonic probe with the tip inside the oil phase.

Ultrasound was also used in the separation of liquids. Sadatshojaie et al. [66] developed an effective and fast as well as cost-effective and environmentally friendly demulsifying process for water-contained crude oil emulsions using ultrasound at 20 kHz frequency and 80–1000 W power.

Meanwhile, Onal et al. [67] used ultrasound on the settlement of clay particles. Ultrasound was applied with an ultrasonic bath at a frequency of 25 kHz and a power of 0.5 kW. Different amounts of flocculants (0, 25, 50, 75 g/t Superfloc A-120) were tested in the experiments. The experimental results of the 25 g/t flocculant dosage are shown in Fig. 10. Similar results were obtained with other reagent dosages and better results were obtained with the ultrasonic application compared to natural conditions in all experimental conditions. Furthermore, there were considerable reductions in the reagent consumption and settling times in the presence of the ultrasonic application.

Ultrasound can also be used to improve the cake filtration process. In the conventional filtration process, the filter is immersed in suspension. When the filter is taken out of the suspension with the cake, the moisture content of the cake is reduced by vacuum. Ultrasound can be applied to the surface of the filter at this stage. Ultrasonic motion supports agglomeration and thus increases the size of the particles. Therefore, the collection efficiency of conventional filters was improved. Figure 11 shows the filter cake solids concentration values depending on the power (10–100 W) for the titanium dioxide (TiO_2) suspension with an initial solid concentration of 30% from the results of Riera-Franco de Sarabia et al. [68]. Figure 11 indicates that the ultrasound application significantly increases the solids concentration of the filter

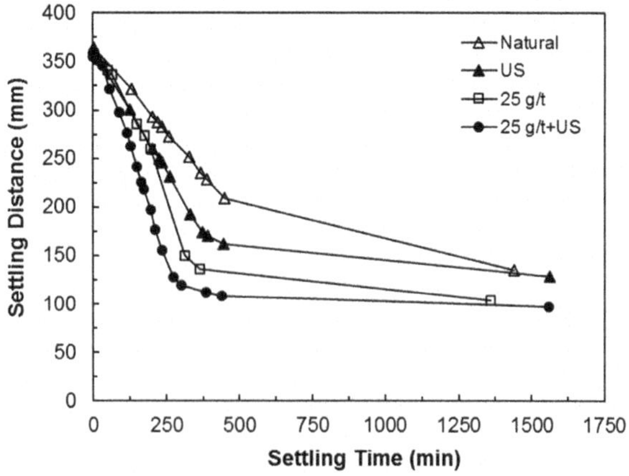

Fig. 10 Comparison of the sedimentation of clay in natural and ultrasonic media without and with flocculant addition (25 g/t Superfloc A-120). Reprinted from Onal et al. [67], with permission from Elsevier. Copyright (2003)

cake, and the power increase has a positive effect on the solids concentration to a certain level [68].

Singh [69] investigated the effect of ultrasound on coal dewatering in vacuum filtration and the presence of surfactant. As a result, it was found that the ultrasonic energy showed a positive effect on dewatering and reduced the concentration of chemicals required as seen in Fig. 12.

In the case of the flocculation process, Zhu et al. [70] investigated the flocculation-sedimentation and thickening of a gold mine concentrator tailings and reported that the sequence among the variables according to their influences is "power > frequency > duration > start time". Demir et al. [71] stated that even if ultrasound might not change the flocculation rate, it could help to reduce the sediment bed height by

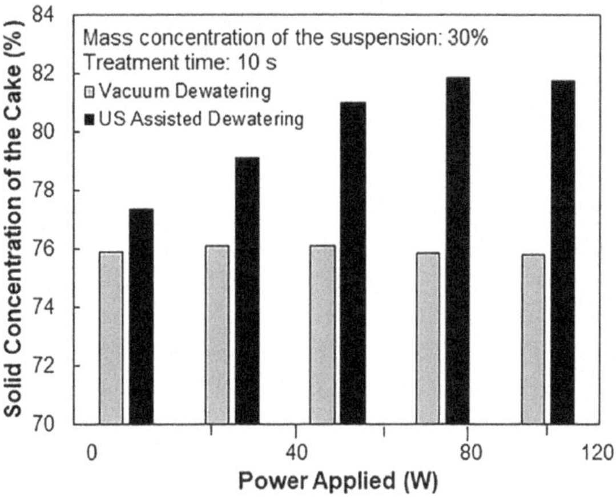

Fig. 11 The effect of power applied to the ultrasonic generator at 30% initial solids concentration on TiO_2 suspension. Reprinted from Riera-Franco de Sarabia et al. [68], with permission from Elsevier. Copyright (2000)

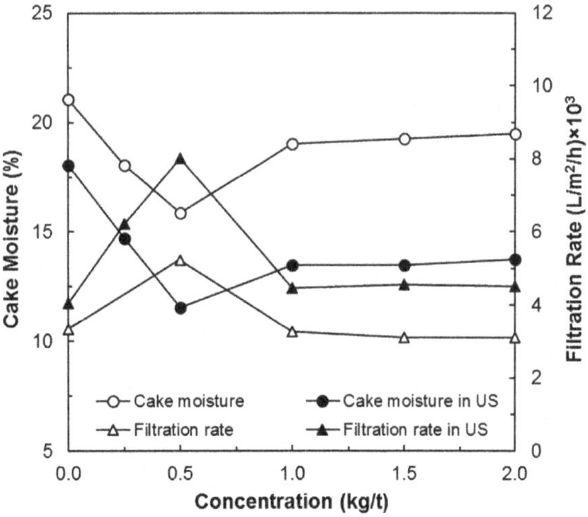

Fig. 12 Effect of ultrasonic pretreatment on filter cake moisture and filtration rate. Reprinted from Singh [69], with permission from Elsevier. Copyright (1999)

creating a more compact sediment bed [71]. In another study on flocculation, Ozkan and Esmeli [72] reported that ultrasonic application increased the surface charge of colemanite mineral, and thus increase the collector adsorption on the mineral surfaces. This made colemanite particles more hydrophobic and improve the flocculation process. Moreover, the results of Burat et al. [73] indicated that the moisture removal from coal fines could be increased by the surface cleaning effect of ultrasound, which increased the adsorption of chemicals on coal surfaces.

Precipitation was another process that can be enhanced by ultrasound. Davies et al. [74] accelerated the precipitation of sulfate with ultrasound at 24 kHz frequency. In clay-boron systems, ultrasound disintegrated the clay flocs and therefore provided the dissolution of boron minerals with the help of temperature. In the subsequent cooling phase, the boron minerals re-crystallized on clay substrate [75].

In the presence of ultrafine solid particles, the cavitation bubbles collapse similarly to a homogeneous medium and therefore liquid jets do not form. Shock waves that occur as a result of homogeneous cavitation create high-velocity inter-particle collisions. The temperature at the collision points can reach the melting point of some types of solid particles. Thus, melting can occur at the collision points, thus the metal particles are welded together. These collisions can alter the material's surface texture, composition, and reactivity. Ultrasound can increase the reactivity of metal powders up to 100,000 times [42].

Suslick [43] used metal powders to calculate the temperatures occurring during the collision. He applied ultrasound at 20 kHz and 50 W/cm^2 to chromium, molybdenum, and tungsten powders of a few micrometers in size. As a result of the study, it was reported that the temperatures in the collision zone increased to around 3000 °C, and the collision speeds increased to 1800 km/h.

5 Effect of Ultrasound on Chemical Mineral Processing Techniques

Ultrasound has been used in both leaching and electrolysis processes as a supporting element.

5.1 Ultrasound in Leaching

Leaching, the process of dissolution and solid-liquid extraction, can be supplemented with auxiliary energies to increase the process recovery and shorten the processing time. In many conditions, ultrasound-assisted leaching is faster, cheaper, and more efficient than conventional leaching. Luque-Garcia and Luque de Castro [38] and Mason et al. [76] compared ultrasound-assisted leaching with conventional leaching. According to their comparison, some additional factors that occur in the presence

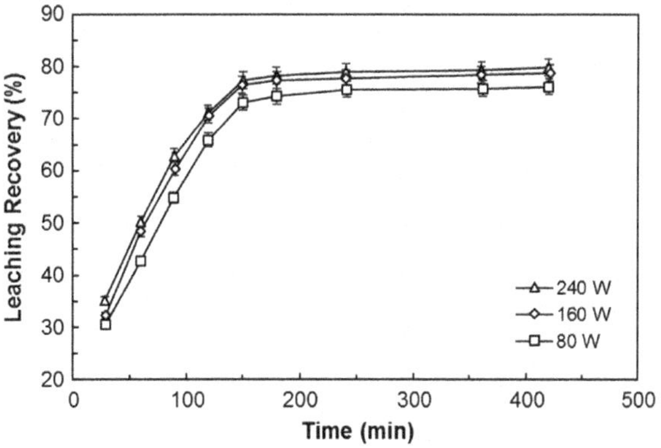

Fig. 13 Effect of ultrasound power on leaching recovery (±2% error). Reprinted from Wang et al. [77], with permission from Elsevier. Copyright (2013)

of ultrasound improve the leaching process. First, high-velocity liquid jets towards the solid surface increase the transportation rate of the chemicals. In addition, abrasion and comminution of the particles increases the total surface area. Furthermore, the shock waves create cracks in the particles and therefore the leaching agent can penetrate the inner regions of the particles by capillary action. Finally, the acoustic stream may disrupt the diffusion layer on the surface. In order to benefit from these advantages, many studies have been conducted on ultrasonic leaching.

Wang et al. [77] subjected 12.31% Zn-containing tailing that contained other metals such as Fe and Pb to sulfuric acid leaching in the presence of ultrasound. Ultrasound-assisted experiments were carried out in a temperature-controlled bath with an accuracy of ±0.1 °C. In the experiments, an ultrasonic generator producing ultrasonic waves with a frequency of 20 kHz and a power of 80, 160, and 240 W was placed in a 1000 cm^3 beaker. As seen from the results in Fig. 13, the leaching recovery increased with the power of ultrasound.

Du et al. [78] studied the removal of iron from silica sand by ultrasound with an aqueous oxalic acid solution. An ultrasonic probe with a diameter of 1 cm, a frequency of 20 kHz, and a power of 0–300 W was used in the experiments. The iron removal processes were carried out at 90, 120, 150, and 180 W ultrasonic power levels, in 4 g/L oxalic acid solution. From the results of this study, it was seen that the ultrasonic application increased the iron removal efficiency, decreased the required acid concentration, and increased the leaching rate compared to conventional leaching.

The effect of ultrasound on the recovery of copper in tailings by ammonium leaching was investigated by Zhang et al. [79]. An ultrasonic generator with a power of 200 W and a frequency of 40 kHz was used in the experiments. Ultrasound was applied for 1 h in each 12 h. The solution was checked every 48 h and ammonia was added to compensate for the decreased amount. Their results indicated that 76% of the residual copper was extracted with 0.2936 g/L Cu after two weeks in the absence of ultrasound. On the other hand, in the ultrasound-assisted experiment, the extraction was 89.5% with 0.3467 g/L Cu.

Oncel et al. [80] investigated the possibility of recovering silver from cyanide leaching plant tailing by using ultrasound at 20 kHz and 0–200 W power. The results showed that almost all of the silver was recovered with the ultrasound-assisted thiourea leaching at 55 μm particle size, 100 g/L pulp density, 7.4 g/L thiourea, 28.7 g/L H_2SO_4 concentration, 24 min leaching time, and 77 °C reaction temperature. They reported the ultrasound energy absorbed by the leaching medium as 80 W/L.

Sayan and Bayramoglu [81] studied TiO_2 leaching with sulfuric acid under ultrasonic effect. In this study, different ultrasonic power, temperature, leaching time, acid concentration, and solid ratios were tested. It was found that the main factors affecting TiO_2 leaching were acid concentration and temperature; other factors showed little effect on the leaching results, and the TiO_2 recovery increased by 20% with the use of ultrasound.

In the study of Bese [82], copper recovery from slag with sulfuric acid-ferric sulfate mixture was investigated in the absence and presence of ultrasound. Under the optimum conditions of 65 °C reaction temperature, 0.2 M sulfuric acid concentration, 0.15 M ferric sulfate concentration, and 180 min reaction time the extraction efficiencies for copper, zinc, cobalt, and iron were 80.41%, 48.28%, 64.52%, and 12.16%, respectively, in the absence of ultrasound. On the other hand, the efficiencies were 89.28%, 51.32, 69.87%, and 13.7%, respectively in the presence of ultrasound.

Carbon cathodes used in aluminum smelting are generally contaminated with F^-, CN^-, and Na^+ ions. When disposing of these cathodes, their impurities must be removed within the scope of environmental regulations. Since the leaching of solid samples in standing water takes a very long time, various physical mixing methods can be used to speed up the process. Saterlay et al. [83] studied the leaching of comminuted carbon samples in the presence and absence of ultrasound. In the study, F^-, CN^-, and Na^+ ions were completely leached with the use of ultrasound for about 20 min. As a result of the experiments, it was determined that the recoveries were reached in 24 h by conventional leaching, and in less than 1 h with the ultrasonic method.

The ultrasound-assisted leaching of low-grade coal was investigated by several researchers. Barma et al. [84, 85] reported that ultrasound improved the leaching process by increasing the particle surface area and extractability of coal impurities. Balakrishnan et al. [86] investigated the ultrasonic leaching of alkali elements from coals. Their results indicated that while similar recovery efficiencies were obtained in the absence of chemicals, the recoveries improved significantly by ultrasound in the presence of ammonium acetate.

In addition to the possibility of adjusting the power of the energy in ultrasonic probes, there is also the option of giving the ultrasound continuously or intermittently in the pulse mode. Pinto and Soares [87] investigated the effect of pulse mode on the leaching of molybdenum. However, their results indicated that the final temperature in the leaching process reached 60 °C, which proved that working in the pulse mode was not sufficient to prevent the rise in the temperature.

The effect of the ultrasound on the process recoveries is directly related to the frequency of the ultrasonic waves. However, since there is no commercial ultrasonic device that allows the testing of various ultrasonic frequencies, it is impossible to perform a frequency optimization. Therefore, multi-frequency ultrasonic transducers can be used to overcome this difficulty, and thus it can be possible to save energy and processing time.

One of the studies in which ultrasonic waves of different frequencies are used together is the study of Swamy and Narayana [88] in which they investigated the effect of ultrasound on copper leaching. In this study, leaching was carried out by designing an experimental setup using dual-frequency ultrasound (20 and 40 kHz). In this study, the effect of ultrasound application at various frequencies (20, 40, 43, and 720 kHz) and intensities (up to 8 W/cm^2) on leaching results was investigated. Figure 14a shows the effect of the ultrasonic irradiation time on copper recovery at various frequencies when the ultrasonic intensity was kept constant at 2 W/cm^2. The metal recovery increased up to approximately 20 min at all frequencies and decreased as the time increased. This indicated that high recoveries could be reached in shorter times. In addition, ultrasound at low frequencies appears to be more effective than at higher frequencies. With the use of 2 W/cm^2 ultrasonic intensity and single-frequency ultrasound at 20 and 40 kHz frequencies for 20 min, 51.5% and 62.5% efficiencies were achieved. When these two frequencies were used together, Cu extraction increased to 92%. Figure 14b shows the effect of ultrasonic intensity on metal recovery for different frequencies when the application time was kept constant at 20 min. In the experiments using a single frequency, the Cu recovery increased with the ultrasonic intensity and reached a maximum of 3 W/cm^2. As the intensity continued to increase, the recovery started to decrease. In experiments using double frequency, better results were obtained compared to both single-frequency experiments.

As well as probes, ultrasonic baths have been used as ultrasound sources in leaching processes. One of these studies was carried out by Xue et al. [89] for nickel ore. As seen from the results of this study in Fig. 15, the leaching under ultrasonic

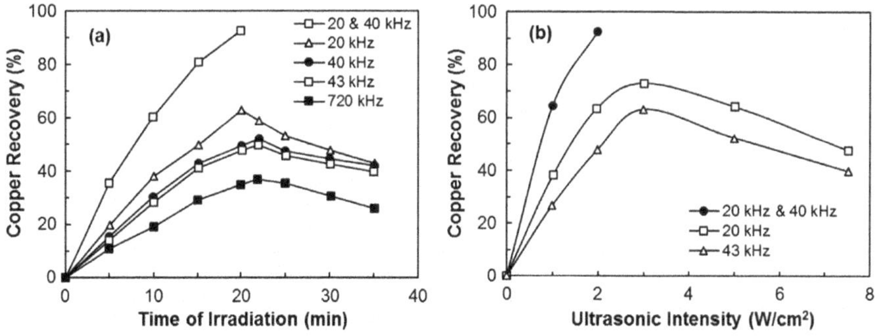

Fig. 14 **a** The effect of ultrasonic frequencies at a constant intensity of 2 W/cm^2 on the copper recovery over time, **b** the effect of single and dual frequencies on Cu recovery as a function of ultrasonic intensity at 20 min sonication time. Reprinted from Swamy and Narayana [88], with permission from Elsevier. Copyright (2001)

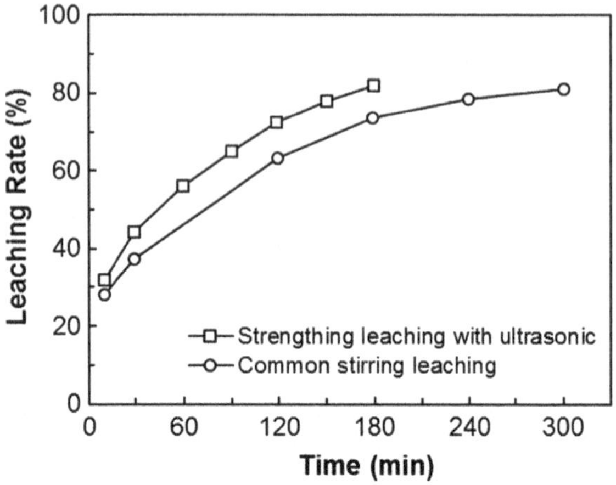

Fig. 15 Comparison of conventional and ultrasonic leaching for nickel ore. Reprinted from Xue et al. [89], with permission from Elsevier. Copyright (2010)

application was more effective than conventional leaching. The use of ultrasound for 30 min increased the leaching rate by 6.7%. It is also seen that while the leaching rate of 83% was reached in 150 min with ultrasound-assisted leaching, the same value was reached in 300 min with conventional leaching.

In another study on nickel, Cetintas and Bingol [90] stated that ultrasound contributed positively to the nickel leaching from nickel laterite in the presence of Na_2SO_4 and Na_2CO_3 by accelerating the reaction and facilitating the mass transfer.

The extraction of metals from sulfides and/or iron-containing ores using microorganisms is called bioleaching. Swamy et al. [91, 92] investigated the supporting bioleaching processes with ultrasound using low-grade nickel ore. In these studies, the bioleaching process combined with ultrasound was called sonobioleaching. According to their studies, the sonobioleaching process increased the advantages and reduced or eliminated the disadvantages of the bioleaching process. Ultrasound application at low power helped the growth and reproduction of microorganisms. Thus, it reduced the leaching time as well as increased the extraction rate, recovery, and selectivity. Furthermore, the leaching process could be controlled by adjusting the intensity of ultrasound. The number of microorganisms could be increased by using ultrasound under convenient conditions by supporting binary fission. Meanwhile, some or all of the microorganisms could be eliminated by giving ultrasound above a limit value. Therefore, the extraction of undesired chemicals could be controlled.

5.2 *Ultrasound in Electrolysis*

In the last decades, an interdisciplinary method called sonoelectrochemistry emerged by combining electrochemistry and sonochemistry. In this research field, Zhang et al. [93, 94] investigated the enhanced desulfurizing flotation of high sulfur coal

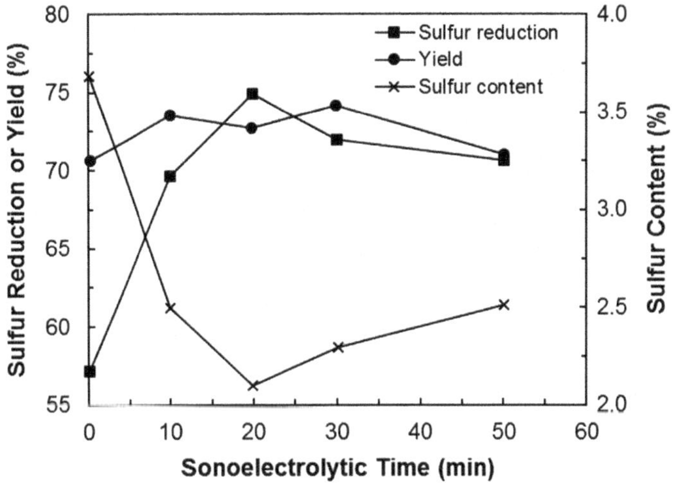

Fig. 16 Effect of sonoelectrolysis time on sulfur removal, clean coal yield, and sulfur content. Reprinted from Zhang et al. [94], with permission from Elsevier. Copyright (2012)

(−0.076 mm) with graphite as an anode and stainless steel as a cathode using ultrasound at 59 kHz frequency and 250 W power. Their results for sulfur reduction, clean coal yield, and sulfur content are given in Fig. 16 as a function of sonoelectrolytic time. As seen in Fig. 16, as the time increases the reduction of sulfur increases abruptly and reaches a maximum at 20 min, then decreases slowly. At the beginning of the sonoelectrolysis period, sulfur-containing groups are adsorbed on the coal surface. In time, the reaction progresses from the coal particles on the surface to the coal particles in the inner parts. At this point, the removal of sulfur becomes more difficult and therefore slows down [94].

Li et al. [95] support nickel sulfate leaching in sulfuric acid and hydrogen peroxide medium with ultrasound. Their results showed that the oxidation film on the nickel surface was peeled off and the reaction area was increased by ultrasound. As a result, while 46.29% nickel leaching rate was obtained in 5 h by conventional leaching, it increased to 60.41% under the optimum conditions of 30% sulfuric acid and 10% hydrogen peroxide concentration, 333 K temperature, 200 W ultrasonic power, and 4 h leaching time. The effect of ultrasound on the leaching rate of nickel is seen in Fig. 17 as a function of leaching time.

The effect of ultrasound (20 kHz) on vanadium leaching recovery from roasted vanadium-bearing shale was investigated by Chen et al. [96]. As seen from their results in Fig. 18a, the leaching recovery of vanadium increased considerably with ultrasonic power. Moreover, Fig. 18b indicated that ultrasound also decreased the leaching time from 240 to 30 min.

John et al. [97] investigated the potential of ultrasound in the extraction of lead from a landfilled metallurgical residue in an acidic saline leaching process. They reported that ultrasound also decreased the leaching time from 240 to 30 min and the

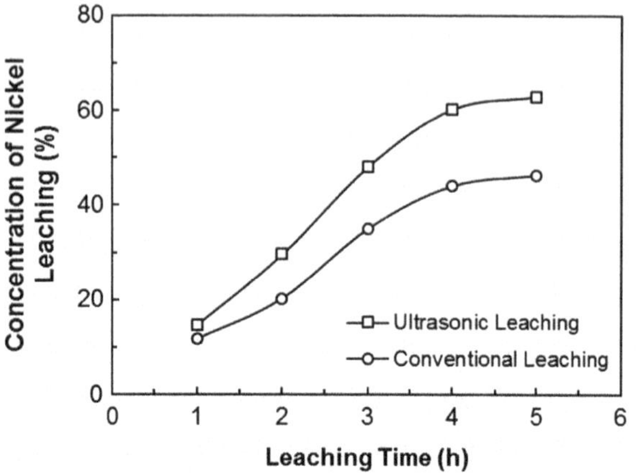

Fig. 17 The effect of ultrasound on the leaching rate of nickel with respect to leaching time. Reprinted from Li et al. [95], with permission from Elsevier. Copyright (2018)

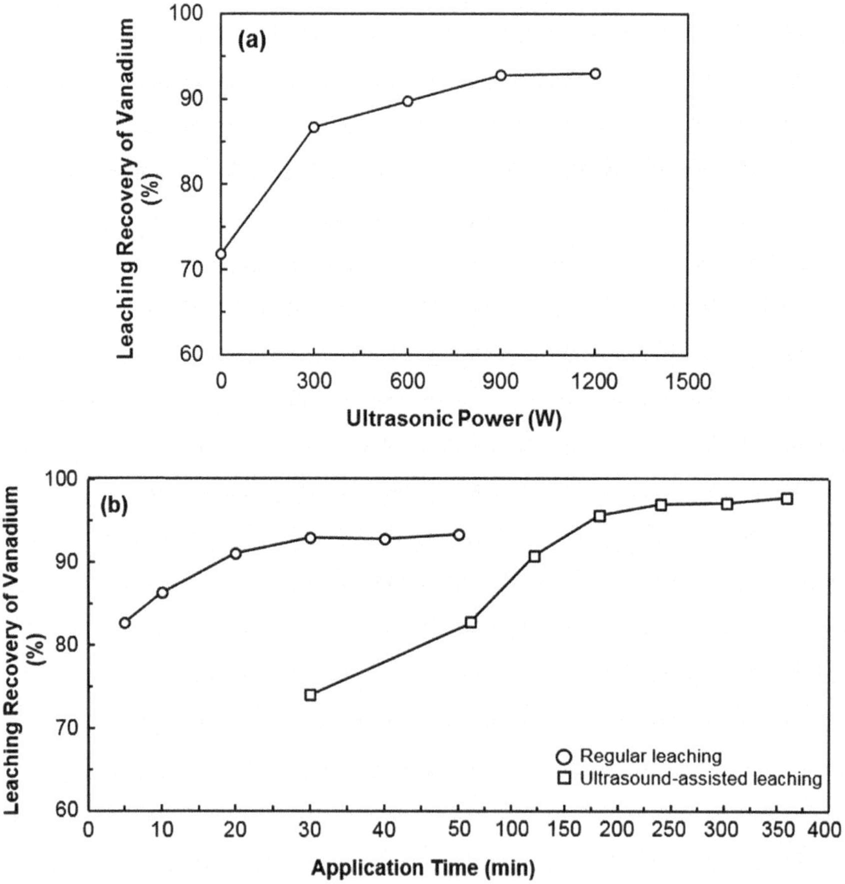

Fig. 18 Effect of **a** ultrasonic power, and **b** leaching time on vanadium leaching recovery. Reprinted from Chen et al. [96], with permission from Elsevier. Copyright (2020)

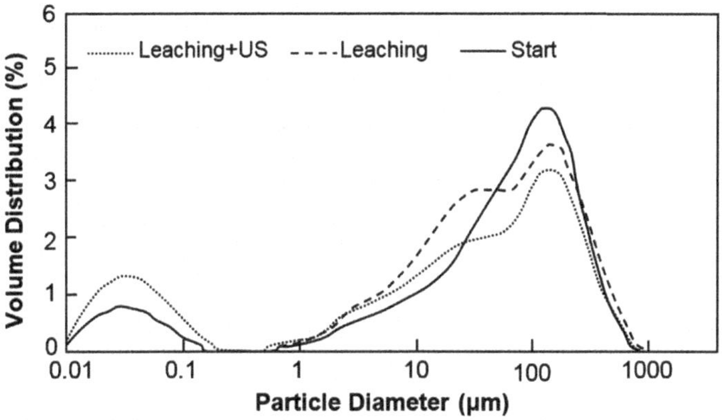

Fig. 19 Volume distribution of the particle size of the 2 mm material (Start) and the material after 2 h of leaching without (Leaching) and with ultrasonic assistance (Leaching + US). Reprinted from John et al. [97], with permission from Elsevier. Copyright (2020)

yield of the process was improved by 19–26%. They stated that one of the reasons for this positive effect was that ultrasound reduced the particle size (Fig. 19), thus creating fresh surfaces, increasing mass transfer, and therefore increasing the solubility of the lead. Furthermore, ultrasound also decreased the dissolution of calcium and therefore improved the selectivity of lead in the process.

Wang et al. [98] used the ultrasound for the copper leaching from chalcopyrite in acidic ferric sulfate media. They reported that copper leaching was enhanced with the ultrasonic power, which was the second strongest parameter after the temperature. They also reported that ultrasound enhanced the leaching rate, decreased the reaction time, and reduced the acid consumption.

In the case of the decopperization from the anode slime, Wang et al. [99] reported that the removal efficiency increased considerably and the decopperization time was significantly shortened in ultrasonic conditions compared with the conventional method (>10%, 24 h).

Yu et al. [100] developed an ultrasound-assisted reactor for cyanide extraction of gold from ore at low temperatures. Ultrasound-assisted extraction increased the extraction rate of gold by 0.6–0.8% in the leaching of gold concentrate. They mention that ultrasound also reduced the NaCN consumption. In another study, on gold leaching, 8 times faster leaching rate and ~25% higher extraction were obtained by Gui et al. [101] with ultrasound compared to conventional conditions.

Ultrasound-assisted leaching of potassium from phosphorus-potassium-associated ore was studied by Zhang et al. [102]. Their results indicated that the dissolution fraction of potassium could be significantly improved with ultrasound and under the optimum reaction conditions; a dissolution fraction of 94% was obtained. They attributed the positive effect of ultrasound to the intra-particle diffusion, increase in the mass transfer rate, the disintegration of large ore particles to small ones, exposing a large surface area, and enabling the easy penetration of the leaching acid into ore particles.

Yin et al. [103] studied the leaching of rare earth ores with $MgSO_4$ instead of $(NH_4)_2SO_4$. They reported that the leaching efficiency was intensified by ultrasound. They obtained 92.3% leaching efficiency at the conditions of 700 W ultrasonic power, 30 min leaching time, and 25 °C temperature.

In the investigation of the ultrasound-assisted HCl-NaCl leaching of Pb-rich and Sb-rich oxidizing slag. Zhang et al. [104] reported that ultrasound treatment elevated the leaching rates of Sb and Pb, and shortened the leaching time significantly.

Johansson et al. [105] attempt to increase the recovery rate and reduce the energy and temperature required for the leaching of tungsten using hydrodynamic and acoustic cavitation (HAC) compared to the conventional autoclave leaching process. They reported that the leaching recovery increased from 36.7% (conventional stirring) to 71.5% in 6 h exposure time and at 80 °C temperature in the presence of HAC.

6 Conclusions and Future Perspectives

As an inaudible sound wave for humans, ultrasound generates unusual conditions in the liquid medium such as acoustic streaming and cavitation. In the presence of cavitation, the local pressure and temperature rise to extremely high values for a very short time. Researchers carried out various investigations to benefit from these features of ultrasound to promote the results of the industrial processes. This chapter presents a summary of the applications of ultrasound on the comminution of the minerals in addition to the physical and chemical beneficiation techniques.

It is seen from the results of this study that ultrasound eases the comminution processes and decreases the average particle size by abrading the surfaces, creating weak spots on the particles in the form of micro-fractures, and increasing the grinding media/particle contact related to several factors including the type, size, and shape of the mineral particle. In the case of physical mineral processing operations, ultrasound provides slime removal from the surfaces, and thus enhances the washing and mechanical cleaning processes. Ultrasound also improves solid/liquid separation by reducing the flocculant dosage and process time in addition to making the sediment more compact. Moreover, ultrasound provides a faster, cheaper, and more efficient leaching process via increasing the total surface area by abrasion and comminution by creating crack formations in the particles. This provides the penetration of the leaching agent to the inner parts of the particles by capillary action, and thus the reagent consumption reduces considerably. Meanwhile, the ultrasonic application conditions such as device, frequency, time, power, and intensity have a crucial role in the process improvement as well as the properties of the medium including purity, viscosity, composition, and temperature.

Therefore, it can be evaluated as a result of this study that ultrasound can be beneficial for many applications of mineral processing to improve efficiencies and the quality of the products. Meanwhile, there is still a need for further studies to improve the benefits of ultrasound. In this context, multidisciplinary studies should be carried out with related research fields such as physics, chemistry, and metallurgy for the modeling of energy consumption for larger capacities, to restrain the noise level created by ultrasound, as well as to produce abrasion-resistant materials to use in the cavitation environment.

Finally, it should be noted that the studies on the use of ultrasound in mineral processing operations are generally at a laboratory or pilot scale. To apply ultrasound in commercial applications at high capacities, the scale-up parameters should be determined with detailed investigations.

Acknowledgements This study was supported by the Research Fund of Istanbul University-Cerrahpasa with Project numbers 34626 and FAB-2017-25658.

References

1. G. Cloutier, F. Destrempes, F. Yu, A. Tang, Quantitative ultrasound imaging of soft biological tissues: a primer for radiologists and medical physicists. Insights Imaging **12**(1), 127–147 (2021). https://doi.org/10.1186/s13244-021-01071-w
2. O. Falou, Ph.D. thesis, Ryerson University, 2011
3. T.G. Leighton, What is ultrasound? Prog. Biophys. Mol. Biol. **93**(1–3), 3–83 (2007). https://doi.org/10.1016/j.pbiomolbio.2006.07.026
4. G. Musielak, D. Mierzwa, J. Kroehnke, Food drying enhancement by ultrasound—a review. Food Sci. Technol. **56**, 126–141 (2016). https://doi.org/10.1016/j.tifs.2016.08.003
5. P.R. Gogate, Cavitation: an auxiliary technique in wastewater treatment schemes. Adv. Environ. Res. **6**, 335–358 (2002). https://doi.org/10.1016/S1093-0191(01)00067-3
6. R.B. Baig, R.S. Varma, Alternative energy input: mechanochemical, microwave and ultrasound-assisted organic synthesis. Chem. Soc. Rev. **41**(4), 1559–1584 (2012). https://doi.org/10.1039/c1cs15204a
7. F.V. de Andrade, R. Augusti, G.M. de Lima, Ultrasound for the remediation of contaminated waters with persistent organic pollutants: a short review. Ultrason. Sonochem. **78**, 105719 (2021). https://doi.org/10.1016/j.ultsonch.2021.105719
8. Y. Zhang, N. Abatzoglou, Review: fundamentals, applications and potentials of ultrasound-assisted drying. Chem. Eng. Res. Des. **154**, 21–46 (2020). https://doi.org/10.1016/j.cherd.2019.11.025
9. Y. Yao, Y. Pan, S. Liu, Power ultrasound and its applications: a state-of-the-art review. Ultrason. Sonochem. **62**, 104722 (2020). https://doi.org/10.1016/j.ultsonch.2019.104722
10. S.G. Ozkan, Further investigations on simultaneous ultrasonic coal flotation. Minerals **7**(10), 1–9 (2017). https://doi.org/10.3390/min7100177
11. S.G. Ozkan, A review of simultaneous ultrasound-assisted coal flotation. JME (2018). https://doi.org/10.22044/jme.2018.6784.1502
12. S.D. Barma, Ultrasonic-assisted coal beneficiation: a review. Ultrason. Sonochem. **50**, 15–35 (2019). https://doi.org/10.1016/j.ultsonch.2018.08.016
13. Y. Mao, W. Xia, Y. Peng, G. Xie, Ultrasonic-assisted flotation of fine coal: a review. Fuel Process. Technol. **195**, 106150 (2019). https://doi.org/10.1016/j.fuproc.2019.106150
14. Y. Mao, Y. Chen, X. Bu, G. Xie, Effects of 20 kHz ultrasound on coal flotation: the roles of cavitation and acoustic radiation force. Fuel **256**, 115938 (2019). https://doi.org/10.1016/j.fuel.2019.115938

15. H. Ebrahimi, M. Karamoozian, Effect of ultrasonic irradiation on particle size, reagents consumption, and feed ash content in coal flotation. Int. J. Coal Sci. Technol. **7**(4), 787–795 (2020). https://doi.org/10.1007/s40789-020-00307-2
16. W. Lv, J. Chen, F. Min, B. Hou, C. Liu, Effect of ultrasonic pre-treatment on coal slime flotation. Physicochem. Probl. Miner. Process. **56**(1), 173–183 (2020). https://doi.org/10.5277/ppmp19095
17. Y. Mao, X. Bu, Y. Peng, F. Tian, G. Xie, Effects of simultaneous ultrasonic treatment on the separation selectivity and flotation kinetics of high-ash lignite. Fuel **259**, 1–10 (2020). https://doi.org/10.1016/j.fuel.2019.116270
18. Y. Mao, G. Xie, X. Qi, Y. Peng, Effects of ultrasonic pretreatment on particle size and surface topography of lignite and its relationship to flotation response. Energy Sources A: Recovery Util. Environ. Eff. **43**(10), 1274–1282 (2021). https://doi.org/10.1080/15567036.2019.1638996
19. M. Kruszelnicki, A. Hassanzadeh, K.J. Legawiec, I. Polowczyk, P.B. Kowalczuk, Effect of ultrasound pre-treatment on carbonaceous copper-bearing shale flotation. Ultrason. Sonochem. **84**, 105962 (2022). https://doi.org/10.1016/j.ultsonch.2022.105962
20. S.D. Barma, P.K. Baskey, D.S. Rao, S.N. Sahu, Ultrasonic-assisted flotation for enhancing the recovery of flaky graphite from low-grade graphite ore. Ultrason. Sonochem. **56**, 386–396 (2019). https://doi.org/10.1016/j.ultsonch.2019.04.033
21. W. Kang, H. Li, Enhancement of flaky graphite cleaning by ultrasonic treatment. R. Soc. Open Sci. **6**(12), 191160 (2019). https://doi.org/10.1098/rsos.191160
22. C. Li, X. Li, M. Xu, H. Zhang, Effect of ultrasonication on the flotation of fine graphite particles: nanobubbles or not? Ultrason. Sonochem. **69**, 105243 (2020). https://doi.org/10.1016/j.ultsonch.2020.105243
23. U. Malayoglu, S.G. Ozkan, Effects of ultrasound on desliming prior to feldspar flotation. Minerals **9**(12), 784 (2019). https://doi.org/10.3390/min9120784
24. C. Gungoren, O. Ozdemir, S.G. Ozkan, Effects of temperature during ultrasonic conditioning in quartz-amine flotation. Physicochem. Probl. Miner. Process. **53**(2), 687–698 (2017). https://doi.org/10.5277/ppmp170201
25. C. Gungoren, O. Ozdemir, X. Wang, S.G. Ozkan, J.D. Miller, Effect of ultrasound on bubble-particle interaction in quartz-amine flotation system. Ultrason. Sonochem. **52**, 446–454 (2019). https://doi.org/10.1016/j.ultsonch.2018.12.023
26. A. Hassanzadeh, H. Gholami, S.G. Ozkan, T. Niedoba, A. Surowiak, Effect of power ultrasound on wettability and collector-less floatability of chalcopyrite, pyrite and quartz. Minerals **11**(1) (2021). https://doi.org/10.3390/min11010048
27. F. Hassani, N. Mohammad, S. Tonkaboni, S. Ziaedin, Enhancement of phosphate flotation by ultrasonic pretreatment. Iran J. Chem. Chem. Eng. **38**(5) (2019)
28. F. Hassani, M. Noaparast, S.Z. Shafaei Tonkaboni, A study on the effect of ultrasound irradiation as pretreatment method on flotation of sedimentary phosphate rock with carbonate–silicate gangue. Iran J. Sci. Technol. Trans. A: Sci. 2787–2798 (2019). https://doi.org/10.1007/s40995-019-00765-7
29. L.O. Filippov, I.V. Filippova, O. Barres, T.P. Lyubimova, O.O. Fattalov, Intensification of the flotation separation of potash ore using ultrasound treatment. Miner. Eng. **171** (2021). https://doi.org/10.1016/j.mineng.2021.107092
30. K. Shu, L. Xu, H. Wu, S. Fang, Z. Wang, Y. Xu, Z. Zhang, Effects of ultrasonic pre-treatment on the flotation of ilmenite and collector adsorption. Miner. Eng. **137**, 124–132 (2019). https://doi.org/10.1016/j.mineng.2019.04.001
31. H. Wu, S. Fang, K. Shu, Y. Xu, Z. Wang, L. Luo, J. Yang, L. Xu, Selective flotation and adsorption of ilmenite from titanaugite by a novel method: ultrasonic treatment. Powder Technol. **363**, 38–47 (2020). https://doi.org/10.1016/j.powtec.2019.12.046
32. U. Ulusoy, H. Kursun, Review: comparison of ultrasonically aided zinc beneficiation by mechanical flotation and column flotation cell. EUREKA Phys. Eng. **1**, 3–13 (2021). https://doi.org/10.21303/2461-4262.2021.001608

33. A. Videla, D. Faúndez, J. Meneses, L. Gaete, Y. Vargas, Enhancement of the sedimentation rate of copper tailings by application of acoustic fields. Miner. Eng. **146** (2020). https://doi.org/10.1016/j.mineng.2019.106096
34. U. Horasan, M. Tanrıverdi, T. Cicek, M. Polat, Investigating the effects of ultrasonic energy on the flotation behavior of pyrite and galena minerals. Physicochem. Probl. Miner. Process. **56**(3), 538–547 (2020). https://doi.org/10.37190/ppmp/120291
35. D. Lu, L. Chen, Y. Ma, X. Zheng, Y. Wang, Effects of ultrasonic pretreatment on the flotation behavior of galena with and without the presence of pyrite. Physicochem. Probl. Miner. Process. **56**(4), 611–624 (2020). https://doi.org/10.37190/ppmp/123406
36. C. Gungoren, Y. Baktarhan, I. Demir, S.G. Ozkan, Enhancement of galena−potassium ethyl xanthate flotation system by low power ultrasound. Trans. Nonferrous Met. Soc. China **30**(4), 1102–1110 (2020). https://doi.org/10.1016/S1003-6326(20)65281-5
37. X. Bu, M. Alheshibri, The effect of ultrasound on bulk and surface nanobubbles: a review of the current status. Ultrason. Sonochem. **76**, 105629 (2021). https://doi.org/10.1016/j.ultsonch.2021.105629
38. J.L. Luque-Garcia, M.D. Luque de Castro, Ultrasound: a powerful tool for leaching. Trends Analyt. Chem. **22**(1), 41–47 (2003). https://doi.org/10.1016/s0165-9936(03)00102-x
39. C. Gungoren, T.M. Erbek, O. Ozdemir, S.G. Ozkan, Effect of simultaneous ultrasonic treatment on Quartz-Amine flotation system, in *Proceedings of the XVI Balkan Mineral Processing Congress*, vol. 1, no. 1, Belgrade, Serbia And Montenegro (2015), pp. 483–490
40. B. Ambedkar, Ph.D. thesis, Indian Institute of Technology Madras, Chennai, India, 2012
41. T.J. Mason, Ultrasound in synthetic organic chemistry. Chem. Soc. Rev. **26**(6), 443–451 (1997). https://doi.org/10.1039/cs9972600443
42. K.S. Suslick, The chemical effects of ultrasound. Sci. Am. 80–86 (1989)
43. K.S. Suslick, The chemistry of ultrasound, in *The Yearbook of Science & the Future* (Encyclopaedia Britannica, Chicago, 1994), pp. 138–155
44. M. Breitbach, D. Bathen, H. Schmidt-Traub, Effect of ultrasound on adsorption and desorption processes. Ind. Eng. Chem. Res. **42**, 5635–5646 (2003). https://doi.org/10.1021/ie030333f
45. M. Ja'fari, S.L. Ebrahimi, M.R. Khosravi-Nikou, Ultrasound-assisted oxidative desulfurization and denitrogenation of liquid hydrocarbon fuels: a critical review. Ultrason. Sonochem. **40**(Pt A), 955–968 (2018). https://doi.org/10.1016/j.ultsonch.2017.09.002
46. L. Raylight, VIII. On the pressure developed in a liquid during the collapse of a spherical cavity. Philos. Mag. Ser. 6 **34**(200), 94–98 (1917). https://doi.org/10.1080/14786440808635681
47. K.S. Suslick, Y. Didenko, M.M. Fang, T. Hyeon, K.J. Kolbeck, W.B. McNamara III., M.M. Mdleleni, M. Wong, Acoustic cavitation and its chemical consequences. Philos. Trans. R. Soc. Lond. A **357**, 335–353 (1999). https://doi.org/10.1098/rsta.1999.0330
48. A.F. Collings, A.D. Farmer, P.B. Gwan, A.P.S. Pintos, C.J. Leo, Processing contaminated soils and sediments by high power ultrasound. Miner. Eng. **19**(5), 450–453 (2006). https://doi.org/10.1016/j.mineng.2005.07.014
49. U. Orthaber, J. Zevnik, R. Petkovsek, M. Dular, Cavitation bubble collapse in a vicinity of a liquid-liquid interface—basic research into emulsification process. Ultrason. Sonochem. **68**, 105224 (2020). https://doi.org/10.1016/j.ultsonch.2020.105224
50. H. Frenzel, H. Schultes, Luminescenz im ultraschallbeschickten Wasser. Z. Phys. Chem. **27B**(1), 421–424 (1934). https://doi.org/10.1515/zpch-1934-2737
51. L.F. Gaete-Garreton, Y.P. Vargas-Hermandez, C. Velasquez-Lambert, Application of ultrasound in comminution. Ultrasonics **38**, 345–352 (2000). https://doi.org/10.1016/S0041-624X(99)00170-5
52. L. Gaete-Garreton, The use of power ultrasound in mining, in *Power Ultrasonics* (2015), pp. 1059–1094
53. G.F. Gao, B. Zhao, D.H. Xiang, Q.H. Kong, Research on the surface characteristics in ultrasonic grinding nano-zirconia ceramics. J. Mater. Process. Technol. **209**(1), 32–37 (2009). https://doi.org/10.1016/j.jmatprotec.2008.01.061

54. V. Singh, R. Venugopal, V.K. Saxena, A.K. Mukherjee, Study of the effect of ultrasonic treatment on mineral materials of different morphologies. Miner. Metall. Process **33**(2), 88–96 (2016). https://doi.org/10.19150/mmp.6626
55. E. Sahinoglu, T. Uslu, Effects of various parameters on ultrasonic comminution of coal in water media. Fuel Process. Technol. **137**, 48–54 (2015). https://doi.org/10.1016/j.fuproc.2015.03.028
56. H. Li, M. Rostamizadeh, K. Mameri, D.C. Boffito, N. Saadatkhah, M.G. Rigamonti, G.S. Patience, Ultrasound assisted wet stirred media mill of high concentration $LiFePO_4$ and catalysts. Can. J. Chem. Eng. **97**(8), 2242–2250 (2018). https://doi.org/10.1002/cjce.23212
57. B. Ambedkar, T.N. Chintala, R. Nagarajan, S. Jayanti, Feasibility of using ultrasound-assisted process for sulfur and ash removal from coal. Chem. Eng. Process. **50**(3), 236–246 (2011). https://doi.org/10.1016/j.cep.2011.02.008
58. J.H. Gibson, H. Hon, R. Farnood, I.G. Droppo, P. Seto, Effects of ultrasound on suspended particles in municipal wastewater. Water Res. **43**(8), 2251–2259 (2009). https://doi.org/10.1016/j.watres.2009.02.024
59. W. Kang, H. Xun, J. Hu, Study of the effect of ultrasonic treatment on the surface composition and the flotation performance of high-sulfur coal. Fuel Process. Technol. **89**(12), 1337–1344 (2008). https://doi.org/10.1016/j.fuproc.2008.06.003
60. V. Raman, A. Abbas, Experimental investigations on ultrasound mediated particle breakage. Ultrason. Sonochem. **15**(1), 55–64 (2008). https://doi.org/10.1016/j.ultsonch.2006.11.009
61. V.M. Fridman, The interaction mechanism between cavitation bubbles and particles of the solid and liquid phases. Ultrasonics 162–165 (1972). https://doi.org/10.1016/0041-624X(72)90357-5
62. I. Ahmad, E.U. Hartge, J. Werther, R. Wischnewski, Ultrasonic processing of bauxite ore to estimate its washing potential. J. Min. Sci. **52**(1), 201–206 (2016). https://doi.org/10.1134/s1062739116010321
63. E. Sonmez, S. Koca, H. Ozdag, H. Ipek, Beneficiation of colemanite concentrates from fine wastes by using ultrasound waves. Miner. Eng. **17**(2), 359–361 (2004). https://doi.org/10.1016/j.mineng.2003.11.008
64. I. Alp, H. Deveci, H. Ozdag, Processing tincal ores using ultrasonic waves. Paper presented at the II international Boron symposium, Eskisehir, Turkey, 23–25 September 2004
65. H. Kyllonen, P. Pirkonen, V. Hintikka, P. Parvinen, A. Gronroos, H. Sekki, Ultrasonically aided mineral processing technique for remediation of soil contaminated by heavy metals. Ultrason. Sonochem. **11**(3–4), 211–216 (2004). https://doi.org/10.1016/j.ultsonch.2004.01.024
66. A. Sadatshojaie, D.A. Wood, S.M. Jokar, M.R. Rahimpour, Applying ultrasonic fields to separate water contained in medium-gravity crude oil emulsions and determining crude oil adhesion coefficients. Ultrason. Sonochem. **70**, 105303 (2021). https://doi.org/10.1016/j.ultsonch.2020.105303
67. G. Onal, M. Ozer, F. Arslan, Sedimentation of clay in ultrasonic medium. Miner. Eng. **16**(2), 129–134 (2003). https://doi.org/10.1016/s0892-6875(02)00309-6
68. E. Riera-Franco de Sarabia, J.A. Gallego-Juarez, G. Rodriguez-Corral, L. Elvira-Segura, I. Gonzalez-Gomez, Application of high-power ultrasound to enhance fluid/solid particle separation processes. Ultrasonics **38**, 642–646 (2000). https://doi.org/10.1016/S0041-624X(99)00129-8
69. B.P. Singh, Technical note-ultrasonically assisted rapid solid-liquid separation of fine clean coal particles. Miner. Eng. **12**(4) (1999). https://doi.org/10.1016/S0892-6875(99)00024-2
70. L. Zhu, W. Lyu, P. Yang, Z. Wang, Effect of ultrasound on the flocculation-sedimentation and thickening of unclassified tailings. Ultrason. Sonochem. 104984 (2020). https://doi.org/10.1016/j.ultsonch.2020.104984
71. I. Demir, C. Gungoren, Y. Baktarhan, M. Yucel, I. Kursun, K. Cinku, S.G. Ozkan, Ultrasound supported flocculation of borate tailings with differently charged flocculants. Boron **6**(3), 348–358 (2021). https://doi.org/10.30728/boron.971892
72. A. Ozkan, K. Esmeli, Improvement of colemanite flocculation with collectors by ultrasound treatment. Part. Sci. Technol. 1–9 (2022). https://doi.org/10.1080/02726351.2021.1937415

73. F. Burat, A.A. Sirkeci, G. Onal, Improved fine coal dewatering by ultrasonic pretreatment and dewatering aids. Min. Proc. Ext. Met. Rev. **36**(2), 129–135 (2014). https://doi.org/10.1080/08827508.2014.898637
74. L.A. Davies, A. Dargue, J.R. Dean, M.E. Deary, Use of 24 kHz ultrasound to improve sulfate precipitation from wastewater. Ultrason. Sonochem. **23**, 424–431 (2015). https://doi.org/10.1016/j.ultsonch.2014.08.017
75. Y. Gursoy, H. Kurama, Ultrasonic treatment and its applicability for the selective treatment of borax clayey waste sludge. Physicochem. Probl. Miner. Process. **57**(5), 80–90 (2021). https://doi.org/10.37190/ppmp/141225
76. T.J. Mason, A. Collings, A. Sumel, Sonic and ultrasonic removal of chemical contaminants from soil in the laboratory and on a large scale. Ultrason. Sonochem. **11**(3–4), 205–210 (2004). https://doi.org/10.1016/j.ultsonch.2004.01.025
77. X. Wang, C. Srinivasakannan, X. Duan, J. Peng, D. Yang, S. Ju, Leaching kinetics of zinc residues augmented with ultrasound. Sep. Purif. Technol. **115**, 66–72 (2013). https://doi.org/10.1016/j.seppur.2013.04.043
78. F. Du, J. Li, X. Li, Z. Zhang, Improvement of iron removal from silica sand using ultrasound-assisted oxalic acid. Ultrason. Sonochem. **18**(1), 389–393 (2011). https://doi.org/10.1016/j.ultsonch.2010.07.006
79. J. Zhang, A. Wu, Y. Wang, X. Chen, Experimental research in leaching of copper-bearing tailings enhanced by ultrasonic treatment. J. China Univ. Min. Technol. **18**(1), 98–102 (2008). https://doi.org/10.1016/s1006-1266(08)60021-8
80. M.S. Oncel, M. Ince, M. Bayramoglu, Leaching of silver from solid waste using ultrasound assisted thiourea method. Ultrason. Sonochem. **12**(3), 237–242 (2005). https://doi.org/10.1016/j.ultsonch.2003.10.007
81. E. Sayan, M. Bayramoglu, Statistical modeling and optimization of ultrasound-assisted sulfuric acid leaching of TiO_2 from red mud. Hydrometallurgy **71**(3–4), 397–401 (2004). https://doi.org/10.1016/s0304-386x(03)00113-0
82. A.V. Bese, Effect of ultrasound on the dissolution of copper from copper converter slag by acid leaching. Ultrason. Sonochem. **14**(6), 790–796 (2007). https://doi.org/10.1016/j.ultsonch.2007.01.007
83. A.J. Saterlay, Q. Hong, R.G. Compton, J. Clarkson, Ultrasonically enhanced leaching: removal and destruction of cyanide and other ions from used carbon cathodes. Ultrason. Sonochem. **7**, 1–6 (2000). https://doi.org/10.1016/S1350-4177(99)00025-5
84. S.D. Barma, R. Sathish, P.K. Baskey, Ultrasonic-assisted cleaning of Indian low-grade coal for clean and sustainable energy. J. Clean. Prod. **195**, 1203–1213 (2018). https://doi.org/10.1016/j.jclepro.2018.06.030
85. S.D. Barma, P.K. Baskey, S.K. Biswal, Chemical beneficiation of high-ash Indian noncoking coal by alkali leaching under low-frequency ultrasonication. Energy Fuels **32**(2), 1309–1319 (2018). https://doi.org/10.1021/acs.energyfuels.7b03291
86. S. Balakrishnan, V.M. Reddy, R. Nagarajan, Ultrasonic coal washing to leach alkali elements from coals. Ultrason. Sonochem. **27**, 235–240 (2015). https://doi.org/10.1016/j.ultsonch.2015.05.014
87. I.S.S. Pinto, H.M.V.M. Soares, Selective leaching of molybdenum from spent hydrodesulphurisation catalysts using ultrasound and microwave methods. Hydrometallurgy **129–130**, 19–25 (2012). https://doi.org/10.1016/j.hydromet.2012.08.008
88. K.M. Swamy, K.L. Narayana, Intensification of leaching process by dual-frequency ultrasound. Ultrason. Sonochem. **8**, 341–346 (2001)
89. J. Xue, X. Lu, Y. Du, W. Mao, Y. Wang, J. Li, Ultrasonic-assisted oxidation leaching of nickel sulfide concentrate. Chin. J. Chem. Eng. **18**(6), 948–953 (2010). https://doi.org/10.1016/s1004-9541(09)60152-x
90. S. Cetintas, D. Bingol, Performance evaluation of leaching processes with and without ultrasound effect combined with reagent-assisted mechanochemical process for nickel recovery from laterite: process optimization and kinetic evaluation. Miner. Eng. **157**, 106562 (2020). https://doi.org/10.1016/j.mineng.2020.106562

91. K.M. Swamy, L.B. Sukla, K.L. Narayana, R.N. Kar, V.V. Panchanadikar, Use of ultrasound in microbial leaching of nickel from laterites. Ultrason. Sonochem. **2**(1), S5–S9 (1995)
92. K.M. Swamy, K.L. Narayana, V.N. Misra, Bioleaching with ultrasound. Ultrason. Sonochem. **12**(4), 301–306 (2005). https://doi.org/10.1016/j.ultsonch.2004.01.035
93. H. Zhang, H. Bai, X. Dong, Z. Wang, Enhanced desulfurizing flotation of different size fractions of high sulfur coal using sonoelectrochemical method. Fuel Process. Technol. **97**, 9–14 (2012). https://doi.org/10.1016/j.fuproc.2012.01.005
94. H. Zhang, X. Ma, X. Dong, Z. Wang, H. Bai, Enhanced desulfurizing flotation of high sulfur coal by sonoelectrochemical method. Fuel Process. Technol. **93**(1), 13–17 (2012). https://doi.org/10.1016/j.fuproc.2011.09.007
95. H. Li, S. Li, J. Peng, C. Srinivasakannan, L. Zhang, S. Yin, Ultrasound augmented leaching of nickel sulfate in sulfuric acid and hydrogen peroxide media. Ultrason. Sonochem. **40**(Pt A), 1021–1030 (2018). https://doi.org/10.1016/j.ultsonch.2017.08.031
96. B. Chen, S. Bao, Y. Zhang, S. Li, A high-efficiency and sustainable leaching process of vanadium from shale in sulfuric acid systems enhanced by ultrasound. Sep. Purif. Technol. **240** (2020). https://doi.org/10.1016/j.seppur.2020.116624
97. J.J. John, V. De Houwer, D. Van Mechelen, T. Van Gerven, Effect of ultrasound on leaching of lead from landfilled metallurgical residues. Ultrason. Sonochem. **69**, 105239 (2020). https://doi.org/10.1016/j.ultsonch.2020.105239
98. J. Wang, F. Faraji, A. Ghahreman, Effect of ultrasound on the oxidative copper leaching from chalcopyrite in acidic ferric sulfate media. Minerals **10**(7) (2020). https://doi.org/10.3390/min10070633
99. S. Wang, W. Cui, G. Zhang, L. Zhang, J. Peng, Ultra fast ultrasound-assisted decopperization from copper anode slime. Ultrason. Sonochem. **36**, 20–26 (2017). https://doi.org/10.1016/j.ultsonch.2016.11.013
100. S. Yu, T. Yu, W. Song, X. Yu, J. Qiao, W. Wang, H. Dong, Z. Wu, L. Dai, T. Li, Ultrasound-assisted cyanide extraction of gold from gold concentrate at low temperature. Ultrason. Sonochem. **64**, 105039 (2020). https://doi.org/10.1016/j.ultsonch.2020.105039
101. Q. Gui, M.I. Khan, S. Wang, L. Zhang, The ultrasound leaching kinetics of gold in the thiosulfate leaching process catalysed by cobalt ammonia. Hydrometallurgy **196** (2020). https://doi.org/10.1016/j.hydromet.2020.105426
102. Y. Zhang, J. Ma, Y. Qin, J. Zhou, L. Yang, Z. Wu, T. Wang, W. Wang, C. Wang, Ultrasound-assisted leaching of potassium from phosphorus-potassium associated ore. Hydrometallurgy **166**, 237–242 (2016). https://doi.org/10.1016/j.hydromet.2016.07.009
103. S. Yin, J. Pei, F. Jiang, S. Li, J. Peng, L. Zhang, S. Ju, C. Srinivasakannan, Ultrasound-assisted leaching of rare earths from the weathered crust elution-deposited ore using magnesium sulfate without ammonia-nitrogen pollution. Ultrason. Sonochem. **41**, 156–162 (2018). https://doi.org/10.1016/j.ultsonch.2017.09.028
104. R.L. Zhang, X.F. Zhang, S.Z. Tang, A.D. Huang, Ultrasound-assisted HCl-NaCl leaching of lead-rich and antimony-rich oxidizing slag. Ultrason. Sonochem. **27**, 187–191 (2015). https://doi.org/10.1016/j.ultsonch.2015.05.020
105. O. Johansson, T. Pamidi, V. Shankar, Extraction of tungsten from scheelite using hydrodynamic and acoustic cavitation. Ultrason. Sonochem. **71**, 105408 (2021). https://doi.org/10.1016/j.ultsonch.2020.105408

Advanced Techniques on Fine and Coarse Particle Flotation

Oktay Şahbaz, Ali Uçar, Çağrı Emer, and Cengiz Karagüzel

Abstract Flotation is a physicochemical process that is used for the separation of hydrophobic particles from hydrophilic ones. This method is widely used in mineral processing for the selective separation of valuable minerals. It is also used in water treatment and recycling processes. In the flotation of minerals, hydrophobic particles form bubble-particle aggregates as a result of micro-events, namely collision, attachment, and detachment. The aggregates float to the top of the flotation cell as a concentrate, while the hydrophilic particles, which are not attached to bubbles, sink to the bottom of the cell and are taken out as tailings in direct flotation. The concentrate and tailing positions in the cell are reversed for reverse flotation applications. To achieve high flotation recoveries, particles shall securely attach to air bubbles and stay attached as they rise to the top of the cell inside the flotation cell. The most critical parameter affecting the stability of bubble-particle aggregates is the size of the particles. The optimum size range for traditional mechanical flotation cells is typically between 20 and 150 μm for metal sulfides. The flotation recovery decreases outside this range due to high turbulence within the cell for coarser particles and low bubble-particle collision probability for the finer end of the particle size spectrum. In recent years, however, many studies have become available that promise improvement in recovery for finer and coarser ends of the particle size spectrum. In this chapter, the development of new technologies to increase the efficiency of the flotation of fine and coarse particles is discussed. General flotation theory and parameters affecting the recovery of particles are explained in detail. Relatively new technologies such as the Jameson cell, Concorde cell, Hydrofloat, Jet Diffuser Flotation Column, NovaCell and Reflux flotation cell have significant potential for both fine and coarse particle sizes.

Keywords Flotation · Coarse particle flotation · Fine particle flotation · Flotation hydrodynamics · Flotation cells

---

O. Şahbaz · A. Uçar · C. Karagüzel (✉)
Mining Engineering Department, Dumlupinar University, 43270 Kütahya, Türkiye
e-mail: cengiz.karaguzel@dpu.edu.tr

Ç. Emer
Metso Maden Teknolojileri A.Ş., Ankara, Türkiye

S. J. Ikhmayies (ed.), *Advances in Minerals Research*, Advances in Material Research and Technology, https://doi.org/10.1007/978-3-031-49175-7_3

1 Introduction

Mineral processing is as old as the first human being, as discussed by Drzymala [1]. In his book, he explained the place of mineral processing within science (Fig. 1) and the meaning of mineral processing as a part of extractive metallurgy (Fig. 2).

From the Stone Age to this century, mineral processing has been used in many different applications. Considerable development of mineral processing and its physicochemical basis occurred within the last hundred years [1]. A breaking point in the history of mineral processing is the invention of flotation.

Flotation is a physicochemical method invented at the end of the nineteenth century [2]. Ores with low grades that cannot be enriched by physical processes can

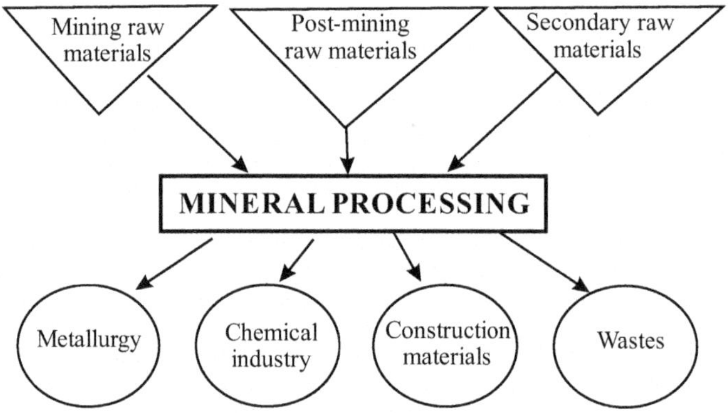

Fig. 1 Place of mineral processing in the science. Reprinted with permission from Drzymala [1]. Copyright by Jan Drzymala, Wroclaw 2007. www.ig.pwr.wroc.pl/minproc

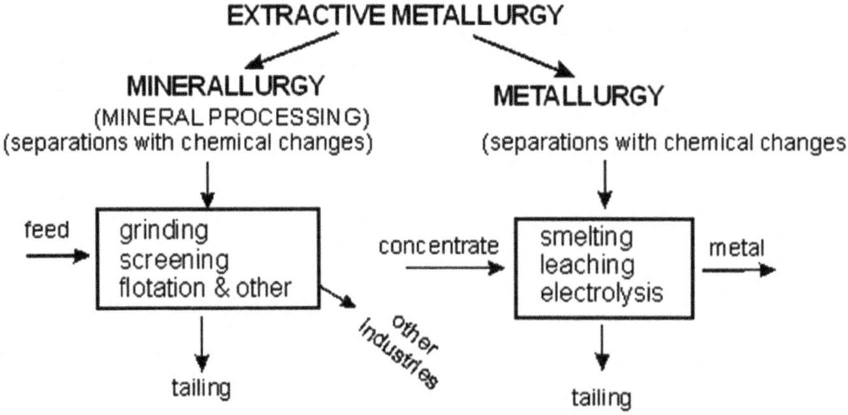

Fig. 2 Mineral processing with the relationship with metallurgy. Reprinted with permission from Drzymala [1]. Copyright by Jan Drzymala, Wroclaw 2007. www.ig.pwr.wroc.pl/minproc

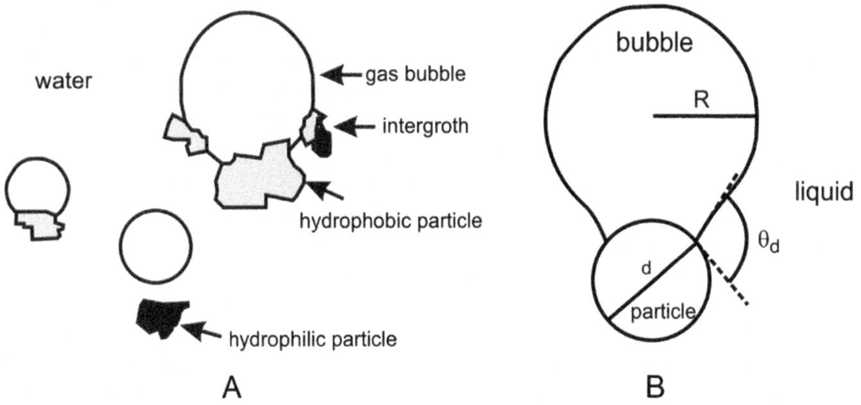

Fig. 3 Schematic explanation of flotation. **a** Different interaction mechanisms of the hydrophobic and hydrophilic particles with bubbles. **b** Detachment mechanism of the bubble-particle aggregate. Reprinted with permission from Drzymala [1]. Copyright by Jan Drzymala, Wroclaw 2007. www.ig.pwr.wroc.pl/minproc

be concentrated by flotation using physicochemical characteristics of particles. Flotation is a method that enables the separation of hydrophobic particles with the help of air bubbles (Fig. 3a). In flotation, the particles adhere to air bubbles depending on solid, liquid, and gas contact phases. The interaction of particles and bubbles includes collision, adhesion, and stability [3–5]. When an attachment occurs between bubbles and particles, hydrophobic particles are taken as float depending on the parameters such as detachment contact angle (θ_d) (Fig. 3b).

The realization of these events, which are known as sub-processes of flotation, is mainly related to the hydrophobicity, size of particles and bubbles and surface properties. These events occur within certain limits in existing flotation technologies [6, 7]. In flotation devices that are being used commercially, while the sizes of particles below 20 μm are removed as slime, the particles of sizes above 150 μm are sent back to the grinding stage depending on the nature of the ore [8] for most metal sulfide applications. While high efficiency is achieved in the stated particle size range, efficiencies of metallic minerals outside the optimum range decrease as shown in Fig. 4 [9–12]. On the other hand, the optimum range is typically 50–500 μm for coal flotation [13].

As shown in Fig. 4, the flotation efficiency decreases for both ends. The reason for low efficiency for fine particles is the decrease in collision probability. For coarse particles, the low efficiency is due to an increase in the probability of detachment as a result of high turbulence. Many studies have been focused on developing new flotation devices to counteract the problems faced for coarse and fine particles. However, it is seen that the industrial applications of the mentioned devices are still limited.

In the present chapter, the flotation efficiency for fine and coarse particles, flotation technologies and techniques have been analysed in detail.

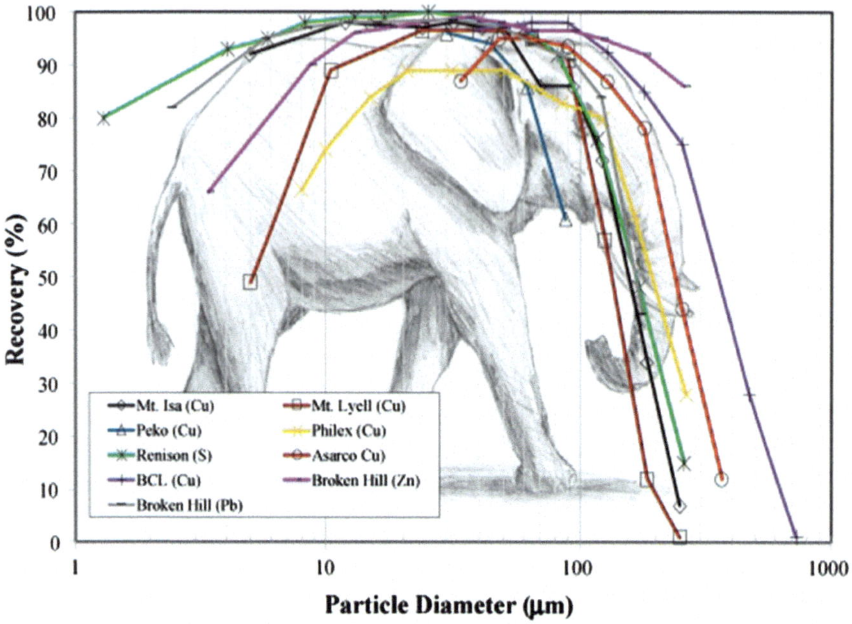

Fig. 4 Effect of particle size on flotation efficiency. Reprinted from Mankosa et al. [14], with permission from Elsevier. Copyrights (2018)

2 Theory of Flotation

The performance of the flotation process depends on more than hundreds of different parameters, mainly including ore characteristics, flotation cell hydrodynamics, and operational parameters (Fig. 5) [1]. This approach is valid mainly for the static condition.

Fig. 5 Main parameters affecting the performance of flotation. Reprinted with permission from Drzymala [1]. Copyright by Jan Drzymala, Wroclaw 2007. www.ig.pwr.wroc.pl/minproc

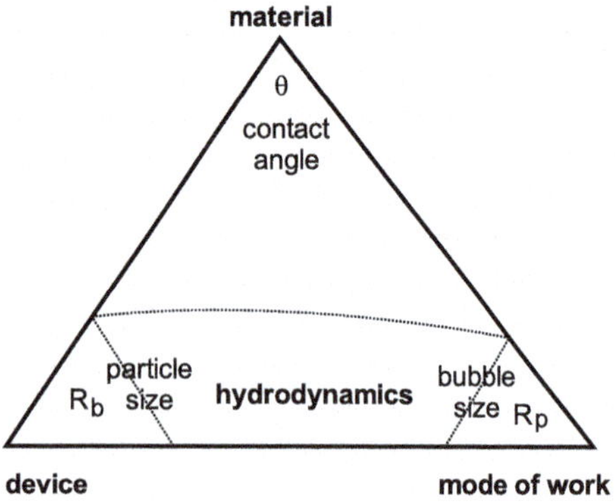

To improve the understanding of the flotation process, Schulze [15–18] divided flotation into three sub-processes; collision, attachment, and stability. The overall probability P of the flotation is given by Eq. (1).

$$P = Pc.Pa.Ps \tag{1}$$

where *Pc* is collision probability, *Pa* is attachment probability, and *Ps* is the probability of non-detachment. Oteyaka and Soto [19] added encountering (*Pe*) as an additional sub-process. Thus, the revised flotation equation can be given as shown below in Eq. (2).

$$P = Pe.Pc.Pa.Ps \tag{2}$$

The probability of encountering depends mainly on air hold-up (ε_g) as mentioned by [20].

$$P_e = 1.209(\varepsilon_g)^{2/3} \tag{3}$$

Schulze [17] took into account the encountering probability with collision probability, and he computed the *Pe* by Eq. (4) that depends on hydrodynamics of liquid flow (ψ), particle velocity (v_p) and bubble rising velocity (v_b).

$$P_e = \frac{2\psi}{\left(1 + \left(v_p/v_b\right)\right)} \tag{4}$$

Collision probability is one of the main sub-processes determining the success of the flotation. Collision is mainly related to particle (d_p) and bubble (d_b) sizes [21, 22] (Fig. 6). The equations improved by many researchers on collision have been benefitted from Sutherland Model.

$$Pc = \frac{3d_p}{d_b} \tag{5}$$

Attachment probability is related to the chemistry and physicochemistry of the bubble surface and bubble-particle interface [18]. There are three steps in the attachment which are "thinning of the film thickness between bubble and particle", "rupturing and formation of three-phase contact line", and "formation of the stable wetting perimeter" [18] as illustrated in Fig. 7.

The last sub-process of the flotation is the non-detachment/stability of bubble-particle aggregate. The probability of aggregate stability strongly depends on the balance between the attachment forces (F_a) and detachment forces (F_{det}). Flotation is only possible when the sum of all attachment forces is greater than the sum of all detachment forces. The attachment forces are capillary force (F_{ca}), the hydrostatic pressure of the liquid (F_{hyd}), and static buoyancy force of the immersed part (F_b). The detachment forces are the force of gravity (Fg), capillary pressure in gas bubble

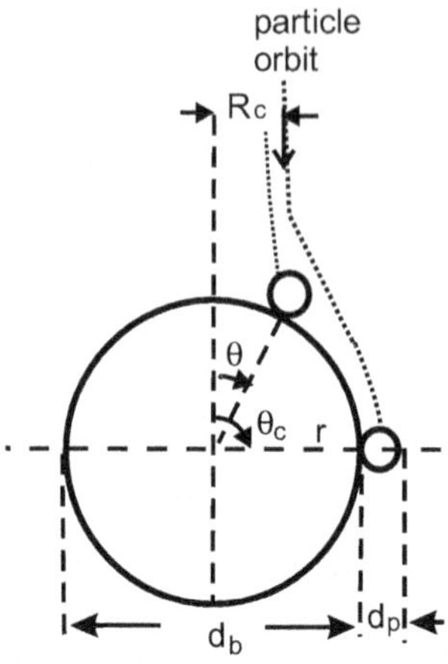

Fig. 6 Particle-bubble collision and particle orbit, where R_c is the collision radius, r is the radial distance, θ is the collision angle, and θ_c is the maximum collision angle. Reprinted from Dai et al. [22] with permission from Elsevier. Copyrights (2000)

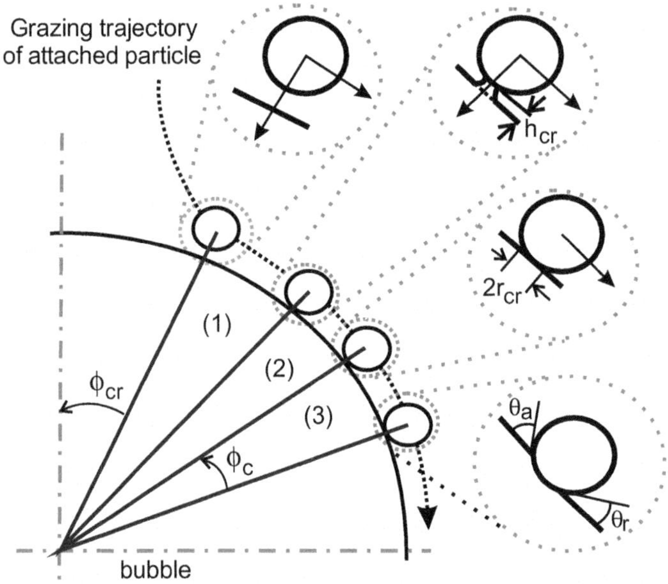

Fig. 7 Three steps of bubble-particle attachment, where (1) is thinning of the intervening liquid film to a critical (cr) film thickness; wetting radius r; (2) is rupture of the intervening liquid film and formation of TPC nuclei of some critical, (3) is expansion of the TPC line to form a stable wetting perimeter, ϕ is polar angle measured from bubble stagnation point, ϕ_{cr} is ϕ beyond which no attachment occurs, h_{cr} is critical value of height, r_{cr} is critical value of r, ϕ is polar angle measured from bubble stagnation point, ϕ_c is ϕ beyond which no collision occurs (collision angle), θ_a is advanced contact angle, and θ_r is receding contact angle. Reprinted from Nguyen et al. [18], with permission from Elsevier. Copyright (1998)

(F_σ) and other detachment occurs due to the particle mass and acceleration in the external field (F_d). Schulze [17] gave the mathematical equation (Eq. (6)) to compute the probability of stability as below:

$$P_S = 1 - \exp[-\frac{F_a - F_{det}}{F_{det}}] \tag{6}$$

There have been studies conducted to determine upper and lower size limits in flotation. Schulze [23] derived an equation, which is related to mainly particle gravity and inertia, to calculate the upper size limits of the flotation process. Kowalczuk et al. [6] improved the equation and implemented for various flotation devices. Kowalczuk et al. [6] derived an expression for the maximum floatable particle size ($d_{\max}$) of flotation (Eq. (7)). In this equation, $d_{\max}$ is related to the surface tension (σ), detachment contact angle (θ_d), particle specific gravity (ρ_p), liquid specific gravity (ρ_l), gravity (g) and inertia (a).

$$d_{\max} = \left(\frac{6\sigma \sin 2(\frac{\theta_d}{2})}{(\rho_p - \rho_1)g + \rho_p a}\right)^{1/2} \tag{7}$$

For coarse particle flotation, low turbulence is needed. On the other hand, lower particle size limits of flotation for the current machines have also been investigated by researchers. The first mathematical approach for fine particle flotation was improved by Scheludko et al. [24]. According to Scheludko et al. [24], there is a critical contact angle which is necessary for the flotation of fine particles. The kinetic energy of fine particles should be larger than the energy needed to provide three-phase contact line [7]. When energies are in balance, a lower particle size limit ($d_{\min}$) can be calculated by Eq. (8) [7].

$$d_{\min} = 2\left(\frac{3\tau^2}{(v_b \gamma_{\text{LV}}.(\rho_p - \rho_l))(1 - \cos\theta_r)}\right)^{1/3} \tag{8}$$

According to Eq. (8), the lower size limit of flotation depends on the solid–liquid–vapour three-phase contact line tension (τ^2), bubble rise velocity (v_b), liquid–vapour surface tension (γ_{LV}), specific gravities of particle and liquid and receding contact angle (θ_r). The lower the bubble size, the higher the fine particle flotation recovery rate.

According to the abovementioned theory, flotation specialists have invented new devices/machines/methods to improve coarse and fine particle flotation. This chapter is aimed to explain the advanced technologies developed for coarse and fine particle flotation.

3 Fine Particle Flotation

Nowadays, fine grinding is required by the processing industry due to the complexity of ores. There is a significant development in the field of tower mills and stirred mills for fine grinding. On the other hand, there is not a substantial change in the flotation technologies. Therefore, researchers have focused on the flotation of fine-ground particles, not only from the perspective of recovery of complex ores but also for the removal of fine particles because of environmental concerns such as acid mine drainage.

Fine particle flotation depends mainly on bubble size, particle-bubble aggregate formation, and flow conditions [25]. The flotation rate of fine particles is relatively slow because of their low collision probability with air bubbles due to the low mass, inertia, high surface area, and the fluid streamlines around the rising bubbles [7, 22, 24, 26–29].

Yoon and Lutrell [30] formulated the collision probability of particles and air bubbles, and they indicated that it is directly proportional to the square of particle size and indirectly to the square of bubble size. Sivomohan [27] stated the problems of the enrichment of fine particles in mineral processing and identified the methods which are based on surface characteristics for future studies. He also showed that a reduction in efficiency is due to fine particles' incapacity of encountering, colliding, and adhering to air bubbles.

Dai et al. [22] compared many models related to the bubble-particle encountering models in their review study and indicated that collision efficiency increased with particle size in all models. Tao [28] explained the decrease in flotation efficiency of fine particles with the low collision probability of bubbles and particles and indicated that detachment effectively decreases flotation efficiency of coarse particles. In the same study, the author also showed that with the background studies he did, the collision probability of fine bubbles and particles would increase, and detachment would decrease. Manouchehri and Farrokhpay [31] stated that the flotation kinetics of fine particles in flotation can be increased by increasing the power input. All these studies show that particle size must be enlarged or bubble size must be minimized for the optimum flotation of fine particles.

One of the techniques used for the selective separation of fine particle minerals is an enlargement of particle size. Polymeric flocculation, coagulation, oil agglomeration, and shear (with the effect of high rupturing power) flocculation are commonly used techniques for optional particle size enlargement [32–36]. Inorganic salt is used for coagulation, while flocculation occurs with anionic or cationic polymers in polymeric flocculation. In the oil agglomeration technique, particle size grows by means of constituting an oil bridge between hydrophobic particles. The polymeric flocculation, coagulation, oil agglomeration, and shear flocculation studies have been conducted for fine particles of raw energy materials [32–36], industrial minerals [36–40], metallic minerals [41–47], and precious minerals [48] in recent years. Additionally, the separation of these minerals, of which particle size is enlarged from accompanying minerals, is achieved with screening, flotation, magnetic separation, and sedimentation. The carrier flotation technique, also known as the flotation of fine minerals with the help of an appropriate carrier mineral, is also used to enrich fine particles [49–51].

The other techniques used for increasing the flotation efficiency of fine particles are dissolved air flotation (DAF), collide gas aphron (CGA), pico bubble flotation, electroflotation, electrostatic air spray, effervescent atomizer techniques in which fine bubbles are generated [27, 29, 52–57]. Most of these techniques have been used for water purification, like clarification of water from oil and heavy metals [58–62]. Rubio et al. [63] investigated several flotation techniques for wastewater treatment in their review. Some of the mentioned techniques have been used with conventional methods to increase flotation efficiency [27, 53, 56].

With the electroflotation technique, 22–50 μm size bubbles are produced by electrolyzing water [54, 55]. While Ketkar et al. [64] worked on the flotation of fine quartz particles with this technique, Lierena et al. [65] worked on the effect of pH on the electroflotation of fine sphalerite particles.

Pico bubble flotation is carried out using pico bubbles (<1 nm), achieved by means of a hydrodynamic cavitation tube. Tao et al. [53] demonstrated coal flotation in the presence of pico bubbles, and the efficiency of column flotation increased, but there was no significant change in the content of coal.

In a similar study by Yalcin et al. [26], the reverse flotation of magnetite ore was tested in a column using the DAF method, which produces 10–100 μm (d_{50}: 40 μm) size bubbles. At the end of the study, the silicate efficiency increased in the column + DAF system while iron grade decreased when compared to conventional and DAF systems. Karagüzel [66], on the other hand, obtained a concentrate containing 11.07% $Na_2O + K_2O$ with the help of positively and negatively charged bubbles from the feldspar wastes of a feldspar plant containing 10.36% $Na_2O + K_2O$ in his study using DAF (Fig. 8).

Another technique by which fine bubble is produced is CGA. In this technique, bubbles are minimized (an average of 50 μm) with the effect of shear force at high speed (>5000 rpm). In a study by Waters et al. [52], flotation was performed in the CGA column and Denver cell using copper oxide and silicon dioxide particles smaller than 10 μm. In this study, bubbles are loaded with sodium dodecyl sulphate (SDS) as an anionic reagent instead of conditioning particles with reagent, and flotation was carried out in the CGA column.

It is seen that modified CGA has benefits compared to the conventional systems. In a study performed by Cilliers and Bradshaw [67] in 1996, pyrite minerals were separated from quartz particles smaller than 38 μm. In this study, CGA and the conventional batch system was compared. Consequently, the sulphur recovery was 50% and 35% for CGA and conventional system, respectively.

Patil and Laskowski [56] enhanced the flotation efficiency by modifying the effervescent atomizer and batch flotation cell. George et al. [68] mentioned the importance of three functions in fine particle flotation; true flotation, entrapment, and entrainment. While entrapment occurs with fine particles forming agglomerate with coarse particles, entrainment is known as the carrying of fine particles with coarse air bubbles

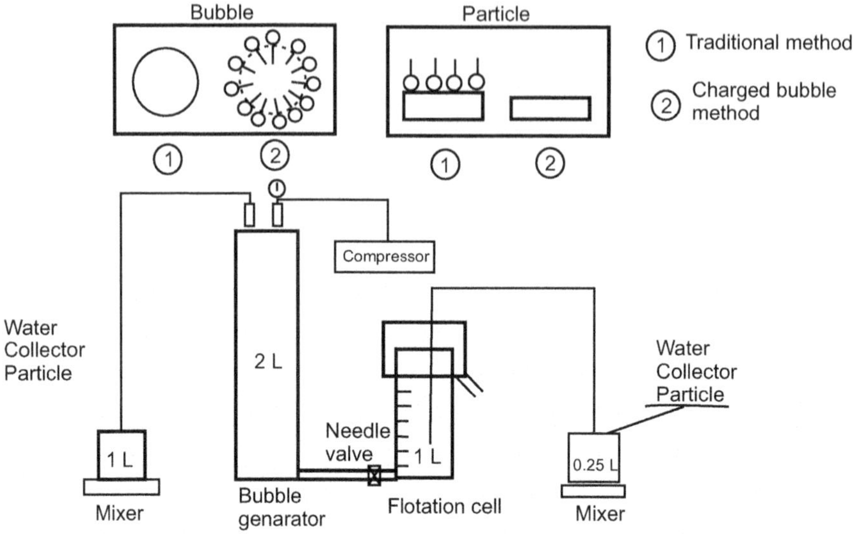

Fig. 8 Application of the charged bubble method in DAF. Reprinted from Karaguzel [66]. Copyright (2010)

in an environment where only fine particles exist. In the same study, it is stated that fine bubbles cannot drag particles such as coarse bubbles because, in such a case, they will be accompanied by a very little slipstream (smaller associated wake).

Almost all the mentioned methods and devices have been performed on a laboratory scale. Having developed through the end of the 1980s, Jameson Cell (Fig. 9) flotation device gained success in fine particle flotation in the industrial dimensions. Nowadays, Jameson Cell is used in more than 400 flotation plants in a wide range of fields such as coal mines, metallic mines, industrial raw materials, and waste cleaning [69]. It is also known that this device has important superiorities in fine particle separation, has been used for oil sand flotation since 2009 [70].

The most crucial reason for Jameson Cell's successful performance in the fine particle flotation is its compact design and generation capability of fine air bubbles compared to conventional flotation devices [69, 71]. Fine air bubbles (400–600 μm) generated in the device enable a superior bubble-particle collision, and the nozzle-downcomer also helps fast collection [69, 72].

Another technology used for fine particle flotation is Imhoflot™ which was commercialized in the early 1980s. The collision between bubble and particle occurs outside the ImhoflotTM cell, producing very fine air bubbles created. This makes the cell very advantageous for the ultrafine particles. According to Battersby et al. [11], the cell in three stages could concentrate an additional 30% Ni from tailings [73]. Imhoflot™ flotation is based on the venturi effect, and Maelgwyn Imhoflot™ technologies have three different types, which are V-cell (vertical fed), G-cell (tangentially fed) and H-cell [74]. Imhoflot™ G-cell is also successful in Au application for ultrafine particles [73].

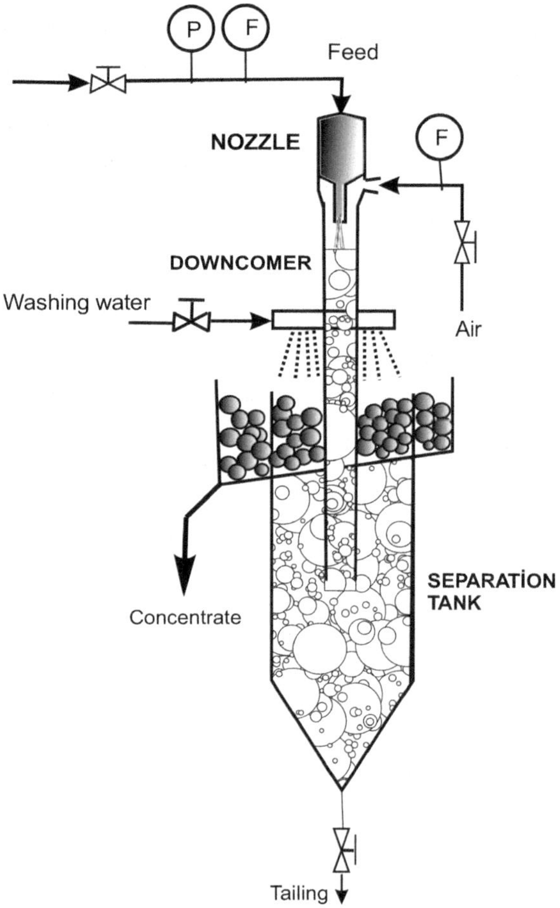

Fig. 9 The Jameson cell (Authors original drawing)

One of the most recent advances in flotation science and technology is Reflux flotation cell™ (RFC) [75]. RFC is a flotation device combining a downcomer, a fluidized bed, and inclined channels. It has a significant potential to improve not only fine but also coarse particle flotation while maintaining a high throughput [76]. Current data shows that the RFC has successful applications for particle size between 25 and 350 μm [77, 78]. A schematic view of the Reflux Flotation Cell is given in Fig. 10.

In the RFC, downcomers having multi-channelled designs create high shear rates. It establishes a high-intensity mixing zone for particles and air which is pressurized through a sparger system. Therefore, a high collision probability is obtained between hydrophobic particles and bubbles. Inclined channels improve segregation of bubbles from the slurry containing hydrophilic particles. The RFC provides fast flotation and can be used in one single machine instead of ten conventional cells [79].

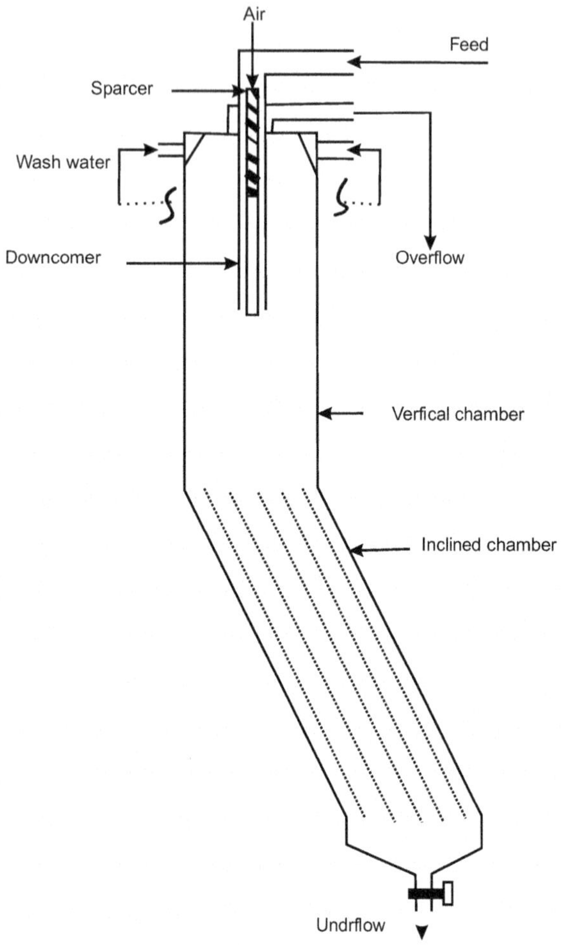

Fig. 10 Schematic view of the Reflux Flotation Cell. Reprinted from Cole et al. [77]. Copyright (2020)

Recently, G.J. Jameson invented the Concorde cell™ to float ultrafine particles [8, 80]. Very fine bubbles are produced in a blast tube under high pressure in the Concorde Cell. The operational procedure was explained by Jameson [8], and according to him, the cell could collect silica particles smaller than 38 μm, which can be called ultrafine. Additionally, the test results showed that the cell was very effective for the flotation of ultrafine coal particles, <20 μm [8]. The schematic view of the Concorde Cell is given in Fig. 11. The Concorde Cell is now commercialised by Metso and it is being used for recovery of fine and ultrafine particles.

An oscillating grid flotation cell (OGC) is another device that can be used in the mineral processing industry for fine particle flotation. The cell was developed at the University of Cape Town and then modified by Massey [81]. Safari et al. carried out the pilot scale implementation of the OGC [82]. The OGC has been applied for the

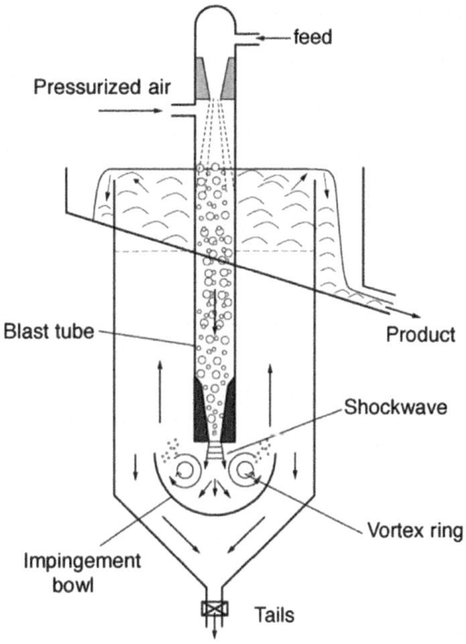

Fig. 11 Concorde cell. Reprinted from Jameson [8], with permission from Elsevier. Copyright (2010)

particle size range up to 650 μm, and the data shows a high potential of the cell for both fine and coarse particle flotation [73].

Battersby et al. [11] used pneumatic Imhoflot G-Cell flotation cells in which pico bubbles are used to obtain fine nickel and zinc. It is stated that when compared to classic flotation cells, an increase at the rate of 20% in zinc obtaining efficiency and a boost at the rate of 30% nickel obtaining efficiency are achieved.

Mankosa et al. [14] conducted a study to show the performance of StackCell™ by Eriez Technology. According to data obtained by the researchers, StackCell™ can be an alternative to conventional technologies, such as column flotation. To substantially increase flotation rate, the StackCell™ incorporates a unique, high-shear, bubble-particle contactor instead of the conventional rotor–stator mechanism historically utilized in mechanical flotation cells. Instead of operating with a large tank volume, the StackCell™ forces the bubbles and particles to contact within a small aeration chamber which is mechanically isolated from the remainder of the tank [14].

4 Coarse Particle Flotation

Coarse particle flotation is generally taken into consideration for the enrichment of liberated particles in coarse sizes to minimize grinding expenses. Only a fraction of the energy used for grinding is utilized as useful energy, and the rest converts into sound energy and heat through friction. Similarly, the energy amount consumed in the size reduction process rises as the final particle size decreases. The difference between the energy amounts that are consumed to reduce the particle size to 0.3 and 0.1 mm from the same top size is between 30–50%. When the energy amount that will be consumed for grinding in enrichment processes of coarse particles is taken into consideration, it is stated that a saving ratio of around 30–50% can be achieved. Energy consumption and slime formation will lead to economic losses and problems related to the process. For this reason, the flotation of coarse particles with less grinding is crucial for the economic enrichment of minerals. The flotation in coarse sizes has attracted many researchers' attention [83–87].

The most critical problem that needs to be solved in the flotation of the coarse particle is preventing the disintegration of bubble-particle aggregate that forms in the process of micro-events. Therefore, the system with high turbulence within the flotation environment is the least favourable. To minimize the turbulence in the flotation environment, flotation columns have been developed, and they are still in use. Flotation columns are preferred in the flotation of fine and coarse particles since turbulence is minimum in these types of cells. Generally, in the enrichment of fine particles, longer columns are used and modified shorter columns are preferred in coarse sizes. For coarse particle flotation, alternative cells, which primarily modify existing commercial cells, are being developed [88]. The most studied alternative is a flotation cell with a fluidised bed [8, 89]. Modified flotation columns also have crucial potential and some companies are developing cells that use methods other than fluidisation as well.

A modified flotation column is a device that is generally preferred in the enrichment of coarse-sized (1–2 mm) ores. Compared to classic flotation columns, the most significant differences can be listed as; their heights are shorter, there is no need for wash water, they work with negative bias and there is no froth zone. Oteyaka et al. [90] subjected the dust coal (<1.16 mm) obtained from Zonguldak washery, Türkiye to an enrichment process with the modified flotation column, and consequently they received a concentrate with 27% ash out of coal that includes 61% feed ash with 51% combustible recovery. Oteyaka and Soto [19] stated that negative bias must be used to enrich the coarse particles with flotation columns so that the feed flow rate must be more than tailings flow rate, and in addition to that, bubble-coarse particle aggregates must leave the column in a very short time and this could be made possible with short modified flotation columns.

In the literature, apart from the techniques for minimizing turbulence, other methods are also available in which mineral surfaces are being modified. These techniques aim to improve bubble-particle interaction, and accordingly, stability of bubble-particle aggregates. The first of these techniques applied is the chemical modification of bubbles and particles with flotation chemicals such as frothers, collectors, oils, etc. The most important subject of physical surface activation for coarse particle flotation is the surface roughness and the flotation systems in which fine bubbles are used as auxiliary bubbles. The cavitation effect of ultrasonic sound waves, hydrodynamic cavitation effect, venturi tubes, and the air dissolved under pressure are utilized to produce micro and nano-sized bubbles (Fig. 12).

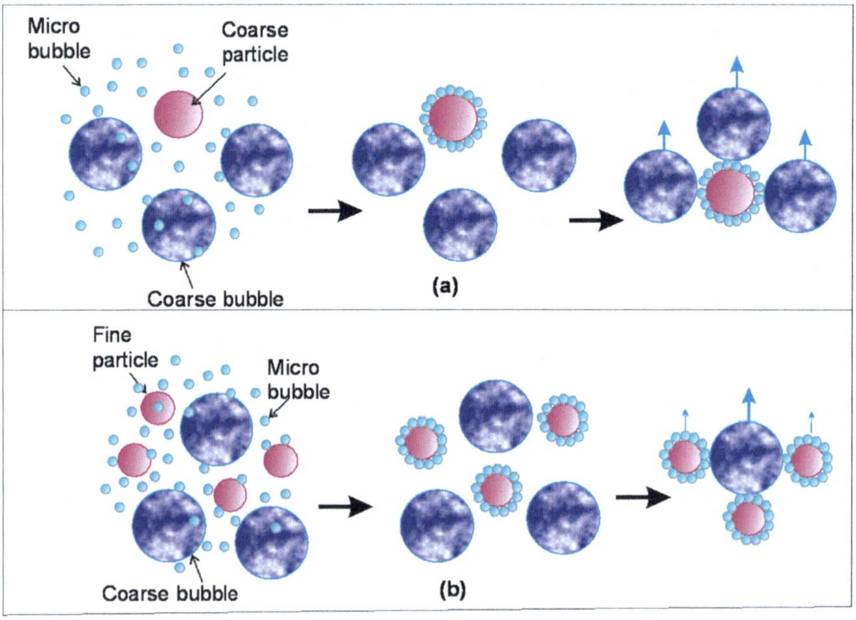

Fig. 12 The interaction of coarse particle (**a**) and fine particle (**b**) micro and coarse bubble (Authors original drawing)

It is known that in these systems, fine bubble producers are installed into these common commercial flotation technologies (Fig. 12). Studies focused on this subject are discussed below.

Tao et al. [53] investigated the effects of pico-bubbles in fine coal flotation, and utilized hydrodynamic cavitation principle to produce pico-bubbles. As a consequence of flotation experiments, 1 nm and smaller-sized pico bubbles were made with the hydrodynamic cavitation principle, and it was also stated that this increased the encounter and collision possibility of particles with pico bubbles and minimized high detachment possibilities. This study mentions an increase of around 10–30% in the efficiency of fine coal concentration is achieved. In another study by Tao et al. [91], the hydrodynamic cavitation principle is similarly utilized to produce pico bubbles. Like the former study, an improvement of around 10–40% in the efficiency of fine coal concentration was reported.

Fan et al. [13] stated that they used nanobubbles. Additionally, they stated that the efficiency of coarse particle flotation with this method is better than the systems in which nanobubbles are not used.

Peng and Yu [92] utilized pico and nanobubbles in the enrichment of fine and coarse-sized coals by the use of flotation, both to increase collision and attachment probability and to increase the non-detachment. They used venturi tubes that were combined with a static mixer to produce pico and nanobubbles. The same authors also stated that with the help of pico and nanobubbles, they obtained clean coal, including 10–12% ash with 85–90% combustible recovery.

Gontijo et al. [7] conducted several experiments to determine the prominent and smallest particle size that could be carried by a bubble in quartz flotation. The study states that a fine bubble size and wide contact angle are required for the flotation of fine and coarse particles.

Yalcin et al. [26] and Yalcin and Byers [93] conducted a study that separates galena which has a hydrophobic surface, from pyrite and quartz with the help of a system they designed via producing microbubbles and the air dissolved under pressure, as it is in DAF. In this study, hydrophobic particles of which microbubbles are absorbed are separated from other minerals.

Sahbaz et al. [69] studied Jameson cells to improve the cell's operable size range. Sahbaz [94] described the velocity gradient (G_t) that can be referred as turbulence for the critical region in the Jameson cell's separation tank. This critical region is the main reason for detachment of the bubble-particle aggregates. Using the mathematical model, Sahbaz [94] computed the velocity gradient change in the turbulent zone of the Jameson cell's separation tank, the so-called critical zone. According to the model (Eq. (9)), the turbulence at the outlet of the downcomer can be calculated and mainly related to downcomer outlet diameter (d_D), density of liquid (ρ_l) and solid (ρ_s); air hold up (ε) and bubble diameter (d_b). Also, the velocity of the bubble (v_k), the velocity of the pulp at the end of the downcomer (v), the total bubble number (Sk), and the volume of the critical region (V) are crucial parameters that have to be taken into account for any research [94].

$$G_t = \frac{\pi.(d_D^2.\rho_l.(1-\varepsilon).v^3 + \rho_s.d_b^2.vk^3.Sk)}{8V\mu} \tag{9}$$

By using the above model (Eq. (9)), Sahbaz [94] modified the cell to improve the flotation performance of coarse particles. Sahbaz [94] stated that the efficiency of coarse particle flotation could be increased by around 10% with a different design of the downcomer of the Jameson cell. Studies on a pilot scale about this emerging device are still in progress, and the first results are promising. Bilir et al. [95] showed that the modified machine, the so-called Jet diffuser Flotation Column (Fig. 13), has 90% talc recovery for the particle size of 350 µm, while it was 70% in the Jameson cell.

One of the advanced coarse particle flotation technologies is Eriez HydroFloat™ technology. It is an aerated conical-shaped fluidized-bed separator. Eriez HydroFloat™ contains flotation and elutriation methods to improve the recovery of coarse particles [96]. According to industrial scale data, the Hydrofloat technology has improved against conventional cells for the coarse sphalerite particle size, 450–850 µm [73, 97]. Awatey et al. [98] compared mechanical cells and HyrdoFloat™ on a lab scale for metallic ores. According to results from Awatey et al. [99], there is a significant recovery difference between the cells for the coarse particle size group of 450–850 µm. The recovery of HydroFloat™ is 44% greater than the mechanical cell

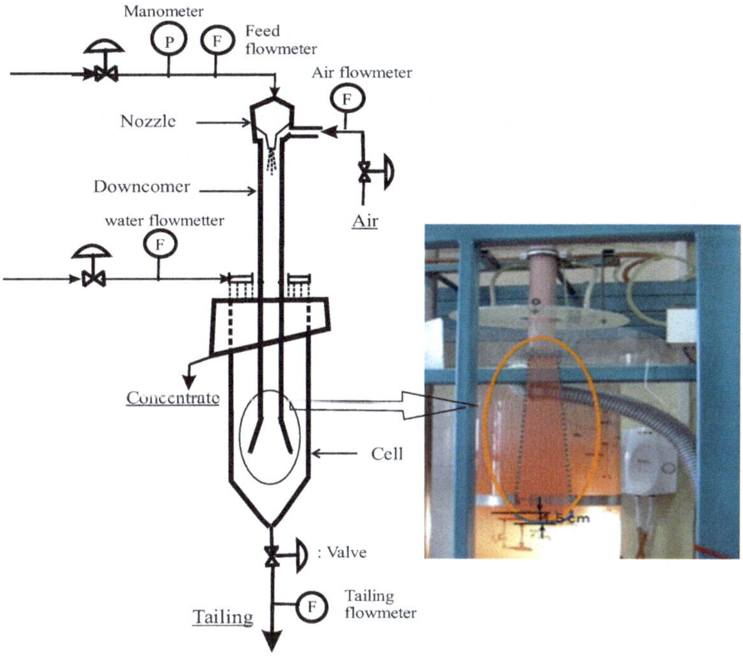

Fig. 13 Schematic view of Jet Diffuser Flotation Column. The right part of the figure was adapted from Bilir et al. [95] under the terms and conditions of the Creative Commons Attribution (CC BY) license. https://%20creativecommons.org/licenses/by/%204.0/

for the size group of 850–1180 μm, while it is 25% for the particle size group of 450–850 μm [73]. Kohmuench et al. [100] indicated that HydroFloat's recovery (93.8%) is higher than the mechanical cell's recovery (78.7%) for phosphate flotation.

Another advanced and new-coming coarse particle flotation technology is Novacell™ which was invented by G. J. Jameson at the University of Newcastle, Australia. The NovaCell can give high recoveries for ultra-fine, mid-range, and very coarse particles [101]. It was tested for copper minerals having a top size of 600 μm. Using batch testing, it was found that it was possible to obtain 100% recovery at large flotation times for particle sizes up to 300 μm [101]. According to Jameson and Emer [101], the NovaCell has an enormous potential to play a significant role in flotation circuit design by rejecting gangue minerals at an early stage of flotation.

Crompton et al. [102] for the first time, reported the CoarseAIR™ which is a novel flotation system. CoarseAIR™ utilized a three-phase fluidized bed as used in Novacell™ and a system of inclined channels as used in Reflux to facilitate coarse particle flotation and internal size classification.

Hasanzadeh et al. [73] summarized the data of various cells, and we have added the bias/bias factor data to create Table 1. We also added the Concorde cell and Jet Diffuser Flotation column which are modifications of the Jameson flotation cell in the industrial testing/commercial phase, to complement the table. The bias factor is the parameter showing the froth zone existence and is mainly connected with the selectivity of the flotation. The negative bias condition is primarily proper for coarse particle operation or rougher application. On the other hand, positive bias conditions are more effective in a cleaner stage and fine particle flotation.

5 Conclusions and Recommendations

Flotation is still the most applied mineral processing technology in the separation of minerals. For high flotation recovery, particles should securely attach to bubbles and stay stable during flotation. The most critical parameter affecting bubble-particle aggregate stability is particle size. It is known that the optimum size range for metal sulfides in conventional flotation devices is between 20 and 150 μm. The recovery diminishes for the coarser and finer particles due to the high turbulence within cells and the low collision probability of bubbles and particles. It is observed that many researchers have worked to develop new technologies and techniques for providing the increase of recovery in coarse and fine particle flotation by exploiting the mentioned shortcomings of traditional devices.

This chapter summarizes developments of new technologies to increase the efficiency of flotation of fine and coarse particles. Parameters affecting the flotation performance and the general flotation theory are explained in detail.

Many new flotation machines are still not being used in mineral processing plants, even though the machines have significant potential to solve the current problems in the plants. For fine particle flotation, the Jameson cell and the Concorde cell are proven technologies. Hydrofloat has a good potential for coarse particle flotation. Also, the Jet Diffuser Flotation Column has the potential to be implemented for coarse particles. Additionally, the NovaCell and Reflux flotation cells have significant potential for both fine and coarse particle sizes. The list is not exhaustive and there are other developments in the industry in addition to the technologies mentioned above.

Table 1 Summarized data for flotation cells considering their key parameters

Cell type	Mechanical cell	Flotation column		Jameson cell		Reflux flotation cell	Concorde cell	Jet diffuser flotation column
First installation	1930s	1960s		1980s		2020s	2021	Testing
Cell residence time (MRT[a]): min	4–20	9–41		1–4		1–5		1–6
Superficial gas velocity (J_g): cm/s	0.7–2.7	2.3–4.8		1–5	0.5–2	0.5–4		0.8–2
Gas hold-up (ε_g): (%)	5–20	5–25		45–70	40–60	40–60		40–65
Bubble size range: (mm)	0.7–3.0	1.5–3.0		0.3–0.7	0.15–0.35	0.3–0.7	0.1–0.4	0.4–0.8
Wash water	Yes (re)-cleaner	Yes	No	Yes	No	Yes	Yes	Yes
Moving parts	Yes	No	No	No	No	No	No	No
Common application	Rougher and Scavenger	Cleaner	Rougher	Cleaner/ Scavenger	Testing	Rougher/ Cleaner	Rougher/ Cleaner	Testing
Throughput	High	Mid	Mid	Mid	High	High	High	High
Footprint	High	Mid	Mid	Small	Small	Small	Small	Small
Froth depth	Low	High	Not	Mid	Mid	Mid	Low/High	Mid
Mixing intensity (turbulence)	Mid	Low	Low	High	High	High	High	Low
Maintenance	Very high	Mid	Mid	Low	Low	Low	Low	Low
Capital costs	High	High	High	Low	Low	Low	Low	Low
Particle size	Fine/Coarse	Fine	Coarse	Fine/Coarse	Coarse	Fine/ Coarse	Fine/ Ultrafine	Coarse
Bias/Bias factor		Positive	Negative	Positive	Negative	Positive	Negative/ Positive	P/N

[a] *MRT* Mean residence time

References

1. J. Drzymala, *Mineral Processing. Foundations of Theory and Practice of Minerallurgy* (Oficyna Wydawnicza PWr, 2007), pp. 381–382
2. B.A. Wills, J. Finch, *Wills' Mineral Processing Technology: An Introduction to the Practical Aspects of Ore Treatment and Mineral Recovery*, 8th edn. (Elsevier, 2016), pp. 265–380
3. A.M. Gaudin, *Flotation*, 2nd edn. (McGraw-Hill, New York, 1957), pp. 24–38
4. A.V. Nguyen, H.J. Schulze, *Colloidal Science of Flotation*, 1st edn. (Marcel Dekker, New York, 2004), pp. 173–196
5. A.V. Nguyen, R.J. Pughb, G. J. Jamesona, in *Colloidal Particles at Liquid Interfaces*, ed. by B.P. Binks, T.S. Horozov (Cambridge, Cambridge University Press, 2006), pp. 328–382
6. P.B. Kowalczuk, O. Sahbaz, J. Drzymala, Miner. Eng. **24**(8), 766 (2011)
7. C.F. Gontijo, D. Fornasiero, J. Ralston, Can. J. Chem. Eng. **85**, 739 (2007)
8. G.J. Jameson, Miner. Eng. **23**, 835 (2010)
9. W.J. Trahar, Int. J. Miner. Process. **8**(4), 289 (1981)
10. C. Karaguzel, G. Çobanoğlu, Sep. Purif. Technol. **74**(1), 100 (2010)
11. M. Battersby, R.M. Battersby, S. Flatman, R. Imhof, H. Sprenger, T. Bragado, in *5th International Flotation Conference (Flotation'11), MEI Conference*, Cape Town, vol. 2 (2011), pp. 779–785
12. J.M. Sokolović, S. Miskovic, Physicochem. Probl. Miner. Process. **54**(4), 1172 (2018)
13. M. Fan, D. Tao, R. Honaker, Z. Luo, Min. Sci. Technol. **20**, 641 (2010)
14. M.J. Mankosa, J.N. Kohmuench, L. Christodoulou, E.S. Yan, Miner. Eng. **121**, 83 (2018)
15. H.J. Schulze, Miner. Process. Extr. Metall. Rev. **5**, 43 (1989)
16. H.J. Schulze, in *Innovations in Flotation Technology. NATO ASI Series*, vol. 208, ed. by P. Mavros, K.A. Matis (Springer, Dordrecht, 1992), pp. 171–180
17. H.J. Schulze, in *Coagulation and Flocculation: Theory and Applications*, ed. by B. Dobias (Marcel Dekker, New York, 1993), pp. 321–353
18. A.V. Nguyen, J. Ralston, H.J. Schulze, Int. J. Miner. Process. **53**(4), 225 (1998)
19. B. Oteyaka, H. Soto, Miner. Eng. **8**, 91 (1995)
20. B. Oteyaka, *Modelisation D'une Colonne De Flottation Sans Zone D'ecume Pour La Separation Des Particules Grossieres*, Ph.D. thesis, Universite Laval, Quebec, Kanada, 1993
21. Z. Dai, S. Dukhin, D. Fornasiero, J. Ralston, J. Colloid Interface Sci. **197**, 275 (1998)
22. Z. Dai, D. Fornasiero, J. Ralston, Adv. Colloid Interface Sci. **85**(2–3), 231 (2000)
23. H.J. Schulze, Int. J. Miner. Process. **4**, 241 (1977)
24. A. Scheludko, B.V. Toshev, D.T. Bojadiev, J. Chem. Soc. Faraday Trans. I **12**, 2815 (1976)
25. S. Farrokhpay, I. Filippova, L. Filippov, A. Picarra, N. Rulyov, D. Fornasiero, Miner. Eng. **155**, 106439 (2020)
26. T. Yalcin, A. Byers, K. Ughadpaga, Miner. Process. Extr. Metall. Rev. **23**, 181 (2002)
27. R. Sivomohan, Int. J. Miner. Process. **28**, 247 (1990)
28. D. Tao, Sep. Sci. Technol. **39**(4), 741 (2004)
29. R.T. Rodrigues, J. Rubio, Int. J. Miner. Process. **82**, 1 (2007)
30. R.H. Yoon, G.H. Luterell, Miner. Process. Extr. Metall. Rev. **5**, 101 (1989)
31. H.R. Manouchehri, S. Farrokhpay, in *IMPC 2016: XXVIII International Mineral Processing Congress Proceedings* (The Canadian Institute of Mining, Metallurgy and Petroleum, Quebec, Canada, 2016)
32. Y.A. Attia, Sep. Sci. Technol. **17**(3), 485 (1982)
33. D.P. Patil, J.R.G. Andrews, P.H.T. Uhlherr, Int. J. Miner. Process. **61**(3), 171 (2001)
34. S. Song, A.L. Valdivieso, Miner. Process. Extr. Metall. Rev. **23**, 101 (2002)
35. J.J. Rosa, J. Rubio, Miner. Eng. **18**, 701 (2005)
36. A. Ozkan, H. Ucbeyiay, S. Duzyol, Comparison of stages in oil agglomeration process of quartz with sodium oleate in the presence of Ca(II) and Mg(II) ions. J. Colloid Interface Sci. **329**, 81 (2009)

37. J.S. Laskowski, in *Mineral Processing on the Verge of the 21st Century: Proceedings of the 8th International Mineral Processing Symposium*, Antalya, Turkey, 16–18 October 2000, ed. by C. Hicyilmaz (2000)
38. Z. Sadowski, I. Polowczyk, Int. J. Miner. Process. **74**, 85 (2004)
39. I. Sonmez, Y. Cebeci, Int. J. Miner. Process. **71**, 219 (2003)
40. Y. Cebeci, I. Sonmez, J. Colloid Interface Sci. **273**, 198 (2004)
41. J.M. Vergouw, A. Difeo, Z. Xu, J.A. Finch, Miner. Eng. **11**(2), 159 (1998)
42. J.M. Vergouw, A. Difeo, Z. Xu, J.A. Finch, Miner. Eng. **11**(7), 605 (1998)
43. S. Song, A. Lopez-Valdivieso, Y. Ding, Powder Technol. **101**, 73 (1999)
44. S. Song, A. Lopez-Valdivieso, J.L. Reyes-Bahena, C. Lara-Valenzuela, Miner. Eng. **14**, 87 (2001)
45. S. Song, S. Lu, A. Lopez-Valdivieso, Miner. Eng. **15**, 415 (2002)
46. Z. Sadowski, Colloids Surf. A **173**, 211 (2000)
47. J. Rubio, F. Capponi, R.T. Rodrigues, E. Matiolo, Int. J. Miner. Process. **84**, 41 (2005)
48. S. Sen, A. Seyrankaya, Y. Cilingir, Miner. Eng. **18**, 1086 (2005)
49. L. Valderrama, J. Rubio, Int. J. Miner. Process. **52**, 273 (1998)
50. R. Liang, Y. Numata, T. Fujita, J. Min. Mater. Process. Inst. Jpn. Jpn. **115**(3), 164 (1999)
51. G. Atesok, F. Boylu, M.S. Çelik, Miner. Eng. **14**(6), 661 (2001)
52. K.E. Waters, K. Hadler, J.J. Cilliers, Miner. Eng. **21**(12–14), 918 (2008)
53. Y. Tao, J. Liu, S. Yu, D. Tao, Sep. Sci. Technol. **41**, 3597 (2006)
54. N.K. Khosla, S. Venkatachalam, P. Somasunduran, J. Appl. Electrochem. **21**, 986 (1991)
55. S.E. Burns, S. Yiacoumi, C. Tsouris, Sep. Purif. Technol. **11**, 221 (1997)
56. D.P. Patil, J.S. Laskowski, Miner. Eng. **21**, 373 (2008)
57. A.H. Englert, R.T. Rodrigues, J. Rubio, Int. J. Miner. Process. **90**(1–4), 27 (2009)
58. N.M. Mostefa, M. Tir, Desalination **161**, 115 (2004)
59. L.A. Feris, A.T. De Leon, M. Santander, J. Rubio, Int. J. Miner. Process. **74**, 1 (2004)
60. G. Chen, Sep. Purif. Technol. **38**, 11 (2004)
61. P. Gao, X. Chen, F. Shen, G. Chen, Sep. Purif. Technol. **43**, 117 (2005)
62. R.G. Casqueira, M.L. Torem, H.M. Kohler, Miner. Eng. **19**, 1388 (2006)
63. J. Rubio, M.L. Souza, S. Smith, Miner. Eng. **15**, 139 (2002)
64. D.R. Ketkar, R. Mallikarjunan, S. Venkatachalam, Int. J. Miner. Process. **31**, 127 (1991)
65. C. Lierena, J.C.K. Ho, D.L. Piran, Chem. Eng. Commun. **155**, 217 (1996)
66. C. Karagüzel, Miner. Eng. **23**(1), 17 (2010)
67. J.J. Cilliers, D.J. Bradshaw, Miner. Eng. **9**(2), 235 (1996)
68. P. George, A.V. Nguyen, G.J. Jameson, Miner. Eng. **17**(7–8), 847 (2004)
69. O. Sahbaz, A. Uçar, B. Oteyaka, Miner. Eng. **41**, 79 (2013)
70. O. Neiman, B. Hilscher, R. Siy, in *Proceedings of the 44th Canadian Mineral Processors Operators Conference*, Ottawa, Ontario, 17–19 January 2012 (Canadian Institute of Mining, Metallurgy and Petroleum), pp. 115–124
71. G.M. Evans, G.J. Jameson, Chem. Eng. Res. Des. **73**(6), 679 (1995)
72. G. Harbort, S. De Bono, D. Carr, V. Lawson, Miner. Eng. **16**, 1091 (2003)
73. A. Hassanzadeh, M. Safari, D.H. Hoang, H. Khoshdast, B. Albijanic, P.B. Kowalczuk, Miner. Eng. **180**, 107509 (2022)
74. A. Hasanzadeh, M. Safari, D.H. Hoang, Fine, coarse and fine-coarse particle flotation in mineral processing with a particular focus on the technological assessments, Paper presented at the 2nd international electronic conference on mineral science: part of the international electronic conference on mineral science series, 1–15 March 2021 (Online)
75. F.L. Smidth, *REFLUX™ Flotation Cell Enhanced Flotation Hydrodynamics*, https://www.flsmidth.com/-/media/brochures/brochures-products/flotation/reflux-flotation-cell-rfc-brochure.pdf. Accessed 19 June 2023
76. J. Chen, W. Chimonyo, Y. Peng, Miner. Eng. **181**, 107519 (2022)
77. M.J. Cole, J.E. Dickinson, K.P. Galvin, Sep. Purif. Technol. **240**, 116641 (2020)
78. J.L. Sutherland, J.E. Dickinson, K.P. Galvin, Miner. Eng. **149**, 106224 (2020)
79. K. Jiang, J.E. Dickinson, K.P. Galvin, Chem. Eng. Sci. **196**, 463 (2019)

80. Metso, Outotec Concorde Cell, https://www.mogroup.com/globalassets/portfolio/brochure-concordecell-4426-05-22.pdf. Accessed 19 June 2023
81. W.T. Massey, *Investigating the Effect of Energy Dissipation on Flotation Kinetics in an Oscillating Grid Flotation Cell*, Master's thesis, University of Cape Town, 2011
82. M. Safari, M. Harris, D. Deglon, L.L. Filho, F. Testa, Int. J. Miner. Process. **156**(10), 108 (2016)
83. E. Tabosa, K. Runge, K. Duffy, Paper presented at the sixth international flotation conference (Flotation '13)—MEI, Cape Town, South Africa, 18–21 November (2013)
84. J.S.J. Van Deventer, W.A. Van Dyk, L. Lorenzen, D. Feng, Miner. Eng. **15**(9), 659 (2002)
85. Xiong, Y., *Bubble Size Effects in Coal Flotation and Phosphate Reverse Flotation Using a Pico-nano Bubble Generator*, Ph.D. thesis, Department of Mining Engineering West Virginia University, 2014, pp. 17–18
86. S. Ata, G. Jameson, Miner. Eng. **45**, 121 (2013)
87. S. Fosu, B. Awatey, W. Skinner, M. Zanin, Miner. Eng. **77**, 137 (2015)
88. R.C. Dunne, in *Proceedings of the XXVI International Mineral Processing Congress-IMPC 2012*, New Delhi, India, 24–28 September (2012), pp. 1259–1268
89. J. Kohmuench, H. Thanasekaran, B. Seaman, Paper presented at metallurgical plant plant design and operating strategies (MetPlant 2013), Perth, Western Australia, 15–17 July (2013)
90. B. Oteyaka, Y. Uçbaş, K. Bilir, H. Özdağ, in *Changing Scopes in Mineral Processing: Proceedings of the 6th International Symposium*, Kusadasi, Turkey, 24–26 September (1996), p. 333
91. D. Tao, S. Yu, X. Zhou, R.Q. Honaker, B.K. Parekh, Int. J. Coal Prep. Util. **28**, 1 (2008)
92. F. Peng, X. Yu, Int. J. Min. Sci. Technol. **25**, 347 (2015)
93. Y. Yalcin, A. Byers, Miner. Proc. Ext. Metall. Rev. **27**(2), 87 (2007)
94. O. Sahbaz, Jameson hücresinde düşeyboru modifikasyonu ve flotasyon performansına etkisi, Doktora tezi, Kütahya Dumlupınar Üniversitesi Fen bilimleri Enstitüsü, Downcowner modification in Jameson cell and its effect on flotation performance, Ph.D. thesis, Kutahya Dumlupınar University Institute of Science, Kütahya, Türkiye, 2010
95. K. Bilir, A. Ucar, O. Sahbaz, Y.H. Gursoy, B. Oteyaka, Physicochem. Probl. Miner. Process **54**(1), 174 (2018)
96. J. Kohmuench, M. Mankosa, H. Thanasekaran, A. Hobert, Miner. Eng. **121**, 137 (2018)
97. M. Mankosa, A new paradigm in sulfide processing. Technical report, Eriez Flotation Division (2017), pp. 1–9
98. B. Awatey, H. Thanasekaran, J.N. Kohmuench, W. Skinner, M. Zanin, Miner. Eng. **50–51**, 99 (2013)
99. B. Awatey, W. Skinner, M. Zanin, Can. Metall. Q. **52**(4), 465 (2013)
100. J.N. Kohmuench, G.H. Luttrell, M.J. Mankosa, Miner. Metall. Process **18**(2), 61 (2001)
101. G.J. Jameson, C. Emer, Miner. Eng. **134**, 118 (2019)
102. L.J. Crompton, Md.T. Islam, K.P. Galvin, Minerals **12**(6), 783 (2022)

Differentiation Between Natural Quartz-Based on Thermoluminescence Properties

Shrouk Farouk, Ahmed Gad, and Nabil El-Faramawy

Abstract Numerous naturally occurring minerals serve as thermoluminescent dosimeter (TLD) materials. The thermoluminescence (TL) characteristics of these minerals are influenced by the specific type and concentration of trace element impurities they contain. Mineral formation and impurity concentration are both influenced by mineral's geological and geographical origin. Silicate minerals, including quartz and feldspar, along with certain oxide minerals like aluminum oxides, distinctly exhibit TL peaks in glow curves resulting from traps formed through thermal treatment and irradiation. Quartz (SiO_2), a particularly abundant silicate mineral in the Earth's crust, forms under various geological conditions, including magmatic, hydrothermal, sedimentary, and metamorphic processes. It plays a crucial role as a rock-forming mineral across all rock types (igneous, sedimentary, and metamorphic). Natural quartz, one of the dosimeters utilized in luminescence investigations, is significantly important for assessing the radiation record of materials in a wide range of applications including evaluating the authenticity of artifacts or retrospective dosimetry (nuclear accident). This chapter discusses TL phenomena and types of luminescence. The kinetic models which describe the TL phenomena were included. Different methods used to analyze TL glow curves and extracting the kinetic parameters were deliberated. The physical properties and the application of quartz were declared in this chapter. Thermoluminescence phenomena and Dosimetric parameters of Quartz collected from various places were shown clearly.

Keywords Quartz · Minerals · Thermoluminescence · Cathodoluminescence · Kinetic parameters · Dosimetric properties · Glow curve · TL analysis techniques

---

S. Farouk · N. El-Faramawy (✉)
Physics Department, Faculty of Science, Ain Shams University, Abbassia, Cairo 11566, Egypt
e-mail: nabil_elfaramawi1@sci.asu.edu.eg

A. Gad
Geology Department, Faculty of Science, Ain Shams University, Abbassia, Cairo 11566, Egypt

S. J. Ikhmayies (ed.), *Advances in Minerals Research*, Advances in Material Research and Technology, https://doi.org/10.1007/978-3-031-49175-7_4

1 Introduction

1.1 Luminescence

Luminescence can be described as the mechanism by which certain formerly irradiated substances emit light of specific wavelength upon stimulation. Though, luminescence could likewise be detected in the absence of previous irradiation (photoluminescence). Materials are classified as insulators, semiconductors, or metals according to their electronic structure [1]. Just insulators and semiconductors demonstrate luminescent characteristics. Such substances are recognized as phosphors. Luminescence is not the same as spontaneous light emission, which occurs when a used material is heated to incandescence [2]. The two primary steps in the luminescence phenomena are excitation and irradiation [3]. During the excitation step, the investigated material was subjected to an external energy source. Ionizing radiation such as nuclear radiation (e.g., γ-rays, and β-particles), and X-rays are widely accepted excitation sources. In addition, the excitation source could be optical (e.g., ultra-violet light) photons [2, 3]. The excitation generates electron–hole pairs that could subsequently radiatively recombine. Following excitation, the luminescence of the material is thermally or optically stimulated. Thermal stimulation is performed by progressively raising the material's temperature, at controlled heating rate, from a certain temperature (T_o) to final temperature (T), a process known as thermoluminescence [4]. When a specimen is subjected to a beam of monochromatic light (for example, blue spectrum of wavelength = 470 nm), the subsequent luminescence is commonly known as optically stimulated luminescence (OSL). Luminescence is generally divided into two types: fluorescence and phosphorescence.

1.2 Types of Luminescence

1.2.1 Fluorescence

If light is emitted after stimulation with a period (τ) $<10^{-8}$ s, the luminescence is recognized as fluorescence. The basic process of fluorescence emission is depicted in Fig. 1. After absorbing energy from an external source, the excited electron goes from the ground state (g) to the excited state (e). After the light emission, the moved electron returns to the initial ground state (g).

1.2.2 Phosphorescence

Phosphorescence occurs when light is emitted after stimulation with period (τ) $>10^{-8}$ s. The steps involved in phosphorescence are illustrated in Fig. 2. The phosphorescence energy level diagram differs from the fluorescence energy level diagram in

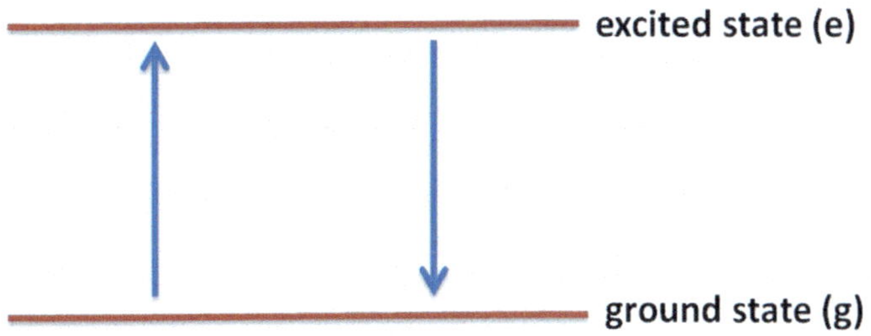

Fig. 1 The energy level diagram of the fluorescence process. Redrawn after McKeever [2]

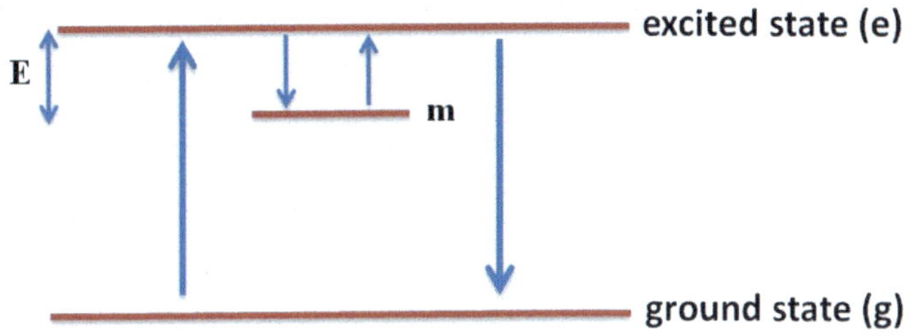

Fig. 2 The phosphorescence process's energy level diagram. Redrawn after McKeever [2]

that it includes an extra state labeled (m). An electron at level (e) enters and exits level (m) prior to actually relaxing to the initial level (g). The amount of time spent at level (m) is responsible for the majority of the time linking stimulation and light emission. Fluorescence is independent of temperature, whereas phosphorescence is dependent on temperature. Phosphorescence's thermal energy assists in the release of the electron from level (m). Therefore, heating energy should consequently be preferably equivalent to the difference in energy that exists between levels (e) and (m).

1.2.3 Thermoluminescence (TL)

TL is the thermally stimulated light emission from a previously irradiated substance [1, 2]. The energy band model of solid states indicates that electrons in crystalline structures can only fill a specific group of energy levels. These permitted energy levels are disconnected by prohibited gaps for which there is no Schrödinger Eigen function [4]. Therefore, localized energy levels should not exist within the forbidden gap in a perfect crystal. Defects or imperfections in natural crystals provide extra energy levels inside the crystal. As a result, electrons are now able to exist at energy levels that are inside the forbidden energy gaps. Insulators and semiconductors, which have wide and narrow gaps, are both suitable examples for luminescence investigations.

For instance, in natural quartz mineral the forbidden energy gap is approximately 9 eV in width [5].

According to the energy band model, the valence band is the lowest possible energy band that electrons are permitted to fill. The conduction band represents the next available energy band. Valence electrons in the substance gain energy and become liberated to move inside the crystal when it is exposed to ionizing radiation, forming electron–hole pairs. A positively charged hole signifies that the valence band is empty. Free electrons move across the conduction band, a few of them become trapped at localized energy levels (electron traps). Likewise, holes may become trapped at hole centers. If the energy from thermal stimulation equals or exceeds (E), the trapped electrons will be liberated from the trap and transferred into the conduction band (the trap's energy depth under the conduction band). Some of the trapped electrons recombine with holes after being re-trapped. If the recombination process is radiative, luminescence is produced. In the circumstances of non-radiative recombination, the energy released by electron–hole recombination is dispersed as phonons.

1.2.4 Cathodoluminescence (CL)

Cathodoluminescence (CL) is the emission produced in cases where excitation is made by the electron-beam generated in the electrical cathodes. This specific type of emission can be observed by the glow on displays of cathode-ray tubes and older versions of TVs to glow. Zinc and cadmium sulfide phosphors are utilized in cathode ray tubes. The manufacturing process of phosphors for TV screens is a highly technical procedure that necessitates a wide range of colors and their proper perseverance to smooth out scan flicker.

1.2.5 Radioluminescence

Radioluminescence is the term used to describe luminescence that results when α, β, and γ rays or X-rays are used as the excitation source. To distinguish it from X-ray fluorescence (XRF), X-ray excited optical luminescence (XEOL) is another term for X-ray luminescence. Both are excited by X-rays; however, in XEOL, optical photons are emitted, whereas in XRF, X-ray photons are emitted. Several inorganic minerals exhibit strong luminescence when excited by X-rays and nuclear particles, but none when excited by UV light. The luminescent dials of antique watches and numerous instruments glow in the dark because of the luminescence stimulated by β-particles of ^3H incorporated into the ZnS phosphor. Because they contain radioactive materials, they are no longer widely used.

1.2.6 Electroluminescence (EL)

Many phosphors can illuminate when electric fields are applied to them. Electroluminescence has a complicated mechanism. Electrons cannot be released from their valence states by electric fields solely. When electric fields are applied, steep voltage gradients are thought to be reproduced across crystal defects. In these locations, electrons are sufficiently pushed to acquire the energy necessary to ionize the atoms. Typically, varying voltages are placed within the phosphors' tinny layers. Injection luminescence, which involves injecting electrons across a semiconductor p-n junction from an external source, is another type of electroluminescence. When a DC voltage is applied throughout the junction, electrons move to the p region, where electron–hole recombination produces luminescence. This principle enables the usage of light emitting diodes (LED) as display devices in many scientific instruments. LEDs are presently being drawn up as general-purpose illuminators, with the goal of eventually replacing mercury-vapor fluorescent light bulbs. Bhushan [6] has provided an extensive overview on electroluminescence.

1.2.7 Chemiluminescence

Luminescence is a result of specific chemical reactions. The best-known case of this is when white phosphorus oxidizes in the air. Certain molecules emit energy as light when they are in high-energy electronic states, although most of the substance is kept at room temperature. On the other hand, chemiluminescence is not a typical companion to many chemical reactions since the quantity of released energy, even in exothermic reactions, is inadequate to generate electronic excitation, that requires multiple energy eV. Second, not all chemical molecules can emit light. Lyoluminescence is a kind of chemiluminescence that occurs after the dissolution of specific compounds that have previously been inundated with X-rays. An example that is well-known for emitting a flash of light when rapidly dissolves in water is X-irradiated NaCl. When released during dissolution, electrons trapped in the NaCl crystal lattice due to irradiation are assumed to stimulate the luminescence.

1.2.8 Bioluminescence

Electrically excited states of biomolecules are generated by biochemical reactions that take place inside live organisms' cells. Fireflies, glow worms, certain kinds of bacteria and fungus, and many marine organisms (such as phytoplankton) exhibit luminescence both at the surface and at extreme depths. Enzymic oxidations are chemical processes. For example, the light of the firefly was described as a result of luciferin oxidation with the enzyme luciferase. Similarly, marine mussel and the luminous bacteria emit light, despite the fact that the molecular composition of the luciferin in each of these is thought to be extremely different.

1.3 *The Mathematical Treatment of TL*

1.3.1 The One-Trap and One-Recombination Center (OTOR) Model

The one trap, one recombination center (OTOR) concept is depicted in Fig. 3. The trap T gradually empties under thermal stimulation as electrons transition to the conduction band (transition c). The probability p of an electron being released from the trap once every unit of time, which is proportional to the Boltzmann factor, estimated by Arrhenius equation:

$$p = s\exp\left(-\frac{E}{kT}\right) \quad (1)$$

where s stands for the frequency factor, also known as the attempt-to-escape factor [4, 7]. S is a constant that is independent of temperature in this simple model. Its value is in the range of the frequency of lattice vibration, namely 10^{12}–10^{14} s^{-1} [8]. The energy required to release a trapped electron into the conduction band is known as the trap depth (E), k is Boltzmann's constant, and T is the absolute temperature. Equation (1) shows that if $E \geq kT_0$, where T_0 is the temperature at the irradiation time, when p is extremely small, the electron remains trapped permanently.

The probability p rises with temperature, allowing trapped electrons to be released into the conduction band. Light is released as a result of the recombination of conduction electrons and holes at the center R. The rate of electrons and holes recombination is proportional to the intensity of the light that is emitted $I(t)$ as stated in Eq. (2):

$$I(t) = -\frac{dm}{dt} \quad (2)$$

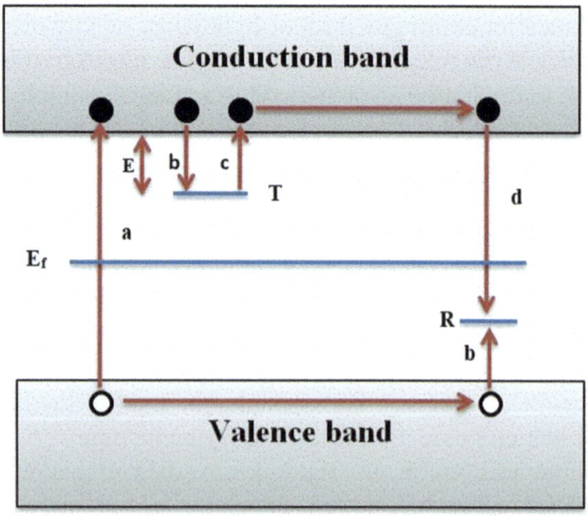

Fig. 3 The energy band model shows electronic transitions in a thermoluminescent (TL) material. **a** Creation of electrons and holes, **b** electron and hole trapping, **c** releasing of electrons due to thermal stimulation, and **d** recombination. Redrawn after Bos [7]

The negative sign denotes a decrease in the concentration of holes (m) because of recombination. But Eq. (2) indicates the recombination rate is proportional to the electron concentration in the conduction band and the hole concentration m at the recombination center R, both of which are thought to be Radiative processes. Equation (2) is now changed to Eq. (3):

$$I(t) = -\frac{dm}{dt} = n_c m A_m \tag{3}$$

where the electron concentration in the conduction band at time t is indicated by n_c. A_m is the assumed temperature-independent recombination probability represented in terms of volume per unit time. The rate of change in the trapped electron concentration n with respect to time t showed in Eq. (4):

$$\frac{dn}{dt} = -np + n_c(N - n)A_n \tag{4}$$

where A_n is the possibility of re-trapping (measured in units of volume per unit time) and N is the concentration of electron traps. Therefore, the following definition applies to the rate of change in the concentration of free electrons as:

$$\frac{dn_c}{dt} = np - n_c(N - n)A_n - n_c m A_m = -\frac{dn}{dt} + \frac{dm}{dt} \tag{5}$$

Also, charge is required to be conserved, Eq. (6) implies that:

$$n_c + n = m \tag{6}$$

Equations (1)–(5) describe the charge transfer resulting in TL. There is no analytical solution to these equations. However, using simplified assumptions, a general solution could be constructed. It is expected that charges do not accumulate in the conduction band at this phase of the derivation. In another means, the concentration of electrons in the conduction band is essentially constant. The quasi-equilibrium (QE) assumption is the name ascribed to this supposition [4]. Mathematically, the QE assumption turns to Eq. (7):

$$\left|\frac{dn_c}{dt}\right| \ll \left|\frac{dn}{dt}\right|; \left|\frac{dn_c}{dt}\right| \ll \left|\frac{dm}{dt}\right| \tag{7}$$

The above assumption indicates that $n_c \approx 0$ which then leads to $n \approx m$ hence Eq. (5) declines to Eq. (8):

$$-dm/dt \approx -dn/dt = I(t) \tag{8}$$

with the QE assumption, Eqs. (3) and (5) lead to Eq. (9):

$$I(t) = \frac{mA_m np}{(N-m)A_n + mA_m} = \frac{mA_m}{(N-n)A_n + mA_m} ns \exp\left(-\frac{E}{kT}\right) \quad (9)$$

Equation (9) is a representation of the general one trap (GOT) equation. The GOT equation can be used to build equations that describe first, second, and general order kinetics under certain conditions.

1.3.2 First-Order Kinetics

First-order kinetics was derived by Randall and Wilkins [9]. They concluded that once electrons are trapped, the possibility of them being re-trapped is negligible, which: $mA_m >> (N - n)\, A_n$. This assumption leads to a reduction of the GOT equation to Eq. (10):

$$I(t) = -\frac{dn}{dt} = sp = ns \exp\left(-\frac{E}{kT}\right) \quad (10)$$

Equation (10) is considered as the first-order kinetics equation since it indicates that the intensity, I is proportional to n. If the temperature is constant during the TL reading, in which case p is constant the intensity follows an exponential decrease of the form as shown in Eq. (11):

$$I(t) = I_0 \exp(-t.p) \quad (11)$$

where I_0 is the initial intensity at the time t. If the recorded temperature changes during the readout process while assuming a linear heating profile the temperature of the sample at time t is obtained as shown in in Eq. (12):

$$T(t) = T_0 + \beta.t \quad (12)$$

where β is the heating rate. By integrating Eq. (10) can determine the concentration of trapped electrons at time t and obtain Eq. (13):

$$n(t) = n_0 \exp\left[-\int_{T_0}^{T} \frac{s}{\beta} e^{-\frac{E}{kT'}} dT'\right] \quad (13)$$

where n_0 is the trapped electrons concentration at time $t = 0$, i.e., before stimulation, and T' is the variable of the integration that reveals the temperature. The luminescence intensity $I(T)$ will be obtained by deriving Eq. (13) with regard to T as illustrated in Eq. (14):

$$I(T) = -\frac{1}{\beta}\frac{dn}{dt} = n_0\frac{s}{\beta}e^{-E/kT}\exp\left[-\int_{T_0}^{T}\frac{s}{\beta}e^{-\frac{E}{kT'}}\mathrm{d}T'\right] \quad (14)$$

When a first-order glow curve (a plot of luminescence intensity as a function of temperature) reaches its maximum intensity, the derivative of Eq. (14) with respect to temperature is equal to zero ($dI/dT = 0$) at maximum $I(T)$, and T_m is the temperature corresponding to the maximum intensity. The equation for the relation between T_m and the other factors is:

$$\frac{\beta E}{kT_m^2} = s\exp\left(-\frac{E}{kT_m}\right) \quad (15)$$

According to Eq. (15), T_m is independent of dose for first-order kinetics and does not depend on n_0, which is an attribute of first-order kinetics [4, 10].

1.3.3 Second-Order Kinetics

Garlick and Gibson [11] discussed about an instance of re-trapping arising i.e., $mA_m << (N - n)\,A_n$ and this leads to Eq. (16):

$$I(t) = -\frac{dn}{dt} = m\frac{A_m}{(N-n)A_n}ns\exp\left(-\frac{E}{kT}\right) \quad (16)$$

Additionally, If the trap is considered to be close from being fully saturated, i.e., $N >> n$ and $n = m$, we obtain Eq. (17):

$$I(t) = -\frac{dn}{dt} = s\frac{A_m}{NA_n}n^2\exp\left(-\frac{E}{kT}\right) \quad (17)$$

In Eq. (17) intensity $I(t)$ is proportional to n^2, it is recognized as the second-order kinetics equation. Assuming a linear heating process that $A_m = A_n$, the luminescence intensity can be obtained as a function of temperature by integrating Eq. (17) to get Eq. (18):

$$I(T) = \frac{n_0^2 s}{N\beta}\exp\left(-\frac{E}{kT}\right)\left[1 + \frac{n_0 s}{N\beta}\int_{T_0}^{T}\exp\left(-\frac{E}{kT'}\right).dT'\right]^{-2} \quad (18)$$

If the glow curve is observed at constant temperature, the phosphorescence in second order kinetics obeys a Becquerel's decay law of a type illustrated by Eq. (19):

$$I(t) = \frac{I_0}{(1 + n_0\alpha t)^2} \quad (19)$$

where $\alpha = s/N \exp(-E/kT)$. By taking the derivative of Eq. (18) with respect to temperature and setting it to zero, it is possible to determine the temperature T_m for second-order kinetics that corresponds to the maximum intensity of a TL glow peak. The relationship between T_m and the other variables is provided by Eq. (20):

$$\frac{\beta E}{kT_m^2} = s\left(1 + \frac{2kT_m}{E}\right)\exp\left(-\frac{E}{kT_m}\right) \tag{20}$$

1.3.4 General-Order Kinetics

The first- and second-order kinetics were derived using the presumptions of May and Partridge [12]. They created the formula for general-order kinetics shown in Eq. (21):

$$I(t) = -\frac{dn}{dt} = n^b s^a \exp\left(-\frac{E}{kT}\right) \tag{21}$$

where s^a has the dimension of $m^{3\,(b-1)}\,S^{-1}$ and b is the order of kinetics. The derived equation of intensity similar to Eq. (13) becomes as Eq. (22):

$$I(T) = \frac{s^*}{\beta} n_0 \exp\left(-\frac{E}{kT}\right)\left[1 + (b-1)\frac{s^*}{\beta}\int_{T_0}^{T} \exp\left(-\frac{E}{kT'}\right) dT'\right]^{-b/(b-1)} \tag{22}$$

where $s^* = s^a\, n_0^{\,(b-1)}$. If b approaches 1 or 2, Eq. (22) can be reduced to the proper equation for either first- or second-order kinetics. General order kinetics using $dI/dT = 0$, we can obtain Eq. (23):

$$\frac{\beta E}{kT_m^2} = s\left(1 + \frac{2kT_m(b-1)}{E}\right)\exp\left(-\frac{E}{kT_m}\right) \tag{23}$$

1.4 Analysis Methods for TL Glow Curve

A TL glow's properties in the context of general kinetics depend on the following kinetic parameters: Peak, glow-curve, thermal activation energy E, frequency factor s, and Kinetics order (KO) b. As a result, the traps of the thermoluminescent material are revealed by the glow curve. Determining trap settings and interpreting the glow curve are therefore important in dosimetry.

1.4.1 Empirical Methods

The stated empirical method involves correlating the glow curve's T_m of the maximum peak temperature with other temperatures. Solving $dI/dt = 0$ yields the activation energy E in Eqs. (10), (14), and (22). For first-order (FO) kinetics use:

$$\frac{s}{\beta} = \frac{E}{kT_m^2} \exp\left(\frac{E}{kT_m}\right) \tag{24}$$

Although this equation cannot be solved directly to provide T_m as a function of E, numerical computations were utilized to determine T_m for various combinations of E and s/β. If we use $s/\beta = 10^{13}$, we obtain $E = T_m/304$. The first method for determining activation energy was developed by Urbach [13], who discovered empirically that $E = Tm/500$ yields a reliable approximation of the trap energy, and that E is measured in electron volts. Halperin and Braner [14] achieved the relationship $E = 38\,kT_m$ for NaCl samples. Miller and Bube [15] attained that $E = 39\,kT_m$ for LiF.

1.4.2 Initial Rise (IR) Method

The term $\exp(-E/kT)$ was observed to be present in the TL Eqs. (14), (18), and (22) in the three models previously described. This factor predominates in the low-temperature interval of the glow peak regardless of the model. This information is used by the IR method to approximately determine the trap depth E. Garlick and Gibson [11] recommended the use of this technique. This method disregards the kinetic order when calculating E. In the low-temperature interval of the glow peak we have Eq. (25):

$$I(t) \propto \exp\left(-\frac{E}{kT}\right) \tag{25}$$

The Arrhenius plot, also known as the $\ln[I(T)]$ versus $(1/T)$ plot, produces a straight line with a slope of $-E/k$, from which the value of E can be derived. If the intensity is greater than 15% of the glow peak I_m maximum intensity, the IR technique shouldn't be employed. The downside of this technique is that it can only be used when the light curves have one or more distinct, non-overlapping peaks.

1.4.3 Various Heating Rate (VHR) Method

Based on the relationship between the maximum peak temperature T_m and the heating rate β, the various heating rate (VHR) approach is used. According to this correlation, it has been shown that T_m shifts as β changes. The relationship was shown in Eq. (15), and it may be expressed as Eq. (26) for the first-order kinetics model:

$$\beta = s\frac{kT_m^2}{E}\exp\left(-\frac{E}{kT_m}\right) \tag{26}$$

As E and s are constants for a given glow peak, T_m increases as β increases.

Equation (15) can be used to get the maximum intensity temperatures T_{m1} and T_{m2} for two distinct heating rates, β_1, and β_2, Eq. (27) can be obtained:

$$E = s\frac{kT_{m1}T_{m2}}{T_{m1} - T_{m2}}\ln\left(\frac{\beta_1}{\beta_2}\left(\frac{T_{m2}}{T_{m1}}\right)^2\right) \tag{27}$$

If T_m can be measured with an accuracy of 1%, this method can be used to determine the value of E with 5% accuracy. Equation (15) can be changed to find the value of s once E has been established.

By rearranging Eq. (20) and taking the natural logarithm of its two sides, Hoogenstraaten [16] proposed an additional technique to determine E using Eq. (28):

$$\ln\left(\frac{T_m^2}{\beta}\right) = \frac{E}{k}\frac{1}{T_m} - \ln\left(\frac{sk}{E}\right) \tag{28}$$

Then using multiple heating rates and plotting $\ln\left(\frac{T_m^2}{\beta}\right)$ against $\frac{1}{T_m}$, a straight line is obtained. The slope of this straight line can be used to compute the activation energy E, and its vertical intercept can be used to get the value of s. While being based on a first-order kinetics model, the various heating rate method has been shown to be a reliable approximation for non-first order kinetics [17, 18].

1.4.4 Isothermal Decay Method

This approach was first presented for single FO glow peaks by Randal and Wilkins [19]. They showed that it is possible to determine the value of E of an isolated FO glow peak by simply charting the peak's isothermal decay at various temperatures. It is possible to express the TL intensity $I(t)$ at a specific temperature T using Eq. (29):

$$I(t) = I_0\exp(-P(T)t) \tag{29}$$

where $P(T)$ is the intensity decay constant determined by Eq. (1) and I_0 is the intensity at time $t = 0$. The value of $P(T)$ for the peak under consideration can be obtained by plotting $\ln[I(t)/I_0]$ versus t. Following that, the value of E can be obtained by evaluating $P(T)$ at two distinct temperatures, let's say T_1 and T_2, and then using $P(T_1)$ and $P(T_2)$ to obtain Eq. (30):

$$E = s\frac{kT_1T_2}{T_1 - T_2}\ln\left(\frac{P(T_1)}{P(T_2)}\right) \tag{30}$$

The value of s may then be calculated by substituting the resulting value of E, T, and $P(T)$ in Eq. (1).

The provided isothermal decay approach can be used in the general situation of any order of kinetics, regardless of the fact that it was initially introduced for FO glow peaks [12, 20, 21]. But in this case, the decline pattern won't be exponential. Equation (21) from the GO kinetics equation can be written as Eq. (31):

$$\frac{dn}{n^b} = s^a \exp\left(-\frac{E}{kT}\right)dt \tag{31}$$

Integrating this previous equation and substituting the value of n into Eq. (21) yields Eq. (32):

$$\left(\frac{I}{I_0}\right)^{\frac{b}{b-1}} = 1 + n_0^{b-1}(\mathrm{b} - 1)s^a \exp\left(-\frac{E}{kT}\right)t \tag{32}$$

This shows that a plot between $\left(\frac{I}{I_0}\right)^{\frac{b}{b-1}}$ versus t at a given temperature T produces a straight line with a slope N defined in Eq. (33):

$$N = n_0^{b-1}(b - 1)s^a \exp\left(-\frac{E}{kT}\right) \tag{33}$$

The correct value of b is necessary to construct a straight line for $\left(\frac{I}{I_0}\right)^{\frac{b}{b-1}}$ versus t. The natural logarithm of the slope N observed for the same dose n_0 at various temperatures T is then plotted against $1/T$ to determine the value of E.

1.4.5 Peak Shape Method

This approach was developed using an approximation of the glow peak depicted in Fig. 4. The approach simply makes use of the three-temperature points T_m, T_1, and T_2. Numerous terms can be deduced to derive the kinetic parameters of the light peaks using these three temperature points [22, 23].

The following geometrical parameters should be determined: The width at half maximum intensity $\omega = T_2 - T_1$, the high-temperature half width $\delta = T_2 - T_m$, and the low-temperature half width $\tau = T_m - T_1$ [23]. Another geometrical parameter μ_g is defined in terms of δ and ω as shown in Eq. (34):

$$\mu_g = \frac{\delta}{\omega} = \frac{T_2 - T_m}{T_2 - T_1} \tag{34}$$

The general equations developed by Chen [22] can be used to compute the activation energy (E), as shown in Eq. (35):

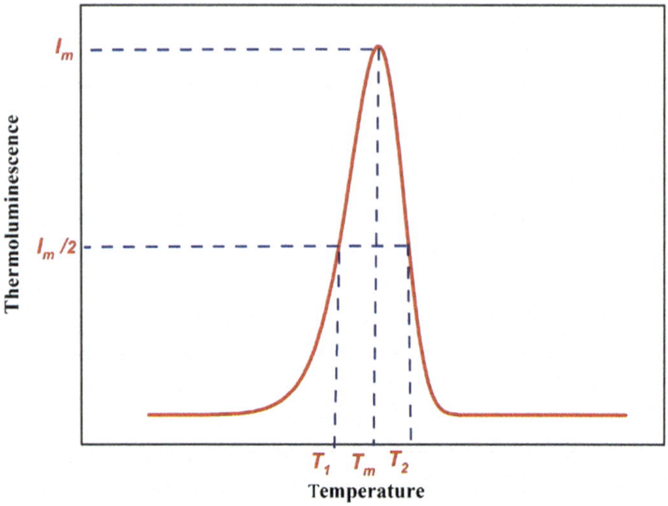

Fig. 4 TL glow curve peak described by T_1, T_2, and T_m. Redrawn after McKeever [2]

$$E = C_\alpha \frac{kT_m^2}{\alpha} - 2kT_m b_\alpha \tag{35}$$

where α stands for τ, δ and ω respectively, C_α, and b_α are attained using the expressions given in Eqs. (36–41):

$$C_\tau = 1.51 + 3(\mu_g - 0.42) \tag{36}$$

$$b_\tau = 1.58 + 4.2(\mu_g - 0.42) \tag{37}$$

$$C_\delta = 0.976 + 7.3(\mu_g - 0.42) \tag{38}$$

$$b_\delta = 0 \tag{39}$$

$$C_\omega = 2.52 + 10.2(\mu_g - 0.42) \tag{40}$$

$$b_\omega = 1 \tag{41}$$

Another parameter proposed by Balarin [23], γ defined by Eq. (42):

$$\gamma = \frac{\delta}{\tau} = \frac{T_2 - T_m}{T_m - T_1} \tag{42}$$

In first- and second-order kinetics, μ_g has values of 0.42 and 0.52 respectively, but in first-order kinetics, γ varies from 0.7 to 0.8 and in second-order kinetics, γ is between 1.05 and 1.2 [23]. The frequency factor s can be computed from the circumstance at the greatest intensity of Eq. (22) to obtain Eq. (43) as the activation energy is obtained:

$$s = \left(\frac{\beta E}{kT_m^2}\right)\frac{1}{1+(b-1)\frac{2kT_m}{E}}\exp\left(\frac{E}{kT_m}\right) \tag{43}$$

1.4.6 Curve Fitting Method

The aforementioned approaches can only examine one peak at a time and need for the peak to be pure, as was explained in the preceding sections on studying glow peaks. Contrarily, curve fitting can be utilized to simultaneously examine each glow peak in a multi-peak glow-curve. The basic method requires the following two steps: (1) choosing the TL equation that best matches the data points. One can choose to fit data using the first-order equation Eq. (14) based on the glow-curve versus dose–response. This would indicate that $b = 1$ is the KO determined by all peaks. E, s, and n_0 serve as the fitting parameters in this instance. But if the GO equation was chosen to match the data, then the number of fitting parameters would increase to four: E, s, b, and n_0. (2) A clear determination of how many glow-peaks are required to create this glow-curve. Numerous approaches were put forth to ascertain the amount of TL components, including initial rise, isothermal decay, various heating rates, T_M-T_{stop} "second derivative of the glow curve," etc. [2, 24, 25]. Once the chosen equation spanning the full glow curve has been selected, the fitting operation is normally carried out by using linear superposition. You can adjust each fitting parameter independently until a favorable match is found.

1.5 *Quartz*

At low temperatures and pressures, quartz mineral (α-quartz), is a stable form of silicon dioxide (SiO_2) [26]. There are known mineral modifications (at least 15), or polymorphs having the chemical composition of SiO_2 but a distinct atomical arrangement and crystal structure. β-quartz, tridymite, cristobalite, coesite, stishovite, and keatite are the well-known forms of polymorphs. The various silica (SiO_2) modifications, particularly quartz, have an immense impact in the earth's geologic materials composition. Because of their plenty and characteristics, silica minerals were used for a variety of applications since the earliest days of time, including tools, weapons, jewelry, and construction materials. Quartz is a common raw material used in high-tech industries (e.g., piezo quartz, optoelectronic appliances, renewable energy sector) or as bulk products in modern industries (e.g., glassmaking, ceramic, refractories, smelting, and metal casting, abrasives, and filler in paints) [27].

Quartz possesses distinctive characteristics (e.g., crystalline structure, color, trace element, isotopes, and luminescence attributes) that vary from the point defects to the macro-appearance and are influenced by their geological context and formation settings. Quartz is commonly colorless, although various colored kinds were identified, including pink (rose quartz), grey (smoky), white (milky), purple (amethyst), and yellow (citrine). The luster is glassy, and no cleavage, therefore conchoidal fracture occurs. Quartz has considerable high hardness (7) on Mohs scale, and moderate density (2.67 g/cm^3). It is positive anisotropic uniaxial mineral and has a maximum birefringence of 9.5×10^{-3} nm/cm. Quartz is composed of corner-sharing tetrahedral arrangement (SiO_4), with each Si atom that is attached to four O atoms and each O atom connected to two Si atoms. Due to this three-dimensional crystallization, quartz is categorized as tectosilicate (framework) [28, 29].

Quartz is a mineral occurring abundantly in igneous, metamorphic, hydrothermal, and sedimentary geologic settings. It is prevalent in acidic and intermediate plutonic igneous rocks but missing at higher mafic crystallizations. It is widespread in acidic volcanic rocks (rhyolites), but rare in silicic vitric tuffs. It is prevalent in pelitic to granitic schists and gneisses in metamorphic rocks. As the main gangue mineral, it is prevalent in hydrothermal mineral assemblages, and in low to high temperature ore deposits. In addition to being the main mineral phase in sandstone rock varieties, it is common in other non-marine sedimentary rocks due to its resistance against chemical weathering [26, 28].

Quartz is usually a homogeneous mixture of pure silica, containing about 0.2% of total impurities. These impurities can be observed as lattice-bound trace elements or inclusions (submicron, nanoscale, mineral, melt, and fluid inclusions). Trace element concentrations are primarily influenced by the melt (fluid of formation) fractionation, temperature, and chemistry. The redistribution of incorporated trace elements is mainly driven by post-crystallization processes such as alteration, metamorphism, and metasomatism, which cause secondary quartz to recrystallize and neocrystallize by replacing primary quartz [28]. Knowledge-based of the links between crystallization conditions as well as different attributes, possibly utilized to reconstruct geological processes as well as for specific technical applications [28, 29].

2 Practical Work

2.1 Preparation of Quartz Samples

Distinct steps are routinely implemented to prepare Quartz samples for experimental work. To ensure the usage of clean samples, distilled water is used as a washing solution [30–32]. If other mineral phases are noticed in the quartz samples such as calcium carbonate ($CaCO_3$) and organic matter (OM) the samples are treated with

HCl (10%) and H_2O_2 (10%), respectively [33]. To obtain the suitable fine grain size fraction samples (63–90 μm) [33] (80–180 μm) [34] samples were pulverized in a ceramic mill and sieved. An alternative approach was employed by Farouk et al. [31, 32] by cutting the samples by rock cutting machine into compact square disks (5 mm) with a thickness of 1 mm.

2.2 *Annealing Process*

In order to prepare a thermoluminescent material (TL) for use in dosimetry applications, it is extremely important to perform a thermal treatment. This treatment is generally called annealing process. The basic aim of this process is to empty the traps from any electron stored on them and stabilize the electron traps to be obtained. The annealing process carried out in oven or/and furnace in which heating up the TL material to a certain temperature is performed, keeping it at that temperature for a certain period before cooling it down to room temperature. Each type of TL materials whether natural or synthetic have a different condition of annealing and cooling to be ready for use. Thomas and Chithambo [35] They enhanced their samples' sensitivity by annealing them at 900 °C for one hour prior to usage. To eliminate the cumulative dose related to natural irradiation, Pedro et al. [36] performed heat-treatment for all of their samples at 400 °C for 1 h before being steadily cooled to room temperature. The Quartz samples studied by Guzzo et al. [37], were annealed using different procedures: heating to 400 °C, annealing for one hour at that temperature, cooling, annealing for two hours at 100 °C, and cooling once more. After each high-dose irradiation increment, this procedure was repeated at least three times. Farouk et al. [31, 32] annealed their quartz samples at 400 °C for one hour followed by slow cooling.

2.3 *TL-Measurements*

The thermoluminescent dosimeter (TLD) reader consists of three parts. One is a heating part, which applies heat to the TL material to release trapped electron. The second part is a photomultiplier tube for the detection of the released light, where the quantity of radiation impacting the phosphor is proportional to the number of photons detected. The third part is a registration system, which records the outgoing signals. Thomas and Chithambo [35] carried out their TL measurements using a RISØ-TL/OSL/DA-20 Luminescence Reader. At room temperature, samples were exposed to radiation from a $^{90}Sr/^{90}Y$ beta source at a nominal rate of 0.1028 Gy/s. Pedro et al. [36] studied the glow curves from 50 to 400 °C with a linear heating rate of 2 k s^{-1} using a Harshaw-Bicron 3500 TLD reader fitted with a 9125-B photomultiplier tube devoid of filtering. Kalita and Wary [30] used a TL reader (TL1009I, NUCLEONIX) to measure TL within 30 s of X-ray irradiation at a constant linear heating of 2 K/s from 300 to 520 K. A quartz filter was incorporated into the

TL reader's design in front of the photomultiplier tube to pass UV–Vis light but absorb infrared wavelengths. Farias and Watanabe [34] studied their glow curves of quartz samples that had been irradiated to 10.000 Gy using the Daybreak model 1100 TL reader. Thermoluminescence properties of Egyptian milky quartz samples were investigated by a Harshaw Model 3500 TLD Reader operated by WinREMS, which operates on a personal computer and is linked to the reader through a serial port, as well as nitrogen cooling system. The glow curves of the samples were recorded from 323 to 673 K at a constant rate of 2 k/s [31, 32].

3 Quartz Thermoluminescence Properties Worldwide

3.1 Natural White Quartz (Gelephu, Bhutan)

The powder sample was subjected to X-ray diffraction (XRD) analysis. The samples' crystalline form is well shown by the sharp and intense peaks. The chemical analysis revealed that the quantity of O and Si were much higher than other elements (Table 1) [30].

The trapping parameters such as *E*, *b*, and frequency factor *s*, etc., were estimated by Kalita and Wary [30] for 15, 30, 45, 60, 75, and 90 mGy irradiation doses and listed in Table 2. After carefully choosing the fitting parameters, it was discovered that the glow curves contained a special arrangement of five overlapping peaks. The deconvolved glow curve of sample irradiated at 60 mGy dose is shown in Fig. 5.

In two cases, the dosage response was investigated. The first comes from the entire glow curve, and the second from the de-convolved peak V. Figure 6 depicts a log–log plot of peak integrals in a function of irradiation dose. The error bars show the standard deviations (st.d.) of averages of the peak integrations obtained from three samples from the same group.

The dose response was non-linear at the time the peak area was calculated based on the entire glow curve, as shown in Fig. 6. Nevertheless, the dose–response determined starting at the peak V is linear within a dose range 10 mGy–10 Gy. The first case's non-linear dose–response is caused by the non-uniform loading of trap charges at

Table 1 Elemental analysis of the sample [30]

Element	wt%	at%
O K_α	55.68	69.08
Si K_α	43.29	30.61
Ca K_α	0.25	0.12
Fe K_α	0.24	0.09
Ni K_α	0.18	0.06
Cd K_α	0.37	0.07
Total	100.00	–

Table 2 TL parameters of quartz evaluated by CGCD analysis [30]

Dose (mGy)	Peak	T_m (K)	b	E (eV)	s (s^{-1})	FOM %[a]
15	I	340.2	1.01	0.68	1.64×10^9	1.21
	II	362.3	1.5	0.9	5.22×10^{11}	
	III	383.2	1.01	0.97	8.92×10^{11}	
	IV	397.4	1.03	1.06	4.64×10^{12}	
	V	426.1	1.1	1.1	1.40×10^{12}	
30	I	341.2	1.02	0.67	1.52×10^9	1.53
	II	360.2	1.54	0.92	6.20×10^{11}	
	III	382.4	1.06	0.96	9.68×10^{11}	
	IV	396.3	1.08	1.05	4.85×10^{12}	
	V	424.2	1.14	1.11	1.69×10^{12}	
45	I	341.3	1.01	0.68	1.56×10^9	1.29
	II	363	1.59	0.91	6.43×10^{11}	
	III	380.1	1.04	0.97	9.83×10^{11}	
	IV	398	1.01	1.06	4.55×10^{12}	
	V	425.7	1.1	1.12	1.45×10^{12}	
60	I	340.9	1.03	0.69	1.73×10^9	1.37
	II	361.3	1.51	0.9	7.11×10^{11}	
	III	381.4	1.02	0.96	7.42×10^{11}	
	IV	397.5	1.07	1.07	5.73×10^{12}	
	V	424.1	1.07	1.11	2.21×10^{12}	
75	I	341.3	1.03	0.67	1.93×10^9	1.59
	II	362.3	1.56	0.91	6.32×10^{11}	
	III	383.4	1.07	0.97	7.44×10^{11}	
	IV	396.6	1.09	1.06	5.21×10^{12}	
	V	426.5	1.09	1.1	2.43×10^{12}	
90	I	341.3	1.02	0.68	1.19×10^9	1.15
	II	361.2	1.6	0.92	7.22×10^{11}	
	III	382.2	1.03	0.96	9.43×10^{11}	
	IV	398.1	1.04	1.07	7.22×10^{12}	
	V	425.7	1.13	1.1	2.33×10^{12}	

[a] *FOM* Figure of merit

trapping sites. Figure 7 depicts how the peak integrals of the whole glow curves and the simulated peak V change over time.

The figure shows that the TL signals faded quite early after irradiation, although the fading rate decreased after 5 days. Nevertheless, the fading rate of peak V was quite low. Within 5 days of irradiation, the entire TL signal faded ~24%, and this rate continued to be ~6% throughout the following 30 days. The TL signal associated

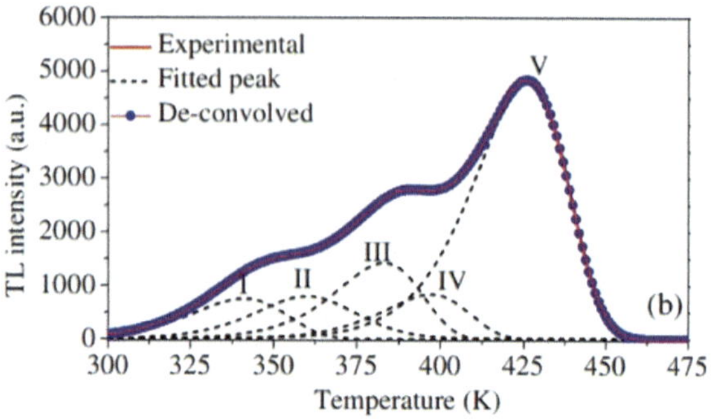

Fig. 5 The 60 mGy dosage irradiated sample's de-convolved glow curves. Reprinted from Kalita and Wary [30], with permission from Elsevier. Copyright (2016)

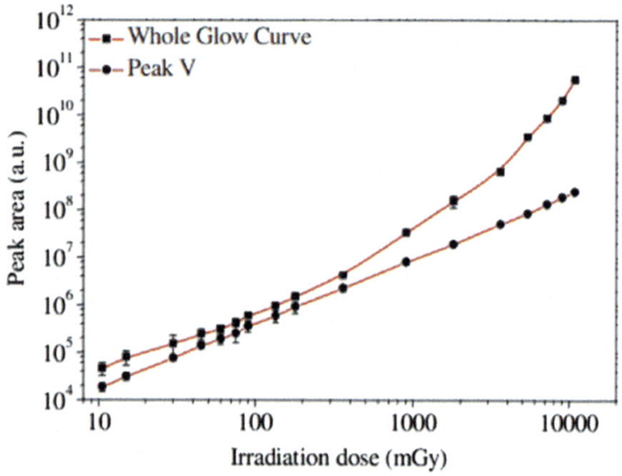

Fig. 6 A plot of peak area versus irradiation dose in log–log scale, derived from the entire glow curve (square) and the peak V (circle). Reprinted from Kalita and Wary [30], with permission from Elsevier. Copyright (2016)

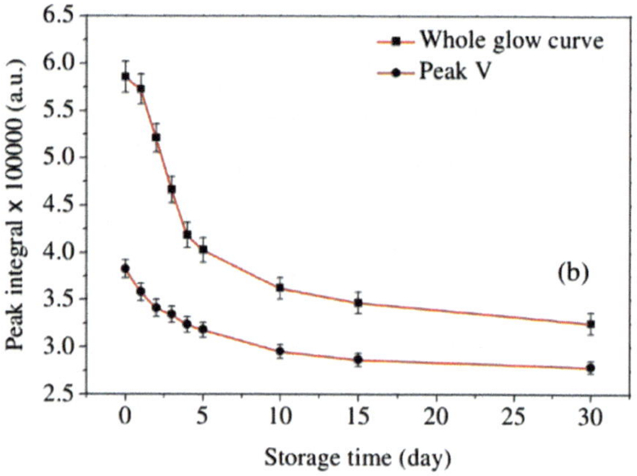

Fig. 7 Peak area changes with storage time as calculated from the whole glow curve (square) and the simulated peak V (circle). Reprinted from Kalita and Wary [30], with permission from Elsevier. Copyright (2016)

with the peak V, on the other hand, faded ~12% within 5 days of irradiation and continued to diminish at a rate of ~4% over the next 30 days.

3.2 *Natural Quartz Annealed at 900 °C*

Figure 8 shows a glow-curve recorded for the sample after 10 Gy irradiation at a heating rate of 1 °C/s. The glow-curve has three peaks: (I), a strong peak at 71 °C, and two weaker peaks (II) and (III), at 125 and 177 °C, respectively. The TL properties were studied solely on the prominent peak (I) [35].

To explore the reproducibility of peak (I), the TL was determined sequentially at a heating rate of 1 °C/s (dose 10 Gy). For 10 measurements, the coefficient of variation (CV %) was determined to be 0.95%. Reproducibility is considered good for dosimetric applications when CV % <5% in TL intensity after 5–10 cycles. As a result, the reproducibility of peak (I) was satisfactory. This could be due to annealing over phase two inversion temperature, as it is known that a major glow-peak (110 °C) of quartz is thermally sensitized after repeated irradiation and measurement cycles. The fading of peak (I) for TL measurements (dose 10 Gy) is shown in Fig. 9. The peak intensity (I_m), dropped with average lifetime (τ) of (2110 s) at room temperature as derived by fitting test data with the function ($I_m = I_o \exp(-t/\tau)$), where t is time and I_o is the initial intensity.

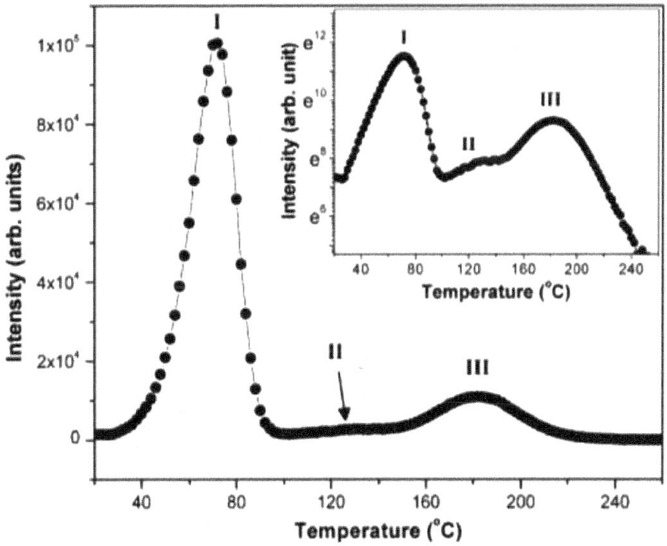

Fig. 8 The main figure represents a TL glow-curve measured at 1 °C/s after irradiation at 10 Gy. The inset figure represents the same data on a semilogarithmic scale. Reprinted from Thomas and Chithambo [35], with permission from Elsevier. Copyright (2018)

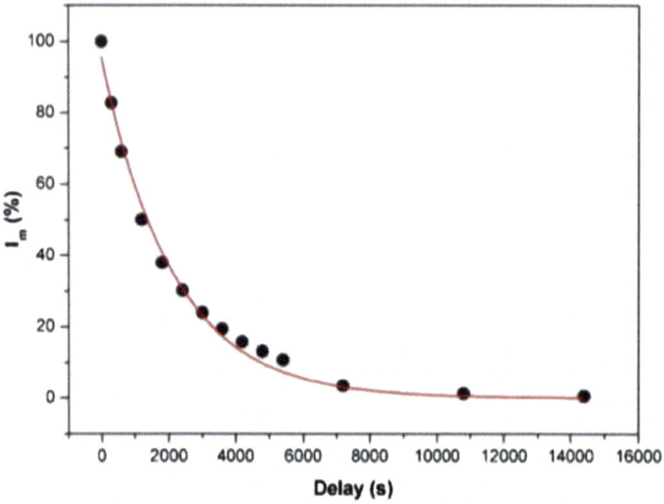

Fig. 9 The Fading of South Africa quartz. Reprinted from Thomas and Chithambo [35], with permission from Elsevier. Copyright (2018)

The dose–response was measured for doses (1:300 Gy). The dose–response curve is illustrated in Fig. 10. The TL intensity increases monotonously (1:40 Gy), then drops as the dosage rises to 130 Gy, with a slight change in intensity, indicating that the intensity saturates at high doses (130 Gy). This saturation occurs because the half-life of the energy level that created the peak (1200 s) is shorter than the irradiation time at high doses (130 Gy). As a result, this peak is an illustration of simultaneous radiation-induced trap filling and fade-induced trap emptying.

It was found that in this study the peak area and intensity increase with the heating rate. This was a thermoluminescent abnormal heating rate effect that has previously been seen in nitrogen-implanted synthetic quartz [38] and $YPO_4:Ce^{3+}$, Sm^{3+} [39]. Using a two-stage model, this anomalous phenomenon has been described as the competitive force between radiative and non-radiative routes [40–42]. This abnormal effect has been described as an inverse thermal quenching brought on by the concurrent thermal release of electrons and holes [43]. Peak area and maximum intensity increased when the heating rate rises due to peak shifting to a higher temperature and extra holes passing through the valence band to the radiative recombination center.

3.3 *Brazilian Natural Quartz [34]*

X-ray diffraction analysis confirmed that diverse kinds of quartz exhibit comparable patterns to typical α-quartz (same crystal structure). The elemental composition (major oxides %) of the studied quartz varieties are listed in Table 3. Figures 11 and 12 demonstrate the TL glow curves of various quartz types. Because the 110 °C

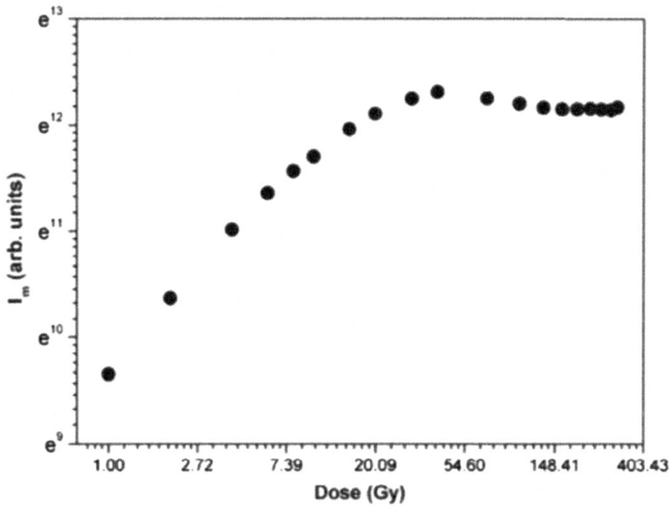

Fig. 10 The dose response of peak I. Reprinted from Thomas and Chithambo [35], with permission from Elsevier. Copyright (2018)

Table 3 Impurities (wt %) in nine quartz types, only high-concentration ones shown [34]

	Alpha	Blue	Green	Black	Pink	Red	Sulphurous	Milky	Synthetic
SiO_2	99.4	97.97	98.1	97.34	98.83	98.35	99.17	98.58	99.2
Al_2O_3	0.22	0.49	0.82	0.3	0.26	0.3	0.07	0.100	0.16
MnO	0.002	0.004	*d*.l.	0.02	0.002	0.004	0.003	0.140	*d*.l.
MgO	0.01	0.01	*d*.l.	0.01	0.01	0.01	0.01	0.010	*d*.l.
CaO	0.05	0.01	0.03	0.19	0.04	0.07	0.01	0.010	*d*.l.
Na_2O	0.02	0.02	*d*.l.	0.02	0.02	0.02	0.02	0.020	0.10
K_2O	0.01	0.13	0.73	0.1	0.01	0.03	0.01	0.010	0.03
TiO_2	0.002	0.241	0.06	0.008	0.005	0.007	0.002	0.001	*d*.l.
P_2O_5	0.01	0.013	0.01	0.064	0.01	0.005	0.003	0.003	*d*.l.
Fe_2O_3	0.02	0.58	0.03	0.18	0.03	0.69	0.01	0.010	0.13

d.l. = determination limits ($\leq$0.001)

peak was so short-lived, all kinds of quartz samples (except synthetic quartz), were pre-heated at 110 °C before TL read-out.

The glow curves were recorded following laboratory irradiation for black, green, pink, sulfurous, and red (doses 1:10 kGy). On the other hand, α and synthetic quartz irradiation was limited to 1 kGy. First, it was discovered that only blue quartz had a glow curve that was comparable to, but not identical to, that of α-quartz, but all the others deviated, some significantly.

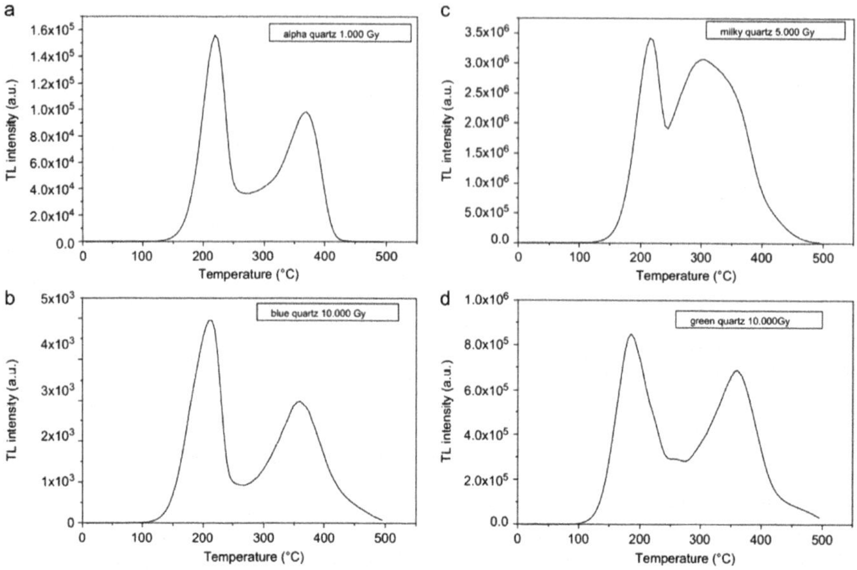

Fig. 11 Glow curve of **a** alpha quartz, **b** blue quartz, **c** milky quartz, and **d** green quartz. Reprinted from Farias and Watanabe [34], with permission from Elsevier. Copyright (2012)

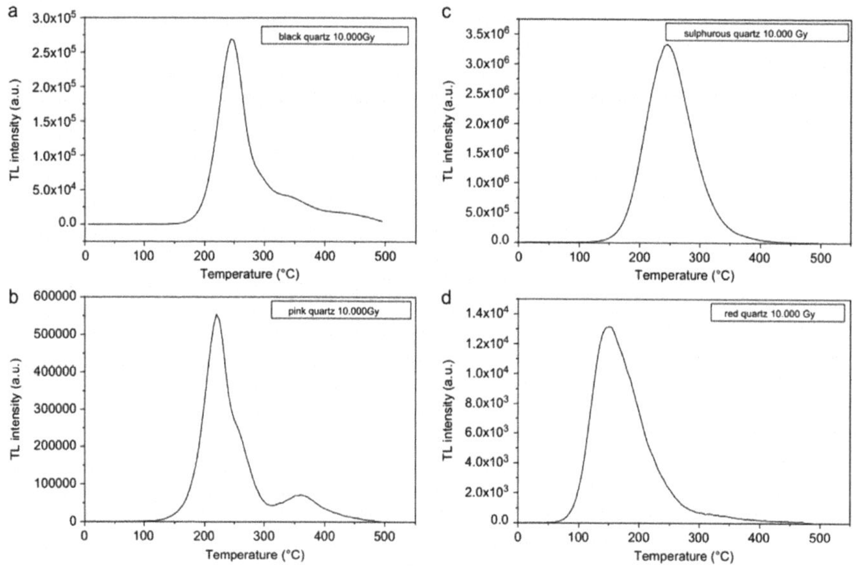

Fig. 12 Glow curve of **a** black quartz, **b** pink quartz, **c** sulfurous quartz, and **d** red quartz. Reprinted from Farias and Watanabe [34], with permission from Elsevier. Copyright (2012)

In fact, as previously stated by McKeever [2] the glow curves of α-quartz exhibit a wide range of patterns. Thus far, a clear overarching explanation has not been pinpointed; however, it is recognized that both intense irradiation and elevated temperature annealing do indeed influence the peaks' position and height. In the study, the peak occurring around 220 °C is noticeable across all quartz types, as depicted in Figs. 11a–d and 12a–d. The corresponding peak temperature ranges from 220 to 230 °C for α, milky, black, sulfurous, and pink quartz, whereas for blue and red quartz, it ranges from 180 to 190 °C. The α-quartz displays characteristic TL peaks at 325 and 375 °C. These peaks are evident in blue, milky, and green varieties, appearing faintly in black quartz. Notably, the 325 °C peak is absent in pink quartz, while sulfurous and red quartz lack high-temperature peaks.

3.4 Chinese Natural Quartz

The major oxides composition (%) of the investigated Chinese quartz are presented in Table 4.

The glow curves presented in Fig. 13 exhibit four distinct TL peaks (labeled as I, II, III, and IV) spanning from room temperature to 500 °C. Notably, these distinct peaks manifest at temperatures of 133, 211, 266, and 405 °C. The lack of an apparent peak below 100 °C was due to the poor thermal stability of the luminescence signal, and the TL signal had totally attenuated prior to the measurements. The baseline was used to analyze the dose–response features of Peak (III), and The TL intensity between 250 and 270 °C has been recorded. Figure 14 a depicts the variation in luminescence intensity with dose. The sensitivity of each aliquot to dose–response varies greatly, possibly due to grinding. Figure 14b shows the luminescence intensity over temperature range 400–420 °C that was recorded for Peak (IV). Nevertheless, the combined luminescence intensity of the highest dose in the figure (1024 Gy) exceeded the statistical software's upper limit. As a result, Fig. 14a captures the summation of intensity from three out of five aliquots, while Fig. 14b records the intensity summation from four aliquots.

The intensity of thermoluminescence increases as radiation dosage increases. The thermoluminescence signal of this quartz has been found to exhibit the necessary TL characteristics for radiation dosimetry and could be utilized for geological dating.

T_m versus T_{stop} is displayed in Fig. 15, and the peaks shown are actual peaks that can only be read after peak identification. This form of glow curve is known as the glow-peak curves or sub-peak. There were at least nine peaks in the glow curve located at (133, 211, 266, 308, 333, 384, 441, 466, and 484 °C).

Table 4 Results of Chinese quartz's XRF analysis (wt%) [44]

SiO_2 (%)	CaO (%)	Al_2O_3 (%)	Fe_2O_3 (%)	Norm. (%)
99.900 0	0.072 9	0.036 1	0.022 1	100.000 0

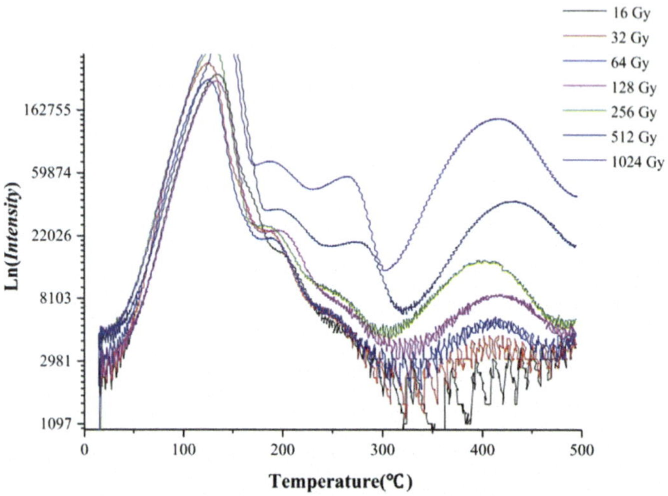

Fig. 13 The glow curves of annealed quartz irradiated with various doses. Reprinted from Zhou et al. [44], with permission from Elsevier. Copyright (2016)

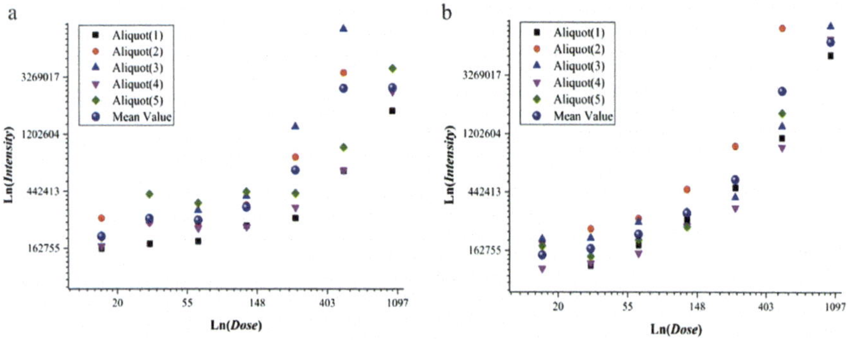

Fig. 14 The dose–response curves of annealed quartz exposed to different radiation doses; **a** peak (III), **b** peak (IV). Reprinted from Zhou et al. [44], with permission from Elsevier. Copyright (2016)

The technique of varying heating rates (VHR) can be employed to analyze trapping parameters across different heating rates. To determine the activation energy E for first-order kinetics, the Hoogenstraaten method employs a range of heating rates [16]. Trapping characteristics of the quartz are investigated using the Hoogenstraaten technique at heating rates of 2, 3, 4, 5, or 6 °C/s. All discs were exposed to a beta dose of 120 Gy for this study. A total of twenty-five aliquots, measured under the same heating rate, were derived from five parallel measurements in this experiment. The outcomes are documented in Table 5.

The fitting results are listed in Table 6. Parameters of Peak (I) inferred from Computer Glow Curve Deconvolution (CGCD) are trustworthy. With CGCD, the

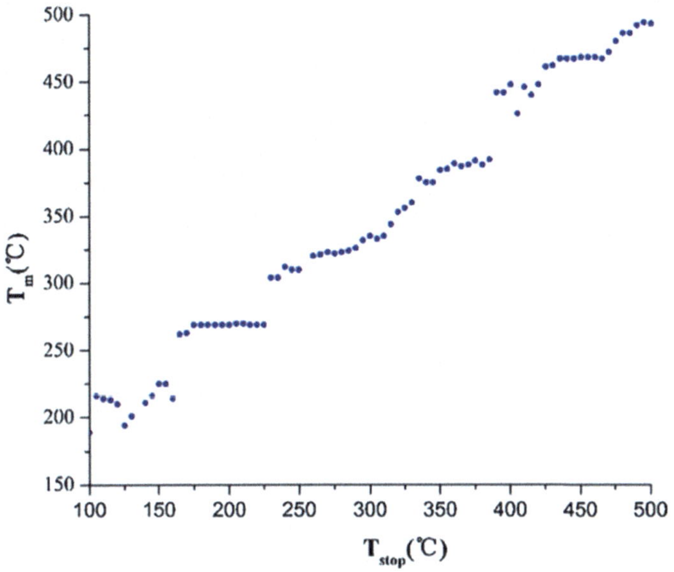

Fig. 15 T_m-T_{stop} plot using a heating rate of 5 °C/s with a beta dose of 120 Gy. Reprinted from Zhou et al. [44], with permission from Elsevier. Copyright (2016)

Table 5 Trapping parameters evaluated by VHR method [44]

Heating rates	Peak (I)	Peak (II)	Peak (III)	Peak (IV)
2 °C s^{-1}	101	180	235	381
3 °C s^{-1}	122	186	256	398
4 °C s^{-1}	129	192	260	400
5 °C s^{-1}	133	204	266	405
6 °C s^{-1}	145	212	275	430
E (eV)	0.56	0.96	1.13	2.53

peaks around 181 and 238 °C are identified. According to their findings, the peaks at 211 and 266 °C may be comprised of two minor peaks. In addition, peaks between 300 and 400 °C can be detected, but their intensity is quite small. The peaks are composed of a sequence of nearly continuous sub-peaks at higher temperatures. The CGCD approach is the best way to retrieve the sub-peaks of Peak (IV).

3.5 French Natural Quartz [46]

Sample purity is ascertained by SEM and XRF techniques to define non-quartz constituents. When examining larger grain sizes (>20 μm), no impurities were

Table 6 Trapping parameters evaluated by CGCD [45]

T_m (°C)	E (eV)	T_m (°C)	E (eV)
139.95	0.79	309.55	1.245
181.35	1	332.85	1.25
209.35	1.14	402.85	1.48
238.05	1.16	442.25	1.69
272.45	1.21	486.15	1.8

detected. However, in the case of smaller grain sizes (<90 μm), they identified slight adherents on the grains, which exhibited enrichment in Cu and Ni. TL curves were recorded for two irradiated quartz samples (150 and 300 Gy). Throughout the temperature range, except around 250 °C, the Fontainebleau (FB) samples exhibited higher TL intensities than the Morar (MR) samples. In the 150–200 °C range, the TL signal of the (MR) sample demonstrated a threefold increase from 150 to 300 Gy (supralinear dose-response), whereas the (FB) sample displayed a sublinear response within the same temperature range. Between 200 and 300 °C, the TL signal of the (FB) sample exhibited a faster dose-dependent increase compared to the (MR) sample, but this trend reversed beyond 300 °C. These observations suggested differing concentrations of defects responsible for distinct electron traps in the (FB) and (MR) samples. The disparities in TL dose-response between the two samples were attributed to an inverse competition between traps and recombination centers. Notably, the final variation in sensitivity of the 110 °C peak with cumulative dose is worth anticipating.

The following procedure was employed to investigate the sensitivity in freshly prepared aliquots of each sample: Achieving a cumulative dose range of 8–600 Gy involved (1) initial heating to 500 °C, (2) irradiation with a dose of 8 Gy, (3) thermoluminescence (TL) measurement up to 250 °C using a heating rate of 2 K/s, and (4) repeating steps 2 and 3 a total of 75 times. Notably, the sensitivity of the (FB) sample experiences a rapid decline with successive doses below 60 Gy. Following a 25% reduction, it exhibits a slight increase before progressively decreasing for doses exceeding 200 Gy, culminating in a value 35% lower than the initial sensitivity at 600 Gy. Conversely, the (MR) sample's sensitivity increases by 120% for doses below 300 Gy, resulting in a plateau phase, and then diminishes to a sensitivity that is 5% lower than the peak observed at 300 Gy.

3.6 *Comparison Between Annealed Natural Quartz at 900 °C with Unannealed Quartz*

Figure 16 depicts the thermoluminescence glow-curve obtained from annealed quartz following a 10 Gy irradiation. Within the glow-curve of the annealed sample, three distinct peaks (labeled as I, II, and III) emerged at temperatures of 71 °C, 125 °C, and 177 °C, respectively. In parallel, the unannealed sample exhibited three peaks

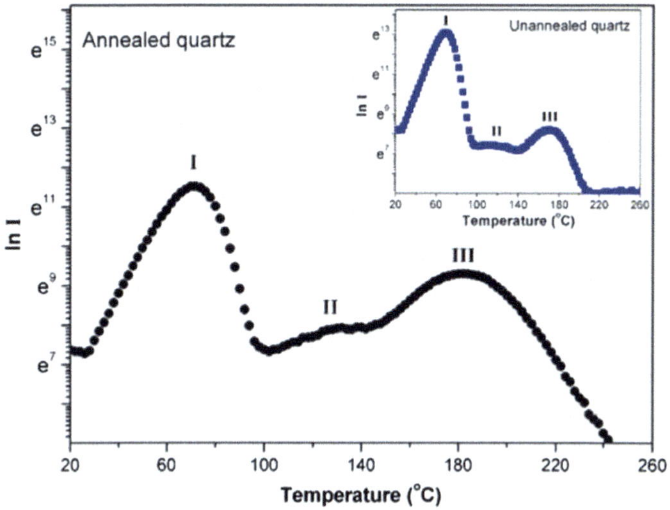

Fig. 16 TL glow-curve of the annealed natural quartz was obtained using a heating rate of 1 °C/s after exposure to a 10 Gy dose of irradiation. The inset presents the glow-curve of an unannealed sample under identical heating rate and dose conditions. Reprinted from Thomas and Chithambo [47], with permission from Elsevier. Copyright (2018)

(I, II, and III) at temperatures of 69 °C, 119 °C, and 170 °C, respectively. Notably, peak (I) predominated within each glow-curve. Peak (II) and (III) maxima for the annealed quartz sample's glow-curve were merely 3% and 11% of peak (I), respectively. Interestingly, the corresponding values for the unannealed quartz sample were significantly lower, at 0.4 and 0.8%. Consequently, annealing at 900 °C notably enhanced the sensitivity of the medium-sized peaks (II) and (III) in comparison to peak (I) [47].

This study looked at the stability of peak (III) by pointing out all the changes in its intensity for different time intervals between irradiation and measurement. Figure 17 compared an annealed quartz glow curve obtained immediately after irradiation to one obtained 20-h after irradiation. Notwithstanding the 20-h delay between irradiation and measurement, peak (III) had a similar intensity and shape. This finding revealed that the electron trap that created peak (III) was stable at room temperature during this period. In contrast, the primary glow-peak at 71 °C had completely dissipated 20 h after irradiation. Figure 17 shows a comparison between the intensities of the annealed and unannealed quartz samples for different delay intervals. Peak (III) is obviously stable in both samples.

To assess reproducibility, TL measurements were conducted ten times using the same dose and heating rate. The resulting consistency of the peaks is satisfactory, a crucial factor for a TL dosimeter. Dose–response analysis was carried out across various doses (1–300 Gy). The growth curves for annealed quartz (depicted by solid squares) and unannealed quartz (represented by blue circles) are presented

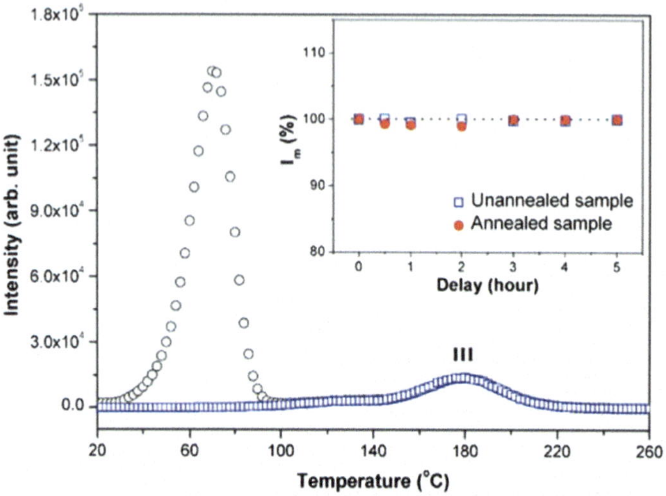

Fig. 17 The glow-curve of the annealed quartz (open squares in the main graph) corresponds to a delay of 20 h between irradiation and measurement. For comparison, the glow-curve of this annealed sample recorded immediately after irradiation (open circles in the main graph) is also provided. The inset illustrates the variation in intensity of peak III over time delay for both annealed and unannealed quartz samples. The dotted line in the inset serves as a visual reference. Reprinted from Thomas and Chithambo [47], with permission from Elsevier. Copyright (2018)

in Fig. 18. Both the annealed and unannealed quartz samples exhibited a sublinear dose–response pattern, aligning with the findings.

Different methodologies were used to study the kinetic analysis of the achieved glow curves for annealed and unannealed quartz samples. Table 7 lists the T_m-T_{stop}, initial rising method, whole glow peak, peak shape method, curve fitting method, and thermal quenching approaches. From the table, it is observed that the calculated kinetic parameters by the different methods are matched well to each other, and that is a good confirmation for their deconvolution method.

3.7 Indian Natural Colorless Quartz

The crystalline nature of the investigated samples was confirmed by the highly intense and sharp peaks of quartz deduced by XRD. The chemical analysis of the investigated natural quartz revealed the inclusion of minor elements such as Al, Na, Fe, Mn, and K. Fourier transform infrared (FTIR) spectra revealed infrared bands of the quartz crystal in the 400–1200 cm^{-1} fingerprint spectrum [45, 49]. Furthermore, the distinctive band at 3438 cm^{-1} is associated with Al–OH/Na^+ defects [50]. After being exposed to ionizing radiations colorless quartz with aluminum (Al) as an impurity, color is changed to dark smokey.

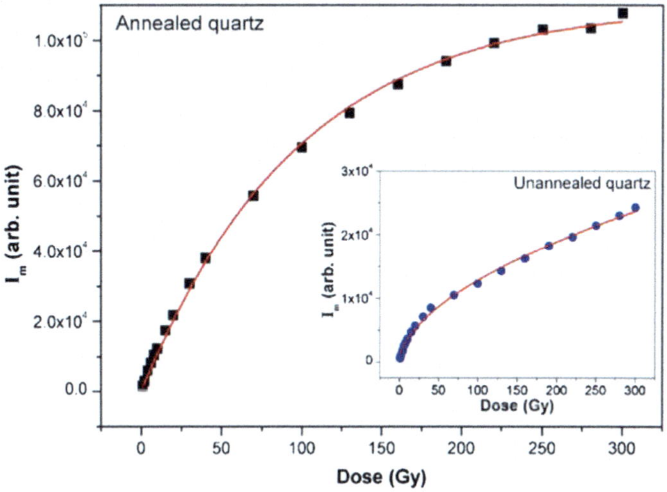

Fig. 18 The dose–response behavior of peak (III) is shown for annealed quartz (solid squares), and the dose–response curve for the unannealed sample is presented in the inset (blue circles). Reprinted from Thomas and Chithambo [47], with permission from Elsevier. Copyright (2018)

Table 7 Comparison of kinetic parameters using different methods [48]

Method	Annealed quartz			Unannealed quartz		
	Activation energy (eV)	s (s^{-1})	b (order of kinetics)	Activation energy (eV)	s (s^{-1})	b (order of kinetics)
T_m versus dose			Non-first-order			First-order
T_m-T_{stop}			Non-first-order			First-order
Initial rise	$E = 1.19 \pm 0.04$			$E = 1.11 \pm 0.07$		
Whole glow-peak	$E = 1.24 \pm 0.01$	4.5×10^{12}	1.5	$E = 1.09 \pm 0.02$	3.85×10^{11}	1.2
Peak shape	$E\tau = 1.32 \pm 0.07$ $E\delta = 1.32 \pm 0.07$ $E\omega = 1.33 \pm 0.07$		General order	$E\tau = 1.12 \pm 0.08$ $E\delta = 1.10 \pm 0.08$ $E\omega = 1.12 \pm 0.08$		First-order
Curve fitting	$E = 1.24 \pm 0.01$	5.0×10^{12}	1.53 ± 0.01	$E = 1.20 \pm 0.02$	2.88×10^{12}	1.21 ± 0.04
Phosphorescence	$W = 0.77 \pm 0.05$			$W = 0.67 \pm 0.02$		

TL glow curves of the gamma ray irradiated Indian natural quartz at doses in the range 0.5–50 kGy are characterized by four TL peaks (I), (II), (III), and (IV), at about 169, 212, 279, and 370 °C, respectively. The intensity of these peaks increased as the gamma radiation increased; however, when a quartz sample was subjected to a high dose (50 kGy), peaks (I), (II), and (III) dropped significantly.

Chen's peak shape and initial rise techniques are employed to derive the kinetic parameters of the identified TL peaks, yielding congruent outcomes. The absence of shallower peaks (around ~110 °C) in the glow curves may be attributed to the thermal fading caused by the surrounding atmosphere.

3.8 Vietnamese Natural Quartz

Ngoc et al. [48] explore the influence of sodium ions (Na^+) on the formation of TL peaks in natural quartz across a temperature range of 77–680 K. They analyze this aspect by integrating spectrally resolved TL and photo transferred thermoluminescence (PTTL) glow curves. To understand the impact of Na^+ on quartz TL properties, they utilized electro-diffusion technology and elevated temperatures to enhance the Na^+ content within the quartz samples.

Figure 19 presents the XRD patterns of quartz before and after Na^+ diffusion, highlighting the emergence of distinct XRD peaks attributed to the Na_2SiO_3 and $Na_2Si_2O_5$ phases in the Na^+-diffused quartz pattern. These new peaks accompany the existing peaks observed in the original quartz pattern. The appearance of the Na_2SiO_3 and $Na_2Si_2O_5$ phases is a consequence of the Na^+ diffusion into the quartz structure. As Na^+ ions diffuse within the quartz crystal lattice, they not only serve as charge compensators at Al^{3+} ion sites but also trigger a chemical reaction leading to the creation of additional lattice components.

In Fig. 20, the three-dimensional (3D) TL emission spectrum is showcased alongside the principal TL attributes of natural quartz subjected to X-ray irradiation for 30 min (2098.4 Gy) at ambient temperature. Employing a heating rate (β) of 5 Ks^{-1},

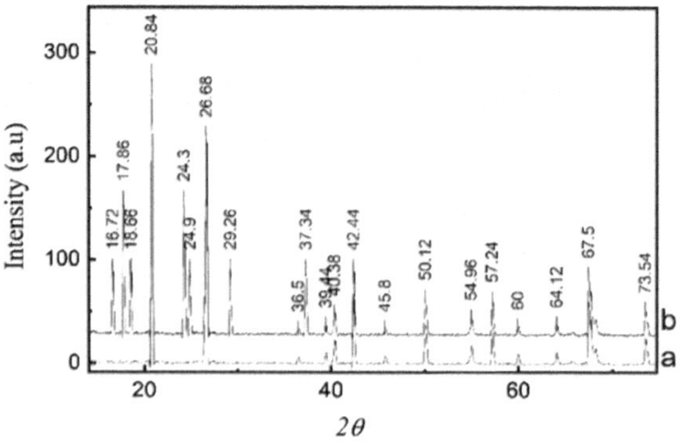

Fig. 19 XRD patterns of the **a** natural, and **b** Na^+-diffused quartz samples. Reprinted from Ngoc et al. [48], with permission from Elsevier. Copyright (2021)

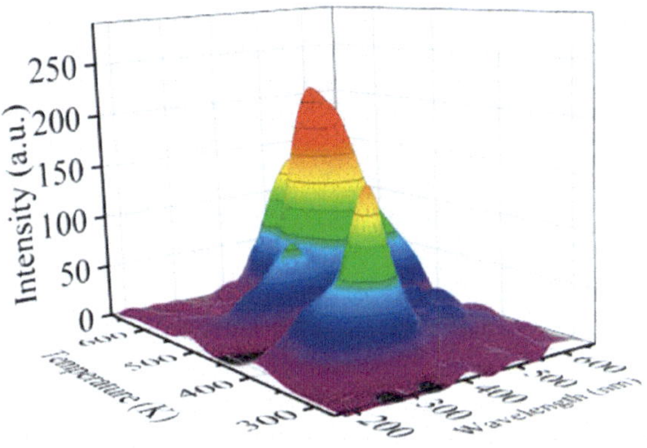

Fig. 20 3D TL spectrum of the irradiated natural quartz emission spectrum. Reprinted from Ngoc et al. [48], with permission from Elsevier. Copyright (2021)

Table 8 The (E_T, τ, s) kinetic parameters of the TL peaks above room temperature [48]

Peak (K)	E_T (eV)	τ (y)	s (Hz)
383	0.97	0.004	8.2×10^{13}
513	1.26	0.72×10^3	2.4×10^{12}
598	1.44	2.6×10^7	1.7×10^{12}
648	1.78	1.4×10^9	2.1×10^{11}

the TL measurements were conducted within the temperature range of 300–680 K and the wavelength range of 270–670 nm. Notably, the TL glow curves unveil the presence of four peaks at 383, 513, 598, and 648 K. The distinctive features of the spectrum arise from four emission bands situated at 380, 470, 590, and 620 nm.

The activation energy, lifetime, and frequency coefficient of the TL glow peaks were determined using the initial rise method, as presented in Table 8. The first group has a comparatively short lifetime at ambient temperature, with the TL peak at 383 K. As a result, neither sample measured after only a few days of irradiation nor those measured without an additional dosage of radiation could produce the peak at 383 K. The intensity of the second group peaked at 513 K and was relatively high between 430 and 550 K. The lifetime of this peak is typically thousands of years. Two TL peaks at 598 and 648 K that were part of the third group had longer lifetimes than the other TL peaks.

3.9 *Egyptian Quartz Thermoluminescence Properties*

Numerous quartz occurrences of exceptional quality can be found in Egypt. They are all dispersed along the Eastern Desert's Red Sea Coast. These deposits manifest as quartz cappings atop plutons, arising from magmatic differentiation subsequent to the solidification and cooling of granitic plutons. Following cap formation, residual solutions infiltrated fractures and crevices, giving rise to vein-like structures or lens-shaped bodies, delineating the genesis of quartz veins. The aggregate quartz reserves in Egypt exceed 20 million tons, distributed across over 50 distinct locations [51].

3.9.1 Quartz Extracted from Wadi Atalla Gold Mine Area

The investigated quartz samples were extracted from the Atalla gold mine location, situated approximately 22 km to the north of the Qift-Quseir asphaltic road within the Central Eastern Desert (CED), as shown in the maps in Fig. 21 [31]. The Wadi Atalla area forms part of Egypt's mountainous country of basement rocks in the CED. For its economic mineral reserves, particularly gold, this region has emerged as one of the extensively mined areas within Egypt's Eastern Desert [52]. The following basement rock units are exposed in this area: Serpentinites, Post-Hammamat felsites, Rhyolite tuffs, Younger granite, Sulfide-bearing quartz veins, and Post-granite dykes [53]. Several quartz veins and veinlets (0.2–1 m thickness), mostly of milky quartz, intrude the granitic rocks in this area.

The Egyptian milky quartz extracted from Wadi Atalla area has a crystal structure with a strong diffraction peak as seen in Fig. 22.

The elemental composition of this quartz variety revealed a concentration sequence of elements as follows: Si > Ti > Al > Na > Fe > Ca > Mg > Mn > Li, as listed in Table 9.

The TL glow curves of this quartz variety, upon exposure to gamma rays, are presented in Fig. 23. Deconvolution was carried out using an innovative TL software

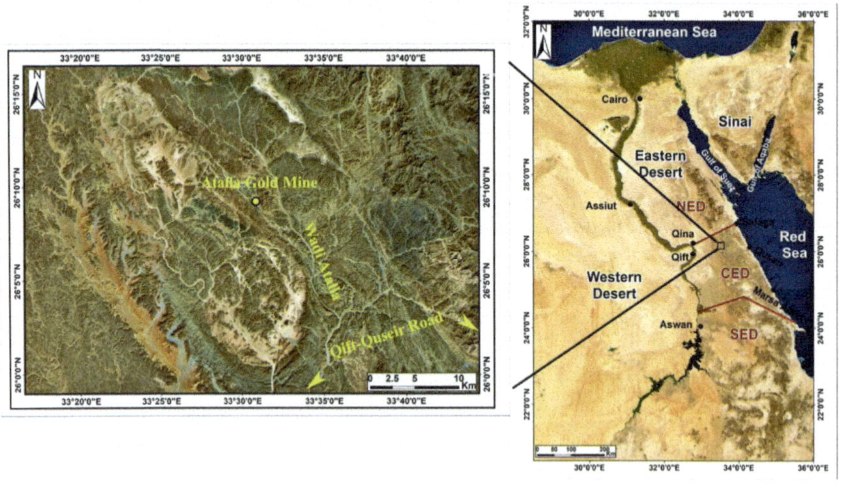

Fig. 21 Satellite image (left), and a map (right) showing Wadi Atalla area from which the investigated samples were taken [31]

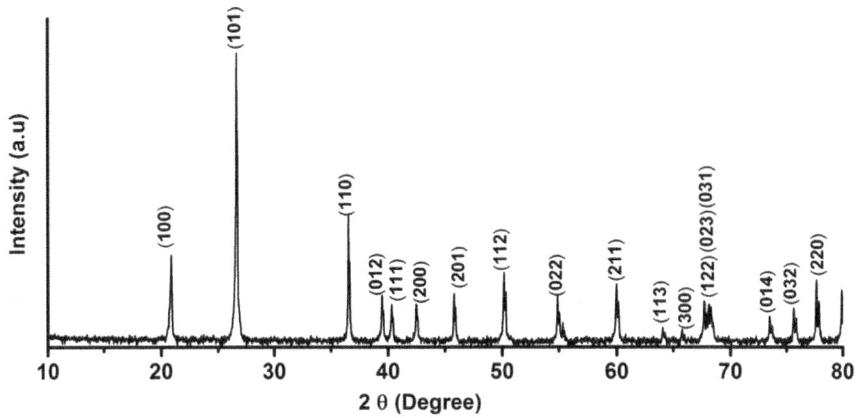

Fig. 22 XRD pattern of Wadi Atalla quartz [31]

Table 9 The elemental analysis of Wadi Atalla quartz [31]

Mg	Ca	Na	Li	Ti	Al	Si	Mn	Fe
8.23	9.88	88.2	0.71	23,520.56	1875.76	139,547.46	2.00	47.04

design [54], yielding crucial trapping parameters such as the activation energy (Ei)—indicating the energy trap's position relative to the conduction band—alongside the frequency factor (s) and kinetic order (b). These parameters for each peak position are enumerated in Table 10.

The resulting glow curves indicated 6 overlapping peaks corresponding to 6 energy trap locations. A direct linear correlation between the intensity of the glow curve and the applied dose across a specific dose range is a significant necessity for TL dosimeters, which will limit the possible applications of the investigated dosimeters. The linearity reflects the material's response to the applied radiation dose. The linearity for this type of quartz sample is divided into three regions as seen in Fig. 24. The first region has 250 mGy and 20 Gy linear dose–response, and the second exhibited a supralinear trend after applying 20 Gy up to 200 Gy doses.

In the third region, as the administered dose exceeded >200 Gy, the TL intensity exhibited an exponential growth pattern with respect to the applied dose, following the equation:

$$\text{TL intensity (a.u)} = Y_0 + A_1.\text{exp. (Dose}/t_1) \tag{44}$$

where Y_0, A_1, and t_1 represent constants. This unexpected behavior can be explained by an augmentation in the sensitivity of milky quartz samples to higher doses, without the concurrent development of additional traps.

As depicted in Fig. 25, the sensitivity of Wadi Atalla quartz exhibited variations aligned with the range of the administered dose. In Fig. 25a, a consistent sensitivity

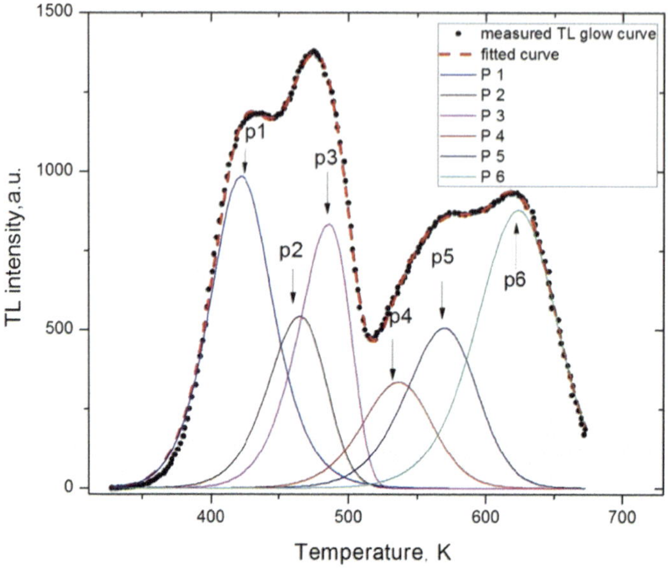

Fig. 23 The deconvoluted glow curve of Wadi Atalla quartz [31]

Table 10 Trapping parameters of Wadi Atalla quartz [31]

Peak	T_m (K)	Activation energy (eV)	Order of kinetics (b)	Frequency factor (s^{-1}) × 10^5
P1	423.5 ± 9.11	0.86 ± 0.031	1.86 ± 0.09	0.66
P2	468.5 ± 10.07	0.97 ± 0.030	1.19 ± 0.03	3.57
P3	498.9 ± 14.48	1.07 ± 0.043	1.18 ± 0.14	10.35
P4	545.1 ± 17.91	1.16 ± 0.022	1.42 ± 0.06	6.56
P5	586.3 ± 14.61	1.22 ± 0.026	1.37 ± 0.04	14.31
P6	639.2 ± 16.05	1.34 ± 0.056	1.43 ± 0.01	8.81

was evident within the linear range (1305.3 ± 10%). However, with the elevation of the applied dose into the supralinear range (20–200 Gy), the sensitivity demonstrated an exponential decrease, as illustrated in Fig. 25b. Notably, the sensitivity re-emerged with a rise in doses >200 Gy. This observation can be attributed to a shift in the disposition of the linearity curve.

The TL measurements of the background dosimeters were collected, yielding a mean and standard deviation (B), which were then employed to assess the realistically achievable minimal detectable dose (MDD) [3, 55] for the prepared dosimeters. The calculated MDD for the examined samples stood at 545 Gy, a correspondence that aligns with the comparatively lower sensitivity values observed in the tested milky quartz samples.

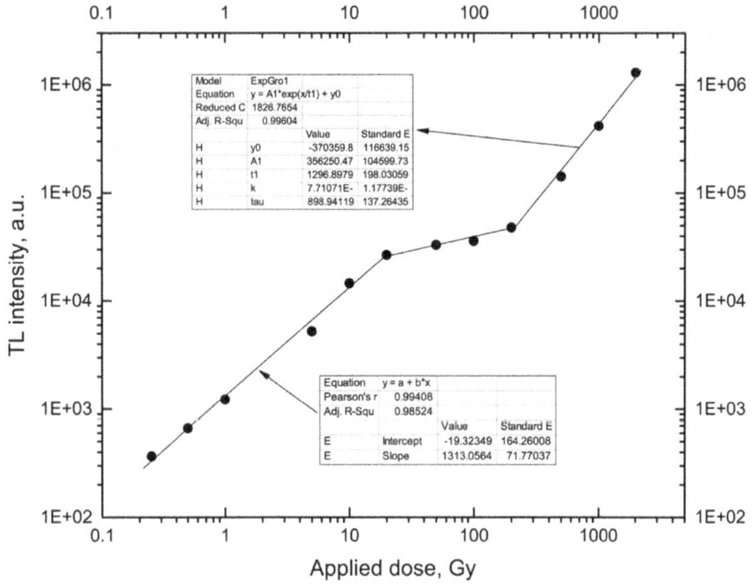

Fig. 24 Linearity of Wadi Atalla quartz [31]

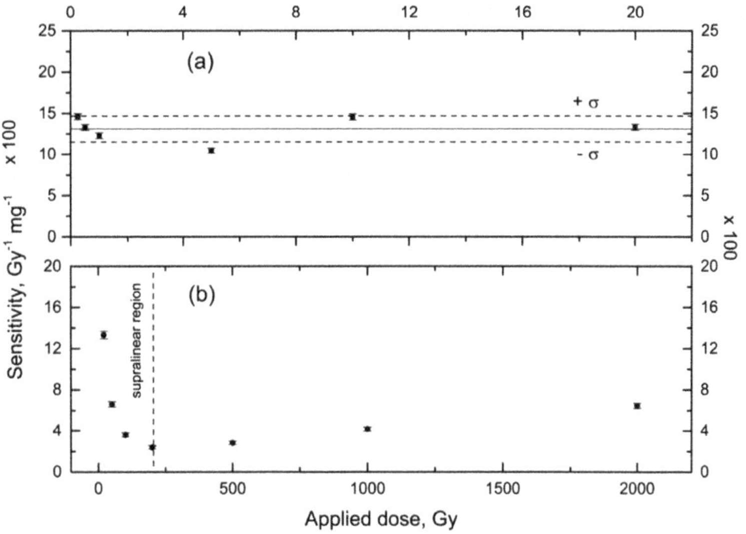

Fig. 25 **a** Shows a Wadi Atalla quartz sensitivity that has been determined across a linear range (0.25–20) Gy. **b** Shows decreasing sensitivity for the supralinear range (20–200 Gy) and rising sensitivity above 200 Gy and [31]

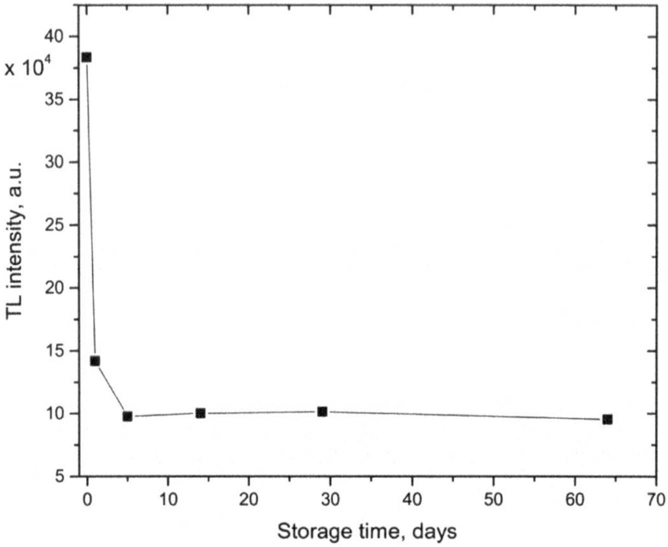

Fig. 26 The fading of Wadi Atalla quartz [31]

As depicted in Fig. 26, the signals from Wadi Atalla quartz samples exhibited rapid fading within the initial 5 days of irradiation. The TL signal diminished to approximately 25% of its initial magnitude. Conversely, no evidence of fading was discernible during extended storage periods of up to 2 months. Consequently, the present study suggests that if researchers aim to employ this quartz for specific applications like dating, obtaining TL measurements after a 5-day interval is advisable.

3.9.2 Quartz Extracted from Gebel Igla Area [32]

This type of quartz was collected from Igla-Tin mine, which is situated in Gebel Igla area, 10 km NW of Marsa Alam, CED as seen in the maps shown in Fig. 27. Gebel Igla is part of Egypt's Pan-African basement [56]. The exposed basement rocks in the area are classified as ophiolitic melange, meta-volcanics, older granitoids, Igla Formation, and younger gabbros and granites. The milky quartz veins represent the youngest event, and are extensively brecciated, with thicknesses ranging from 0.25 to 3 m [57].

The second type of Egyptian milky quartz extracted from Gebel Igla area has a crystal structure with strong diffraction peaks similar to the first type, which was extracted from Wadi Attala. The XRD pattern of Gebel Igla quartz is shown in Fig. 28. The elemental concentrations are listed in Table 11. It is observed that the higher impurities in Gebel Igla quartz include titanium (Ti) and aluminum (Al) elements, the same as Wadi Attala Quartz, but with different values.

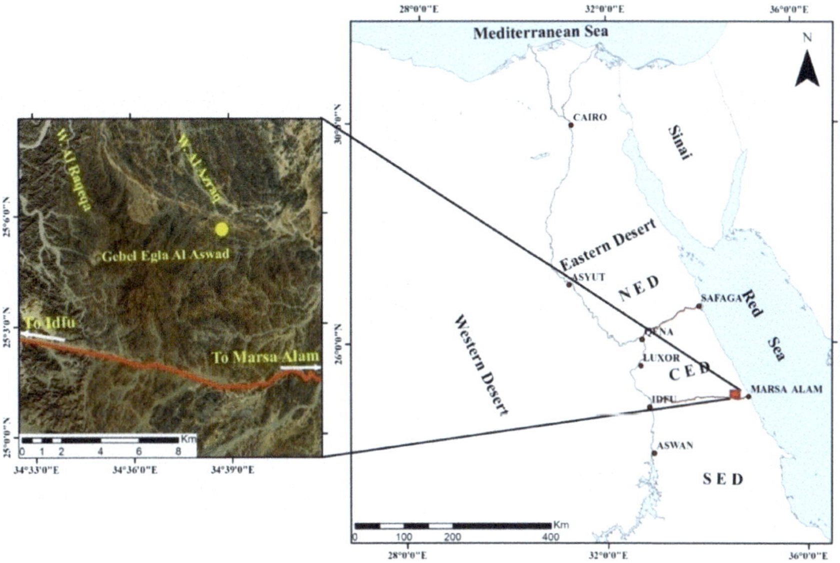

Fig. 27 Satellite image (left) and a map (right) of Gebel Igla area showing from which the investigated samples were taken [32]

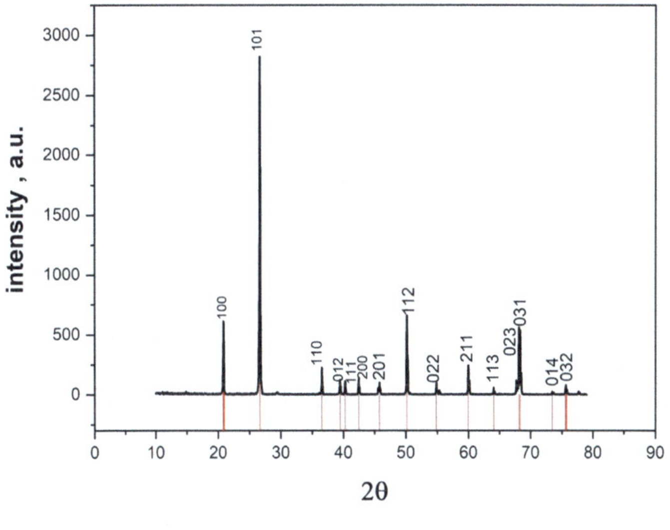

Fig. 28 XRD of Gebel Igla quartz [32]

Table 11 The elemental analysis of Gebel Igla quartz [32]

Mg	Ca	Na	Li	Ti	Al	Si	Mn	Fe
34.54 ± 5.9	74.84 ± 8.6	74.84 ± 8.7	2.3 ± 1.5	12,412.49 ± 111.4	1260.82 ± 35.5	102,754.24 ± 320.6	0.99 ± 1.0	34.54 ± 5.9

The glow curves of Gebel Igla quartz irradiated with gamma rays are shown in Fig. 29 at four different doses, and they were deconvoluted using the new TL software design [54]. The obtained trapping parameters are listed in Table 12 for positions of all peaks, where these trapping parameters are different from those listed in the previous type extracted from Wadi Atalla.

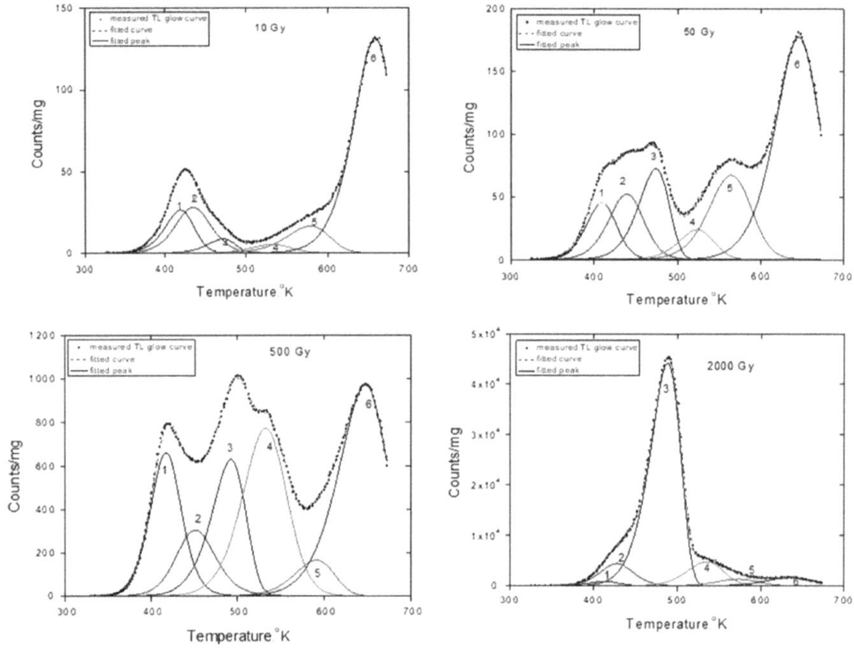

Fig. 29 The deconvoluted glow curve of Gebel Igla quartz irradiated by gamma ray's doses at 10, 50, 500, and 2000 Gy [32]

Table 12 The kinetic parameters of Gebel Igla quartz [32]

		CGCD (5 Gy–2 kGy)		Peak shape method (5 Gy–2 kGy)	
Peak	T_m (K)	E (eV)	s $(s^{-1}) \times 10^{10}$	E (eV)	s $(s^{-1}) \times 10^{10}$
P1	415.1	0.92	1.61	0.925	2.045
P2	440.8	0.99	3.54	0.999	3.015
P3	484.6	1.08	4.68	1.092	2.445
P4	530.6	1.22	2.75	1.222	3.983
P5	573.5	1.32	4.32	1.340	5.540
P6	647.0	1.48	3.61	1.483	2.857

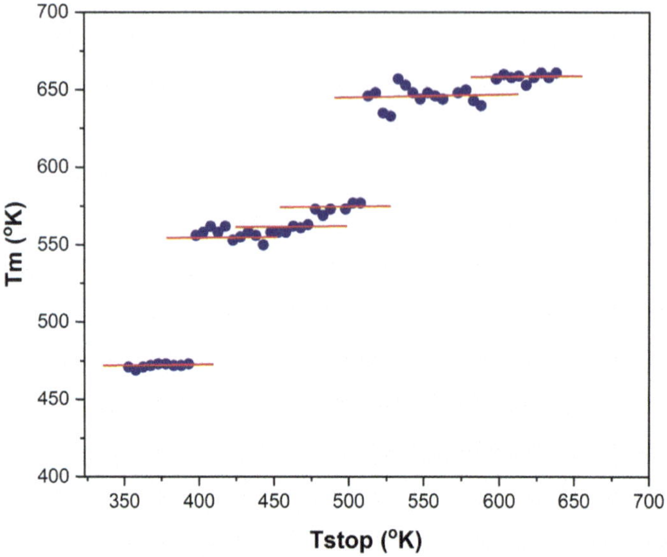

Fig. 30 T_m-T_{stop} results of Gebel Igla quartz [32]

An alternative approach for glow curve analysis is the T_m-T_{stop} method [58]. Peak identification indicates the presence of six distinct peaks within the glow curve, corroborating the findings of the deconvolution method, as illustrated in Fig. 30.

Figure 31 illustrates a linear direct correlation between the intensity of the glow curve and the applied dose within a specific dose range for Gebel Igla quartz. Linearity was established individually for each deconvoluted peak of this type. All deconvoluted peaks within Gebel Igla quartz consistently exhibited dose–response patterns that increased as the gamma-ray dose increased. Peak (1) displayed linearity up to approximately 200 Gy before reaching saturation at around 2000 Gy. Notably, peaks (2), (3), (4), (5), and (6) demonstrated linear behavior with dose increments up to approximately 2000 Gy without encountering saturation. Among these, peaks (3) and (4) showcased superior stability, possessing steeper linear slopes (1.3–1.4) compared to the other peaks. Peaks (1) and (2) exhibited comparatively shallower slopes of linearity (0.9–1.0), whereas peaks (5) and (6) demonstrated the gentlest slope (0.6–0.8) among the high-temperature peaks. Consequently, the linearity of lower peaks up to peak (4) (at 530 K) proves beneficial for TL responses up to 2000 Gy, a substantial value in dosimetry studies. This characteristic enables the examined material to be utilized in high-dose fields beyond medical applications.

The MDD for Gebel Igla quartz is determined to be 627 μGy, surpassing the MDD of Wadi Atalla quartz. The sensitivity of this type of quartz exhibited variability across the applied dose range, with a sensitivity of 293.27 Gy^{-1} mg^{-1} observed within the linear range.

Another vital dosimetric attribute to consider is Reproducibility [59]. Reproducibility pertains to a dosimeter's capacity to consistently reproduce its response

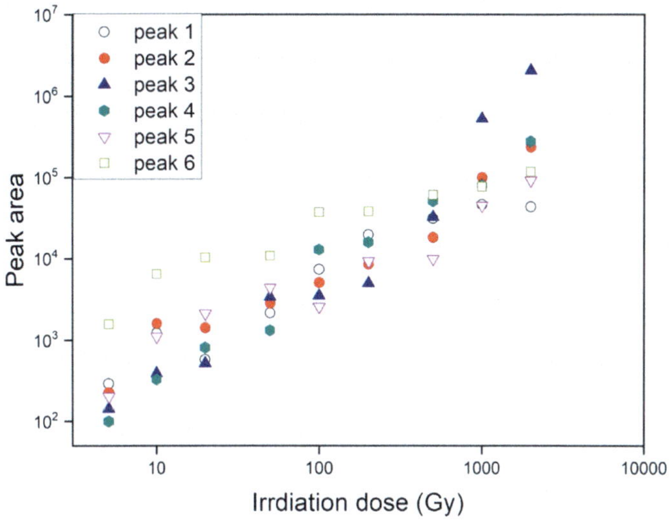

Fig. 31 Linearity of Gebel Igla quartz [32]

under identical experimental conditions. In dosimetric contexts, a dosimeter is deemed to exhibit commendable reproducibility when the coefficient of variation among results after 5–10 cycles remains below 5% [59]. Figure 32 illustrates the reproducibility of Gebel Igla quartz after 50 Gy irradiation to gamma radiation by plotting each deconvoluted peak area versus the number of reuse cycles. The values of the coefficient of variations (%) for peaks (1), (2), (3), (4), (5), and (6) are approximately 1.29%, 0.949%, 3.05%, 5.02%, 1.08%, and 3.55%, respectively. According to their coefficient of variations (%), peaks (2), (1), (3), (6), and (5) have strong stability; peak (4) has the greatest coefficient of variations than the other peaks. The plot clearly illustrates the robust reproducibility of Gebel Igla quartz by peak, rendering it well-suited for applications as a TL dosimeter.

Fading behavior observed in Gebel Igla quartz differs from that in Wadi Atalla, as determined by employing an alternative method that involves measuring fading for each deconvoluted peak. As depicted in Fig. 33, the overall glow curve exhibited a reduction of approximately 45% in its initial intensity within 24 h. Specifically, the intensity of peak (1) at 471.1 K completely diminished within a span of about two weeks. Meanwhile, peak (2) at 433.9 K declined to half of its original value within one month and 14% within two months. A somewhat similar pattern was observed for peak (3). Notably, the TL signals of peaks (4), (5), and (6) remained unchanged after 10 days.

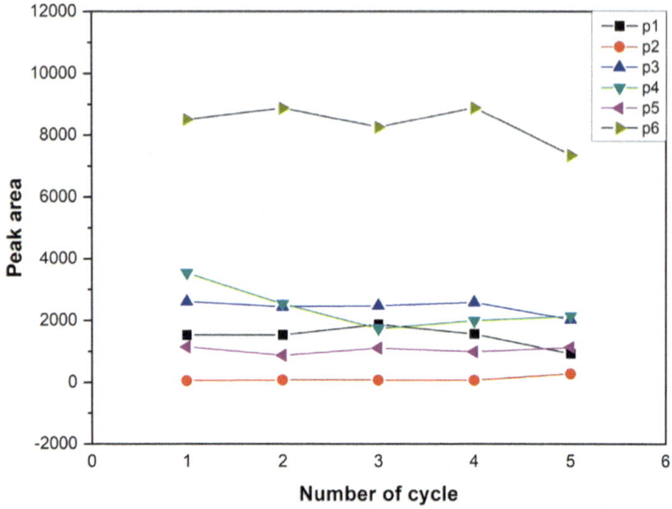

Fig. 32 The reproducibility of Gebel Igla quartz [32]

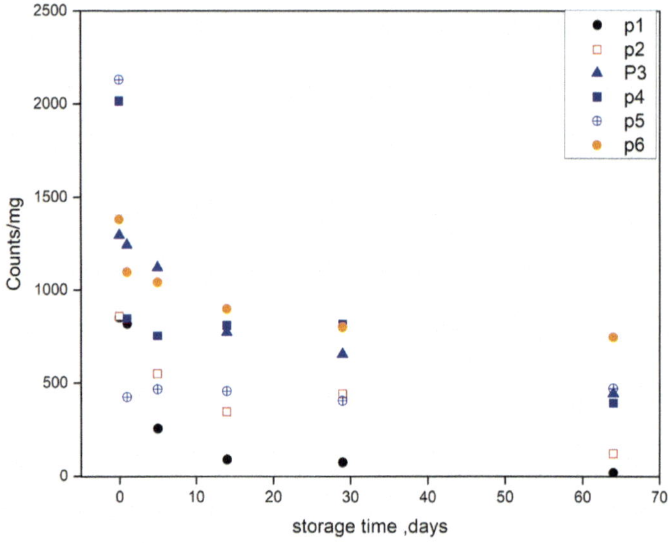

Fig. 33 Fading of Gebel Igla quartz [32]

3.9.3 Quartz Collected from Gebel Rod El Barram [60]

Gebel Rod El Barram quartz is found in conjunction with metamorphic rocks within Egypt's Eastern Desert, as illustrated in the maps provided in Fig. 34. The samples are

Fig. 34 Satellite image (left), and map (right) showing the the location of the investigated quartz from Gebel Rod El Barram [60]

linked to Talc-Schist and the Tectonostratigraphic Unit, representing infrastructural rocks.

The X-ray diffraction pattern of Gebel Rod El Barram quartz is the same as the other two types of Egyptian quartz which represents a completely crystalline structure as shown in Fig. 35. The elemental concentrations listed in Table 13 uncovered that the higher impurities with SiO_2 were found to be titanium (Ti) and aluminum (Al) elements as Wadi Atalla and Gebel Igla quartz. In addition, Gebel Rod El Barram quartz has larger concentrations of calcium (Ca) and magnesium (Mg) than the previous two types.

The glow curves of Gebel Rod El Barram quartz were obtained after exposure to four beta-radiation doses, and then deconvoluted revealing ten overlapping thermoluminescence peaks as shown in Fig. 36. The glow curve peaks of Gebel Rod El Barram quartz were more than those of the obtained from Wadi Atalla and Gebel Igla quartz due to the high concentrations of Ca and Mg. The trapping parameters of Gebel Rod El Barram quartz are listed in Table 14 for each peak position.

The analysis by T_m-T_{stop} method was applied on Gebel Rod El Barram quartz as shown in Fig. 37. The peak detection showed that there are ten peaks within the glow curve. T_m-T_{stop} obtained from Gebel Igla quartz showed six peak positions, and that confirms the dependence of the dosimetric characteristics on the origin of quartz.

The VHR method was implemented, resulting in the generation of the glow curves depicted in Fig. 38a. These glow curves were subsequently subjected to deconvolution using the CGCD method, facilitated by the innovative TL software design. This deconvolution process aimed to isolate the inherent TL glow peaks and ascertain their respective kinetic parameters. Activation energy values were derived from the

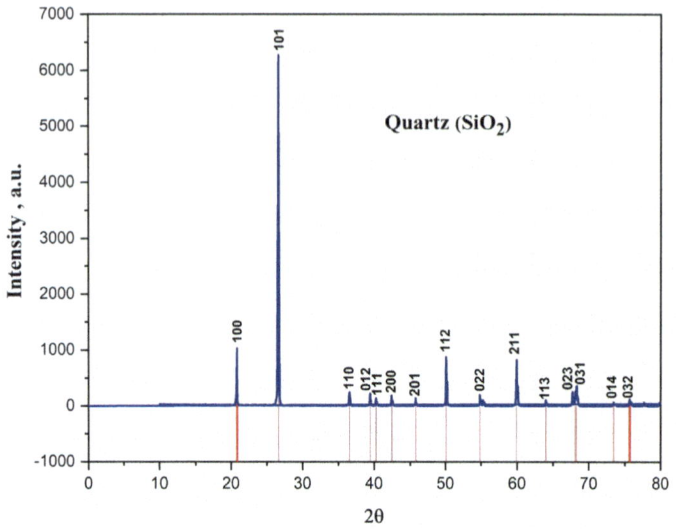

Fig. 35 XRD of Gebel Rod El Barram quartz [60]

Table 13 The elemental analysis of Gebel Rod El Barram quartz [60]

Mg	Ca	Na	Li	Ti	Al	Si	Mn	Fe
32.97	60.46	76.94	1.97	16,949.52	1220.12	1,222,754	1.97	82.43

slopes of linear plots of $\ln\left[I_M^{b-1}\left(\frac{T_M^2}{\beta}\right)^b\right]$ versus $1/T_M$ as illustrated in Fig. 38b, and are enumerated in Table 15. Remarkably, these calculated activation energy values closely align with the deconvoluted dataset.

Figure 39 illustrates the dose–response (linearity) of Gebel Rod El Barram quartz across all ten deconvoluted peaks. Each peak displays a distinct pattern, revealing varying linear dose responses. Peaks (3), (5), (6), and (8) exhibit a linear range from 2.2 to 220 Gy with R^2 values of 0.983, 0.999, 0.987, and 0.992, respectively. Peak (7) demonstrates linearity within the range of 2.2 to 110 Gy ($R^2 = 0.979$), followed by a supralinear trend extending up to 220 Gy ($R^2 = 0.999$). Peaks (2), (4), and (10) exhibit a consistent linear trend within the range of 2.2 to 110 Gy ($R^2 = 0.981, 0.966$, and 0.981, respectively). Peak (1) displays an exponential trend with an R^2 value of 0.989 within the linear range of 2.2–55 Gy.

The MDD of Gebel Rod El Barram quartz was 24 μGy. Also, the sensitivity was determined giving an increasing trend, then decreasing during supralinear rang. This phenomenon can be attributed to the shifts observed in the linearity curve's behavior, as elaborated in Fig. 39.

Figure 40 displays the TL glow curves of reproducibility for Gebel Rod El Barram quartz. The plot clearly shows that the examined sample demonstrates strong repeatability, which is required for TL dosimeter applications.

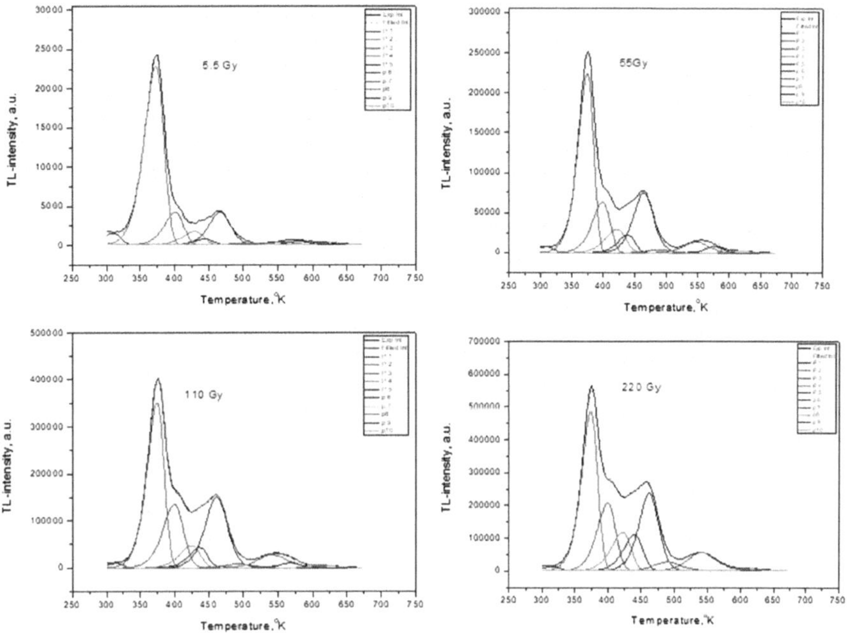

Fig. 36 The deconvoluted glow curves of Gebel Rod El Barram irradiated to beta at 5.5, 55, 110, and 220 Gy [60]

Table 14 The trapping parameters of Gebel Rod El Barram quartz [60]

Peak	T_m (K)	Activation energy (eV)	Order of kinetics (b)	Frequency factor (s^{-1})
P1	310.2 ± 2.85	0.623 ± 0.031	1.029 ± 0.01	3.4×10^{10}
P2	373.7 ± 0.662	0.936 ± 0.060	1.023 ± 0.02	5.77×10^{12}
P3	399.8 ± 1.12	1.02 ± 0.030	1.09 ± 0.06	5.05×10^{12}
P4	425.3 ± 3.17	1.06 ± 0.062	1.12 ± 0.05	2.49×10^{12}
P5	440.3 ± 2.3	1.31 ± 0.038	1.17 ± 0.05	9.67×10^{14}
P6	463.8 ± 1.75	1.47 ± 0.038	1.52 ± 0.06	5.42×10^{15}
P7	496.5 ± 6.2	1.58 ± 0.077	1.82 ± 0.13	1.5×10^{16}
P8	550.6 ± 10.2	1.75 ± 0.046	1.82 ± 0.11	3.91×10^{15}
P9	576.5 ± 9.8	2.49 ± 0.150	1.44 ± 0.09	1.98×10^{23}
P10	622.8 ± 8.6	2.95 ± 0.160	1.59 ± 0.03	2.32×10^{25}

The fading of Gebel Rod El Barram quartz is elucidated in Fig. 41. In this study, samples were exposed to a beta source dose of 11 Gy and subsequently stored for 0, 1, 3, 7, 15, 20, and 30 days, at room temperature within black containers to prevent optical fading. It becomes evident that the entire glow curve experienced an initial

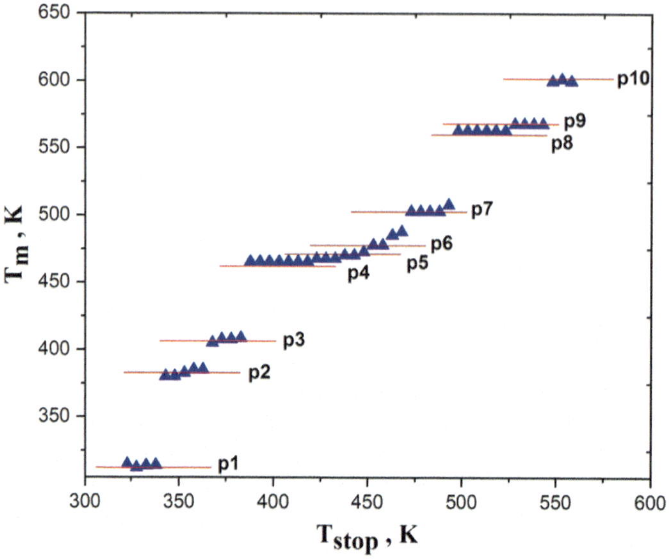

Fig. 37 T_m-T_{stop} results of Gebel Rod El Barram quartz [60]

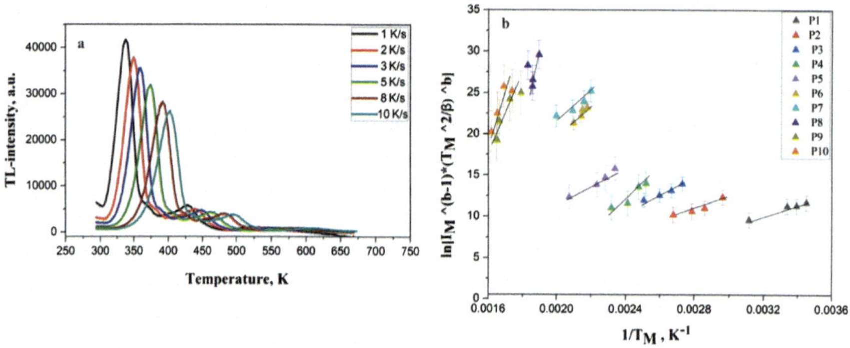

Fig. 38 **a** Various heating rates of Gebel Rod El Barram quartz. **b** Plot of $\ln\left[I_M^{b-1}\left(\frac{T_M^2}{\beta}\right)^b\right]$ versus $1/T_M$ for the detected ten peaks [60]

reduction of approximately 80% from its original magnitude within the first day. Over the subsequent 3 days, the reduction reached around 20% of the remaining signal. Between 7 and 15 days, a marginal fading of about 6% in the TL signal was observed. Subsequently, after 20 and 30 days, the TL signal diminished by roughly 49% relative to the remaining signal.

Table 15 The activation energy of Gebel Rod El Barram quartz by various heating rates method [60]

Peak	E (eV)
P1	0.6961
P2	0.8357
P3	0.951
P4	1.121
P5	1.280
P6	1.354
P7	1.49
P8	1.6378
P9	2.586
P10	2.984

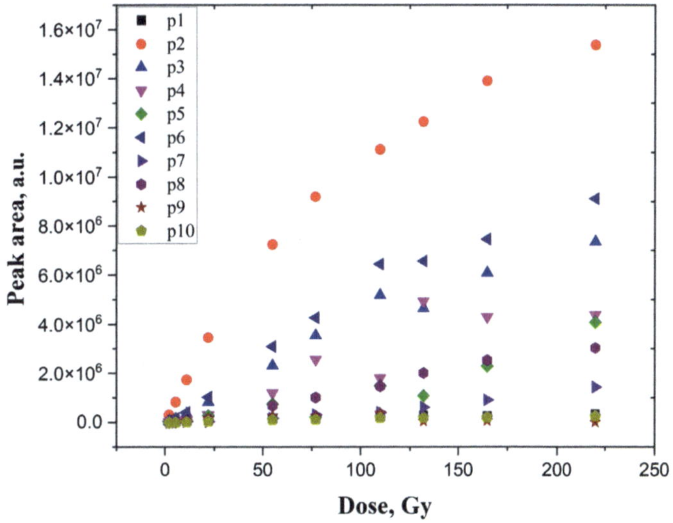

Fig. 39 The dose responses of Gebel Rod El Barram quartz showing linearity regions [60]

4 Applications of Isothermal Decay Method

Randal and Willikans [9] introduced an isothermal decay technique for analyzing glow curves. The partial area (Φ) beneath a segment of a phosphorescence decay curve is given by $(\Phi) = (\kappa)\exp(-E/kT)$, where ($\kappa$) represents a constant, (k) signifies the Boltzmann constant, and (E) is the activation energy. This approach involves plotting $\ln(\Phi)$ versus ($1/kT$), yielding a slope equivalent to the activation energy (E). This methodology was applied by Dawam and Chithambo [61] to their quartz samples' glow curves, which were annealed at 900 °C, as shown in Fig. 42. The resulting activation energy values were determined as 0.85 ± 0.02, 0.79 ± 0.12, and

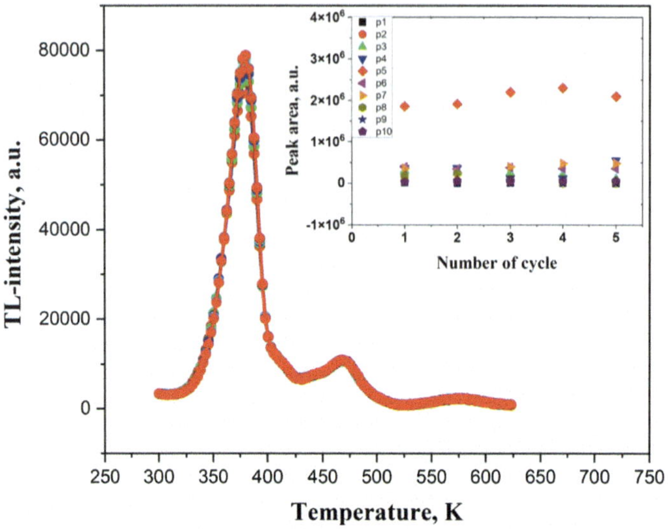

Fig. 40 The main figure displays the TL glow curves of reproducibility for Gebel Rod El Barram quartz. The inset figure shows reproducibility response by each of the ten deconvoluted peaks [60]

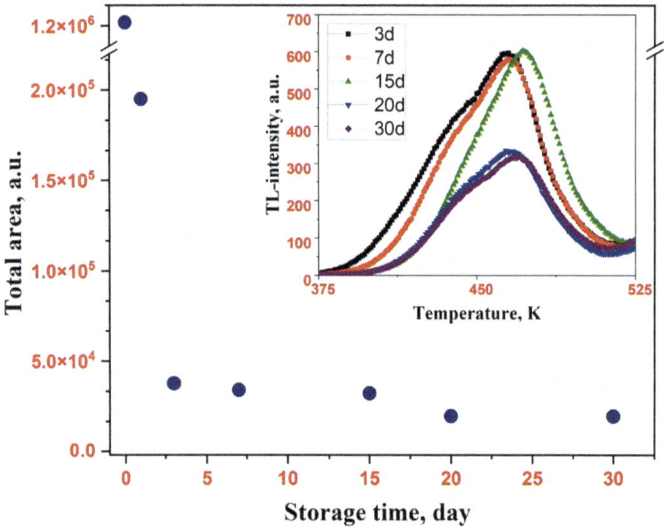

Fig. 41 The main figure shows the relationship between TL intensity with different storage times using the total area under the curve. The inset figure represents the fading TL glow curves of Gebel Rod El Barram quartz at each storage time [60]

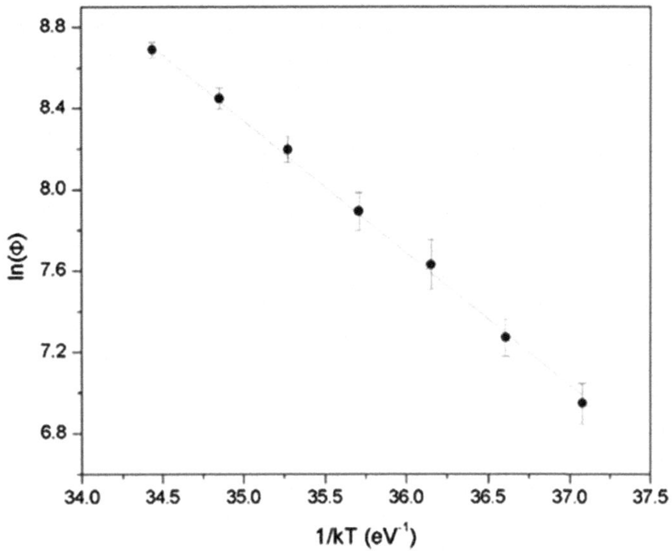

Fig. 42 The relationship between ln(ϕ) and *1/kT* for phosphorescence measurements taken within the temperature range of 40–60 °C from a sample that annealed at 900 °C. Reprinted from Thomas and Chithambo [61], with permission from Elsevier. Copyright (2018)

0.62 ± 0.03 eV for the un-annealed sample, and samples annealed at 500 °C and 900 °C, respectively. Evidently, annealing to 900 °C caused the activation energy to decrease.

In employing the isothermal method, Thomas and Chithambo [47] conducted phosphorescence measurements at different temperatures: 150, 155, 160, 165, 170, and 220 °C for annealed quartz, and 140, 145, 150, 155, 160, and 200 °C for unannealed quartz. The outcomes of their investigation are illustrated in Fig. 43. The figure provides an illustrative instance of phosphorescence, with a focus on determining the area beneath a brief interval of an isothermal decay curve over a 2-s time interval. The inset of Fig. 43 shows a semi logarithmic plot of (Φ_q/Φ_u) versus 1/*kT*. In the case of annealed quartz, the activation energy for thermal quenching was determined to be $E = 0.77 \pm 0.05$ eV, whereas for unannealed quartz, the activation energy was measured as $E = 0.67 \pm 0.02$ eV for the unannealed quartz.

The work conducted by Dawam et al. [62] reported activation energies for thermal quenching of 0.78 ± 0.06 eV and 0.62 ± 0.02 eV for samples annealed for 10 and 60 min, respectively. A comparison of these thermal quenching activation energy values with those presented by Dawam and Chithambo [61] for the same primary peak of the material indicates a favorable alignment, implying a shared recombination center. Interestingly, it seems that the activation energy for thermal quenching tends to decrease as the duration of annealing increases.

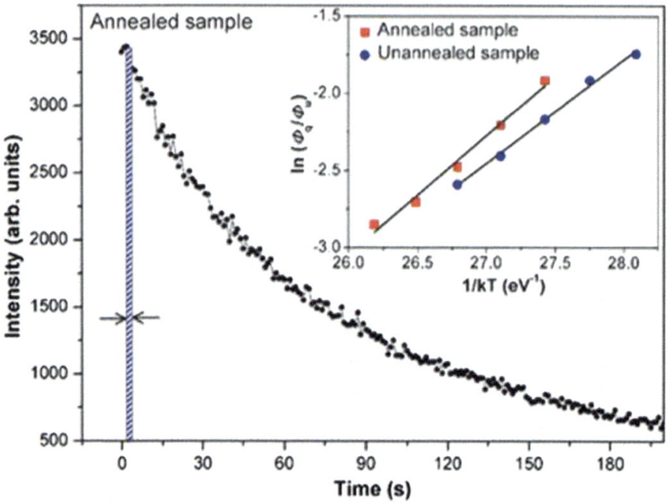

Fig. 43 An illustration of the optimal area under a segment of an isothermal decay curve, for a sample of annealed quartz where an isothermal decay curve was measured at 160 °C. The inset presents (Φ_q/Φ_u) against $1/kT$ for the annealed and unannealed quartz. Reprinted from Thomas and Chithambo [61], with permission from Elsevier. Copyright (2018)

5 Summary and Conclusions

This chapter provided a comprehensive review of the TL properties of natural quartz minerals from several countries, including Egypt. This review has further highlighted the importance of quartz as a luminescence dosimeter. Through kinetic analyses of the glow curves, it became clear that distinct glow curve shapes and kinetic parameters characterize quartz samples from different countries. The result emphasizes the inherent variation in TL characteristics within quartz, which is based on factors such as the mineral's composition, different genetic settings, and the geological contexts of the quartz. The complex interaction of all these factors leads to a wide range of luminescent behaviors that have practical and scientific significance across a wide range of applications.

References

1. C. Kittel, *Introduction to Solid State Physics* (Wiley Inc., 2005), pp. 163–164
2. S.W.S. McKeever, *Thermoluminescence of Solids* (Cambridge University Press, 1988), pp. 5–7
3. S.W.S McKeever, M. Moscovitch, P.D. Townsend, *Thermoluminescence Dosimetry Materials: Properties and Uses* (Nucl. Tech. Pub., 1995), pp. 64–65
4. R. Chen, S.W.S McKeever, *Theory of Thermoluminescence and Related Phenomena* (World Scientific, 1997), pp. 2–21
5. R.B. Laughlin, Phys. Rev. B **22**, 3021 (1980)

6. S. Bhushan, Nucl. Trac. **10**, 215 (1985)
7. A.J.J. Bos, Nucl. Instrum. Methods Phys. Res. Sect. B: Beam Interact. Mater. Atoms **184**(1–2), 3–28 (2001)
8. A.H. Carter, *Classical and Statistical Thermodynamics* (Prentice Hall, 2000), pp. 361–364
9. J.T. Randall, M.H.F. Wilkins, Proc. R. Soc. Lond. **A184**, 366 (1945)
10. V. Pagonis, G. Kitis, C. Furetta, *Numerical and Pratical Exercises in Thermoluminescence* (Springer, 2006), pp. 7–8
11. G.F.J. Garlick, A.F. Gibson, Proc. Phys. Soc. **60**, 574 (1948)
12. C.E. May, J.A. Partridge, J. Chem. Phys. **40**, 1401 (1964)
13. N. Chandrasekhar, K. Bishal Singh, R.K. Gartia, J. Rare Earths **35**, 733 (2017).
14. A. Halperin, A. Braner, Phys. Rev. **117**, 408 (1960)
15. L.D. Miller, R.H. Bube, J. Appl. Phys. **41**, 3687 (1970)
16. W. Hoogenstraaten, Philips. Res. Rep. **13**, 515 (1958)
17. R. Chen, S.A.A. Winer, J. Appl. Phys. **41**, 5227 (1970)
18. T.S.C. Singh, P.S. Mazumdar, R.K. Gartia, J. Appl. Phys. **23**, 562 (1990)
19. J.T. Randall, M.H.F. Wilkins, Proc. R. Soc. A: Math. Phys. Eng. Sci. **184**, 347 (1945)
20. S.V. Moharil, S.P. Kathuria, J. Appl. Phys. **16**, 425 (1983)
21. N. Takeuchi, K. Inabe, H. Nanto, J. Mater. Sci. **10**, 159 (1975)
22. R. Chen, J. Elect. Chem. Soc. **116**, 1254 (1969)
23. M. Balarin, Phys. Stat. Sol. A **31**, 111 (1975)
24. P.L. Land, J. Phys. Chem. Sol. **30**, 1693 (1969)
25. S.D. Singh, M. Bhattacharya, W.S. Singh, W.G. Devi, A.K.M. Singh, P.S. Mazumdar, Ind. J. Phys. **73A**, 471 (1999)
26. J. Götze, Y. Pan, A. Müller, Miner. Mag. **85**(5), 639–664 (2021)
27. J. Götze, Miner. Mag. **73**, 645 (2009)
28. J. Götze, R. Möckel, *Quartz Deposits Mineralogy and Analytics* (Springer, Berlin Heidelberg, 2012), pp. 307–347
29. W.A. Deer, R.A. Howie, J. Zussman, *Rock-Forming Minerals*, vol. 4 (Longmans, London, 1963), pp. 435–436
30. J.M. Kalita, G. Wary, Nucl. Instr. Meth. Phys. Res. B **383**, 177 (2016)
31. S. Farouk, H. El-Azab, A. Gad, H. El-Nashar, N. El-Faramawy, Lumin. **35**, 586 (2020)
32. S. Farouk, A. Gad, H. El-Azab, H. El-Nashar, N. El-Faramawy, Radiat. Phys. Chem. **18**, 109333 (2021)
33. O. Antohi-Trandafir, A. Timar-Gabor, A. Vulpoi, R. Bălc, J. Longman, D. Veres, S. Simon, Radiat. Meas. **109**, 1 (2018)
34. T.M. Farias, S. Watanabe, J. Lumin. **132**, 2684 (2012)
35. S. Thomas, M.L. Chithambo, J. Lumin. Lumin. **197**, 406 (2018)
36. P.L. Guzzo, L.B.F. Souza, V.S.M. Barros, H.J. Khoury, J. Lumin. Lumin. **188**, 118 (2017)
37. P.L. Guzzo, L.B.F. Souza, H.J. Khoury, Radiat. Meas. **46**, 1421 (2011)
38. S. Nsengiyumva, M.L. Chithambo, L. Pichon, Radiat. Eff. Def. Sol. **169**, 919 (2014)
39. A.J.J. Bos, N.R.J. Poolton, J. Wallinga, A. Bessiere, P. Dorenbos, Radiat. Meas. **45**, 343 (2010)
40. A. Mandowski, A.J.J. Bos, Radiat. Meas. **46**, 1376 (2011)
41. R. Chen, J.L. Lawless, V. Pagonis, Radiat. Meas. **47**, 809 (2012)
42. V. Pagonis, L. Blohm, M. Brengle, G. Mayonado, P. Woglam, Radiat. Meas. **51–52**, 40 (2013)
43. R. Chen, V. Pagonis, Radiat. Meas. **106**, 20 (2017)
44. R. Zhou, M. Wei, B. Song, Y. Zhang, Q. Zhao, B. Pan, T. Li, Nucl. Instr. Meth. Phys. Res. B **375**, 32 (2016)
45. J.A. Gadsden, *Infrared Spectra of Minerals and Related Inorganiccompounds* (Butterworths, London, 1975), pp. 291–292
46. C. Schmidt, A. Chruscinska, M. Fasoli, M. Biernacka, S. Kreutzer, G.S. Polymeris, D.C.W. Sanderson, A. Cresswell, G. Adamiec, M. Martini, J. Lumin. **250**, 119070 (2022)
47. S. Thomas, M.L. Chithambo, J. Lumin. **204**, 603 (2018)
48. T. Ngoc, H.V. Tuyen, L.A. Thi, L.X. Hung, N.X. Ca, L.D. Thanh, P.V. Do, N.M. Son, N.T. Thanh, V.X. Quang, Radiat. Meas. **141**, 106539 (2021)

49. A.K. Sandhu, O.P. Pandey, J. Mat. Sci.: Mat. Elec. **32**, 20767 (2021)
50. F.S. Lameiras, in *Infrared Radiation,* ed. by V. Morozhenko (IntechOpen, 2012), pp. 41–56
51. M. EzzEl Din, A.M. Abouzeid, Kh. El maadawy, A.M. Khalid, R.E. El Sherif, J. Mining World Express (MWE) **5**, 9 (2016)
52. I.A. El Kassas, F.S. Bakhit, Qatar Univ. Sci. Bull. **9**, 227 (1989)
53. A. Osman, H. Kucha, A. Piestrzynski, Mieral. Polonica **28**, 87 (1997)
54. M. El-Kinawy, H. El-Nashar, N. El-Faramawy, SN. Appl. Sci. **1**, 834 (2019)
55. C. Furetta, F. Santopietro, C. Sanipoli, G. Kitis, Appl. Radiat. Isot. **55**, 533 (2001)
56. S.El Gaby, F.K. List and R. Tehrani, in *The Pan-African Belt of Northeast Africa and Adjacent Area: Tectonic Evolution and Economic Aspects of a Late Proterozoic Orogen*, ed. by S. El Gaby, R.O. Greiling (Earth Evolution Science Viewing, 1988), pp. 17–70
57. M.S. Amin, Econ. Geol. **42**, 637 (1947)
58. S.W.S. Mckeever, Nucl. Instr. Meth. **175**, 19 (1980)
59. C. Furetta, *Handbok of Thermoluminescence* (World Scientific, Singapore, 2003), pp. 444–447
60. N. El-Faramawy, A. Gad, H. El-Azab, S. Farouk, J. Mater. Res. **37**, 3784 (2022)
61. R.R. Dawam, M.L. Chithambo, Rad. Meas. **120**, 47 (2018)
62. R.R. Dawam, F.B. Masok, S.B. Fierkwap, J. Lumin. **233**, 117918 (2021)

Silicones and Their Applications

Rukhmah Riaz, Maqzia Bashir, Kainat Imtiaz, Abbas Rahdar, Muhammad Faizan Nazar, Sajjad Hussain Sumrra, Leili Mohammadi, and Muhammad Nadeem Zafar

Abstract Silicones have great importance in everyday life due to their extensive framework, permeable films, water resistance, and uniform skin-conditioning effects. Based on comparison with other organic materials, silicones have been proved to be very beneficial in the treatment of scars. This family of high-performing materials is classified into different types based on their structure and properties. A brief discussion on the behavior of silicones is also observed with respect to heat and light along with their mechanical, chemical, and physical properties. Whether it's quarts, masonry work, or a building constructed from cement, all these things are silicones. It depicts those silicones are metallo-organic compounds of significant value and research.

Keywords Silicate rubber · Quartz · Mica · Silicone elastomers · Silicone fluids · Silicone gel · Silicone adhesives

Abbreviations

PDMS	Polydimethylsiloxanes
XRD	X-ray diffraction
LSR	Liquid silicone rubber

R. Riaz · M. Bashir · K. Imtiaz · S. H. Sumrra · M. N. Zafar (✉)
Department of Chemistry, University of Gujrat, Gujrat 50700, Pakistan
e-mail: znadeempk@gmail.com

A. Rahdar
Department of Physics, University of Zabol, 98615-538 Zabol, Iran

M. F. Nazar
Department of Chemistry, University of Education Lahore, Multan Campus, Multan 60700, Pakistan

L. Mohammadi
Infectious Diseases and Tropical Medicine Research Center, Resistant Tuberculosis Institute, Zahedan University of Medical Sciences, 98167-43463 Zahedan, Iran

S. J. Ikhmayies (ed.), *Advances in Minerals Research*, Advances in Material Research and Technology, https://doi.org/10.1007/978-3-031-49175-7_5

SEM	Scanning electron microscopy
nm	Nanometer
IR	Infra-red
3D	3-Dimensional
HCR	High Consistency Rubber
M.P	Melting point
FRM	Fiber reinforced mortars
GGBS	Ground granulated blast furnace slag
Fig	Figure

1 Introduction

In the periodic table, silicon is positioned as 14th element. Silicon (Si) ranks second among all discovered elements, making up about 28% of the Earth's crust and 0.03% of the biosphere. Plants have accumulated up to 10% by weight and this amount is higher than that of some trace elements. It is not present in the free form, and it is obtained naturally from earth's crust in combined form. When silicone combines with other organic compounds or atoms, it results in creation of unique properties in compounds that enable their usage in wide range of temperature. It depicts that manufacturing of silicone-based products is less sensitive than those of organic surfactants. All these properties of silicon make its bonds flexible, and strong as in Si–O bond. In silicones, bonds have freedom of rotation, bringing them to attain the lowest energy configuration. Silicones are also present in almost all the minerals of the earth's crust. These minerals are mostly present in the form of crystal lattice. The basic unit of crystal lattice is its tetrahedron structure [1]. Tetrahedron of silicone comprises four oxygen atoms in which every one of oxygen atom is located at the apex of regular tetrahedron structure. Only a single silicon atom is located in the central position of tetrahedron with the valence charge equal to four [2].

Silicon atom is capable of sharing its four electrons present in the valence shell with oxygen atoms. On the other hand, oxygen atom contributes its two valence shell electrons in bonding with Si for silicones formation. When four oxygen atoms present at the apex of **tetrahedron** contribute in bonding, the valence shell of silicon atom gets completely filled and stable due to its complete octet. This resulting stable tetrahedral arranged molecule is silicate, which is then arranged in the form of different layers, rings, sheets, and frameworks [3].

History

With the scientific research of decades on clay and earth's crust, it has been found that almost every ceramic, agricultural products (croton oil, peanut oil, olive oil), and other composites contain silicones in them. After more thorough investigation of silicones for almost 80 years, colloidal clay chemistry has been observed to be intensely similar with the trademarks of crystal chemistry, particularly of clay minerals. Histologically, signs of chronic inflammation, vascular destruction, and giant cell infiltration characterize silicone granulomas. Silicone embolism syndrome (SES), pneumonitis, sepsis, and pulmonary edema have been reported systemically with associated mortality [4].

Although it was very hard to investigate all the phases and components of colloids in the past when investigation was only limited with the use of petrographic microscope. Later, the advancement took place in the early 1930s and late 1920s when X-ray diffraction technique (XRD) outlined the main components of clay minerals. Afterwards, classification of constituent materials also took place. Mineralogists found that all types of clays, rocks, ceramics, and hydrothermal land deposits contain silicate crystals.

Up till now, silicates are used in every field of life. **Silicone rubber** is the most flexible, thermally stable polymer of extensive research. After the preparation of various silicones, researchers have classified them into various types based on sheets formation and basic frameworks, which are of greater importance. In this chapter, we will study how silicones are classified to enjoy different practical applications of silicones [5].

2 Classification

There are many silicones having an inorganic backbone chain formed by silicon-oxygen (–Si–O–Si–O–Si–O–). Such compounds are called as polysiloxanes or polymerized siloxanes in which two oxygen atoms and two organic groups are attached to the central tetrahedral silicon atom [6]. Due to the presence of organic groups, silicones can be both polymeric and cyclic. Due to the presence of organic groups, polymers contain similarity with ketones, especially because, in most cases, one oxygen and two methyl groups account for one silicon atom on average. The main repeated unit came to be known as "siloxane" and the most common silicones are polydimethylsiloxane, abbreviated as PDMS. When linear siloxanes are catalyzed with acid or base, they result in cyclic structure formation. The extent of cyclic product depends upon the proportion of monomers, type of solvent used in reaction, and temperature conditions. Moreover, the difference in crosslinking, side groups, and Si–O chain length results in the formation of different compounds having different names and compositions (Schemes 1 and 2).

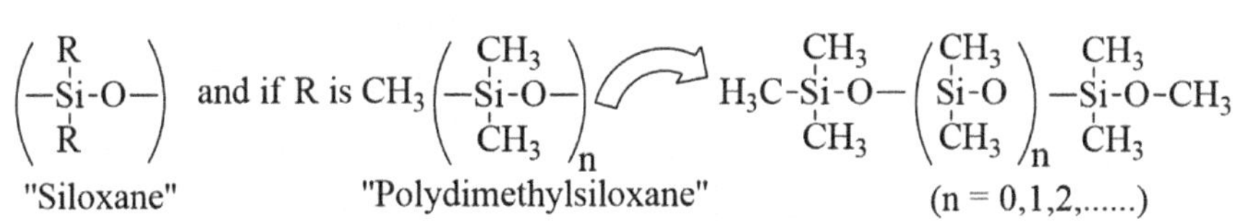

Scheme 1 Formation of polymeric silicones

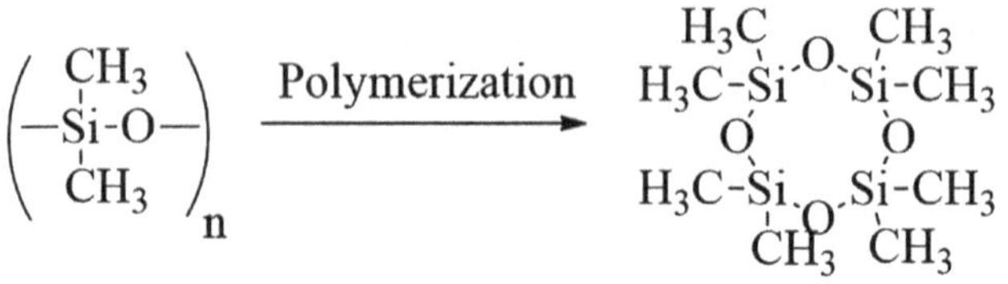

"Polydimethylsiloxane" D_4

Scheme 2 Formation of cyclic silicones

Scheme 3 Structure of PDMS

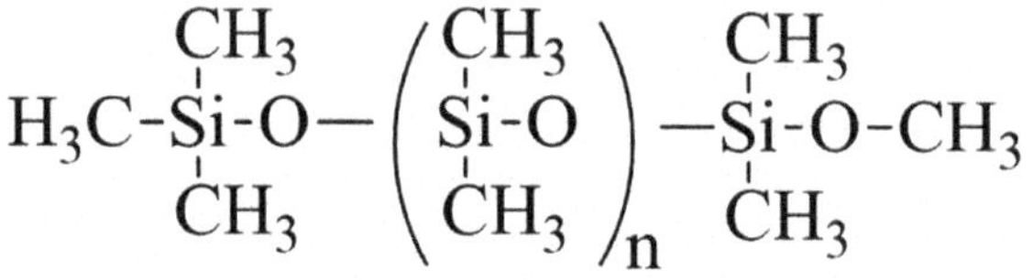

(n = 0,1,2,......)

Silicones are named due to their similarity with ketones as they contain one silicon atom bonded to one oxygen and two methyl groups. However, the main repeating unit is siloxane with the basic framework of –Si(2R)–O. This R can be methyl, ethyl, propyl, or long chain hydrocarbon. Precisely, the most common silicone having a linear structure is polydimethylsiloxanes (PDMS) having main chain unit of a $(Si(CH_3)_2O)_2$. It can also be represented as $(CH_3)_2SiO_{2/2,}$ which is capable of polymerizing with other silicon atoms connected to two oxygen atoms. This polymerization can take place at both ends as shown below [7] (Scheme 3).

Another largest silicone material is of resins group that are branched and known as oligosiloxanes. In this way, the siloxanes are named depending upon the organic group and length of backbone silicone chain [8].

3 Structure of Silicones

Just like other compounds forming ionic, covalent, or coordinate covalent bond (to show the strength of atoms) silicones can also be arranged differently. By looking at their structure, there are two main groups that is, cationic and anionic group. The arrangement of such ions and dense packaging is a determinant of the structural representation. Moreover, their structure can also vary whether atoms are layered, or not. Through packaging and arrangement of ions, they can be identified as liquids, crystalline, or amorphous solid.

When silicones are in liquid and vapor state, there is no repeating unit like **siloxane**. That's why silicone liquids and gases show continuous breaking and rebuilding of bonds between atoms [9]. This continuous bond breakage and formation also affect properties of silicates such as viscosity, reactivity, and strength. To overcome these shortcomings in structural interpretation of silicates, there are two basic models to formulate structural representation of silicones.

3.1 Dense Packaging of Ions

In terms of ions, silicons have oxygen ions O^{-2} with 1.40 Å ionic radius and small cations including Si^{+4} with ionic radius of 0.26 Å [10]. These ions can be packed loosely as well as densely. Mostly cationic and anionic parts are held together through dense packaging of ions either in the form of a cubical structure or in the hexagonal shape. In between every two cations or Si^{+4} ions, interstices are filled with O^{-2} ions.

As the size of oxygen ions remains same, it's the cationic ion size that decides the structural arrangement of silicones. For instance, when there will be small cationic ions, they will perfectly fit into the interstices and holes between O^{-2} ions, leading to octahedral and tetrahedral arrangement. Such strongly packaged structure experiences little distortions. It depicts that the greater the size of cations, the more oxygen ions undergo displacement from their positions, causing reduced density and decreased symmetry packaging [11].

Due to the influence of size and other factors, it often becomes hard to study the three-dimensional structure of silicones. In many silicates, only two-dimensional images of ions overlapping can be extrapolated as shown in Fig. 1. Here Forsterite is represented in different packaging with different radii [12]. In Fig. 1a, packaging of sphere is directly proportional to their ionic radii, having small silicon ($r_{Si}^{+4} = 0.26$ Å), large oxygen sphere ($r_O^{-2} = 1.40$ Å), and magnesium ion ($r_{Mg}^{+2} = 0.72$ Å). In Fig. 1b, dense packaging is proportional to atomic radii instead of ionic radii, having atomic radii of small oxygen spheres, medium silicon spheres, and large magnesium spheres equal to 0.60 Å, 1.10 Å, and 1.50 Å respectively. In the structure shown in Fig. 1c, Ball and stick model is used for representation of silicone structure with broken line Si–O bonds and solid lines Mg–O bonds. However, Fig. 1d forms tetrahedral silicone structure due to packaging of (SiO_4) and cations. Moreover, silicones can also be represented by octahedral due to packaging of (MgO_6) and silicon ions [13].

3.2 Dense Packaging of Atoms

According to Slater, structure of silicone solids is composed of neutral atoms, irrespective of the nature of bond present among their atoms. In account of atomic packaging, atomic radii of silicon and metal atoms are much larger than those of oxygen atoms. ($r_O = 0.60$ Å, $r_{Si} = 1.10$ Å). Only few silicates like olivine structure can be drawn with different atomic sizes and strong packing as shown in Fig. 2.

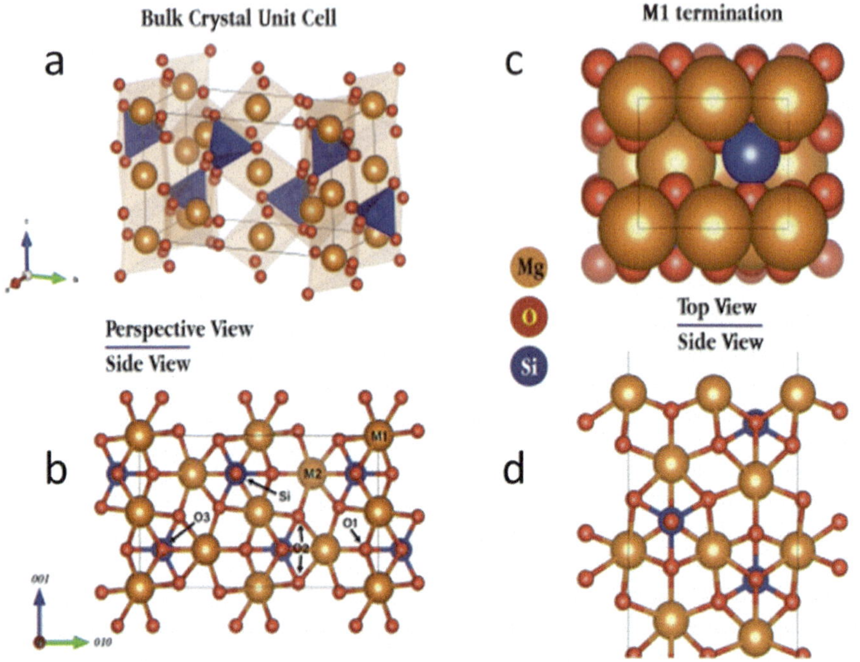

Fig. 1 Different structural representations of Forsterite ($Mg_2(SiO_4)$). **a** Perspective view of bulk crystal unit cell of Forsterite, **b** side view of bulk crystal unit cell of Forsterite, **c** top view of M1 termination of Forsterite, **d** Side view of M1 termination of Forsterite. Reprinted from Geng and Jónsson [11] with permission from American Chemical Society. Copyright (2019)

Fig. 2 Structure of olivine $(Mg_{0.9}\ Fe_{0.1})_2SiO_4$ showing dense atomic packaging

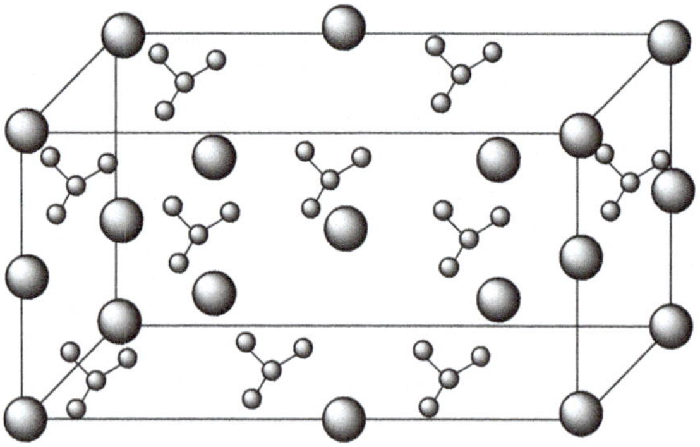

4 Classification of Silicones

Due to the greater versatility and diversity in the formation of silicones, the siloxane unit undergoes different linkage, and cure chemistries to process new material. When the basic tetrahedral structures of silicone unit link together, the formation of different

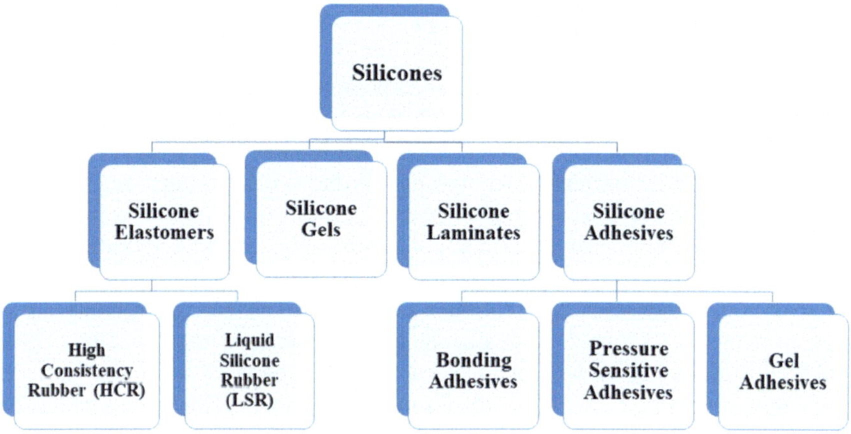

Scheme 4 Classification of silicones

silicates takes place. The incorporation of zinc, aluminum, iron, calcium, magnesium, manganese, titanium, beryllium, and sodium into tetrahedral unit leads to different silicates [14]. When two tetrahedral structures are present isolated, that is the two units are not connected directly, they form a material known as nesosilicate. When two tetrahedral groups are linked directly with each other, this leads to the classification of sorosilicate. When more than one tetrahedral links with one another by back-to-back linkage for the formation of a close ring, it leads to cyclosilicate formation [15]. Upon the basis of different arrangements, materials, and processing techniques, silicones are commonly classified into four major classes.

1. Silicone Elastomers
2. Silicone Gels
3. Silicone Adhesives
4. Silicone Laminates (Scheme 4)

4.1 *Silicone Elastomers*

Majority of **silicone elastomers** are formed by the 3D (3-Dimensional) networking or cross-linking of different silicone units. This cross-linking process of elastomers is known as "curing" that leads to the chemical bond formation. The elastomers are mainly cross-linked either with radicals or by addition and condensation reactions. Which cure strategy to opt for an elastomer is dependent upon the application of that elastomer. For instance, acetoxy cure (condensation cross-linkage) method is used by silicone medical bonding adhesives while addition cure is used to fragment silicon into its parts [16].

4.1.1 Processing of Silicone Elastomers

Silicone elastomers are processed by the incorporation of filler in it. The filler works as the connector to reinforce interstices of cross-linked matrices to impart greater strength. Moreover, filler reinforcement is very helpful in reduction of tackiness and for improvement in its mechanical strength and hardness. Besides imparting strength, elastomer fillers are also added in a silicone to achieve some particular properties [17]. For example, barium sulfate is incorporated in a silicate to boost its radioactivity and carbon black is effective to increase electrical conductivity of a material [16].

Use of fumed silica is also very common as reinforcement material by doing vapor phase hydrolysis of a tetrachloride silicone in the presence of oxygen or hydrogen flame. This reinforcement will take place when hydrogen bond is formed between the hydroxyl groups of silicon polymer and filler matrices. This incorporation of filler takes place before cross linking, for the sake of boosting polymer's viscosity high in the range of 10,000–100,000 mPas [18].

After completing blending process of silica filler with silicone polymer, a catalyst is used to achieve the completion of elastomer cure. There is a need to keep the components separate to avoid premature curing [19]. Unlike thermoplastic material, thermosetting silicone elastomers cannot be reformed, and reprocessed by re-melting. Its proper shape is formed after appropriate linkage. Some methods are also used to shape silicone such as extrusion, molding, and casting. However, which method is appropriate depends upon the feedstock viscosity and desired shape [17, 18].

Here are two basic forms of rubber obtained by processing of silicone elastomers.

- **High Consistency Rubber (HCR)**
  When a silicone polymer of high molecular weight is used, it leads to the formation of high consistency rubber (HCR). There is a need of high shearing technique for processing of HCR of high tensile and tear strengths. The extrusion and bio-pharm processing are used for its tubing. Despite this, low processing methods like injection molding and compression can also be used for HCR [16].
- **Liquid Silicone Rubber (LSR)**
  When a silicone polymer of low molecular weight is mixed with silica, liquid silicone rubber (LSR) is formed. It's less viscous as compared to HCR. The production of LSR occurs by metering, pumping, and injection molding cavity [20]. LSR blend is formed by incurring controlling temperature of molds to lower injection pressure and viscosity of this rubber [16]. The comparison of HCR and LSR is illustrated in Table 1.

Table 1 Comparison of HCR and LSR [16]

HCR (High Consistency Rubber)	LSR (Liquid Silicone Rubber)
• 1-component system • High viscosity • Reduced mixing and batch consistency • Suitable for compression molding or injection molding	• 2-component system • Less viscosity • High batch-to-batch consistency • Appropriate for flash less and automatic molding

4.2 *Silicone Gel*

Silicone gel is another type of silicone polymers in which elastomer chains are tightly cross-linked with either silicone fluids or with non-silicone fluids like mineral oils [16]. This gel undergoes cross-linkage of addition curing reaction such as platinum-catalyzed reactions. This gel is synthesized on the basis of cross-links between ploy (dimethylsiloxane) chain, resulting in the formation of a dense three-dimensional framework [21]. In silicone gel, two-part fluid system comes into existence by mixing both parts A and B in appropriate ratio. That's why cross-linking is usually done after warming the intermediate that is formed in the consequence of pouring mixtures in molds. This technique of cross-linking is very useful in providing protection to sensitive electronic equipment [22].

When silicone gel is applied over any surface, it gets dried within few minutes. According to researchers, silicone gel is effective in the treatment of scars and wounds because of its 86% reducing texture. How much silicone gel will be effective in treating scars depends upon the consistency of silicone gel. In general, a thin application of silicone gel boosts hydration of stratum corneum, increasing fibroblast production that, in turn, reduces collagen making the wounded skin softer. Moreover, it also provides protection against bacterial invasion by reducing redness, itching, and discomfort. However, gel consistency depends upon the quantity, and type of swelling fluid along with its degree of cross-linkage. When both gel components are mixed, the liquid material is casted to mold in various shapes of silicone gel substances [16].

4.3 *Silicone Adhesives*

Silicone adhesives are also practically applicable silicones due to their exclusive bonding, high pressure sensitivity, and gel texture. Silicone adhesives are further divided into these 3 types depending upon their property.

4.3.1 Bonding Adhesives

Silicone bonding adhesives are also components of room temperature vulcanizing elastomers. These are exclusive in attachment and sealing of components and junctions seamlessly altogether. In terms of their practical application, silicone bonding adhesives are used to encapsulate and insulate different electrical components [16].

4.3.2 Pressure Sensitive Silicone Adhesives

Silicon pressure-sensitive adhesives are mainly solvent formulations that undergo polycondensation reactions in the presence of silicate resins, and ammonia catalyst. The main purpose of using ammonia is its stripping with heat to exchange solvent. This class of silicone adhesive has come into existence by the incorporation of ≡SiOH groups into $\equiv SiOSi(CH_3)_3$ groups through the reaction of hexamhethyldisilazane. In practical applications, pressure-sensitive adhesives can be used with or without solvent [23]. However, the resulting substance is permeable to moisture the skin in the presence of vapors along with adhesiveness [24].

4.3.3 Silicone Gel Adhesives

Silicone gel adhesives are two-component silicone systems without any solvent. Such systems came into cure at high temperature in the presence of platinum-catalyzed addition reactions. Just like pressure-sensitive adhesives, these gel adhesives are also skin sensitive and cause gentle traumatic effect on it [25]. It is widely used as a coating material on backing substrates just like polyurethane [24].

4.4 Silicone Laminates

Silicone laminates are a special type of silicone extensively used in apartments, homes, offices, and hotels to give an outstanding style to the buildings. The silicone laminate sheets can bear severe temperature conditions and ultraviolet radiations. In the fabrication of such laminates, solvent or solution dispersions are utilized. The laminates consist of sheets of solid silicone rubber that are reinforced with the glass cloth. The fiberglass is coated from both sides. The fabrics with the coating of silicone are dimensionally very flexible and stable. They are also very tough and thin [8].

Silicone laminates are widely used in press pads, diaphragms, and vacuum blankets. Laminates can even show extra breaking strength and rigidity if almost three piles of fiberglass are utilized in its fabrication. Such properties of silicone laminates are very helpful in very high-pressure operations [26].

We can also enhance the properties of laminates by varying the rubber formulation and glass style. In special construction, coated from one side, and alternate base fabrics are fabricated. Its most amazing property is to withstand extreme temperature conditions. If we compare silicone rubber with other organic rubbers at 25 °C, then it is a quite weak rubber. Under a normal operating condition, at temperature as low as 274 °C and high as 260 °C, the silicone rubbers remain flexible and elastomeric. Moreover, silicone rubbers can bear intermittent exposure to higher temperature as well. Silicon has left far behind the other elastomers by showing more resistance toward the thermal degradation [27]. It has also surpassed the other elastomers in

Table 2 Processing techniques of different silicones materials [16]

Types of silicone material	Primary processing technique
High consistency rubber	High shear mixing followed by transfer, compression, and injection molding
Liquid silicone rubber	Liquid injection molding
Room temperature vulcanizing elastomers	Casting, de-airing, and hand-mixing
Silicone bonding adhesives	Applied on the surface to get adhered
Silicone pressure-sensitive adhesives	Optical heating for removal of solvent, and substrate coating
Silicone gel	Envelope shell filling and mixing
Silicone film-in-place, fast cure elastomers	Spraying or spreading of parts on the surface to cure at ambient temperature
Silicone gel adhesives	Two components are mixed and heated to cure after substrate coating

compression set resistance, non-sticky properties, electrical strength, and general life. They also have a good fluid and chemical resistance.

Amazing electrical insulating properties are also present in inherently silicone polymers. Due to their chemical nature theory are insulators. By compounding the silicone polymers with proper additives and fillers their electrical properties can be enhanced. In this way, they can be used in multiple applications of electrical insulation. Generally, the silicone rubbers well when these are immersed in different liquids, but their property of solvent resistance can be enhanced by increasing temperature and curing time [28] (Table 2).

5 Properties of Silicates

Solid silicate minerals are well characterized and are generally very stable. Silicates with chain-like or small anions and alkali cations such as metasilicate and sodium orthosilicates are impartially soluble in water. When they are crystallized from solution, they form many solid hydrates. Therefore, soluble sodium silicates and their mixtures are known as **water glass**. These are very important household and industrial chemicals [29]. Silicates with three-dimensional polymeric cations or with non-alkali cations are generally very less soluble in water at the normal conditions [30].

The differences in thermodynamic conditions cause chemical weathering of silicate minerals. These differences exist at time of ambient conditions and mineral formation at earth's crust. Generally, solid state properties of silicates are established at high pressure or/and temperature, and these are continuing to be reflected through mineral compositions, textures of rock, the petrographic fabric, and crystallographic structures during weathering [30].

Clay minerals are silicates. They look like mica in chemical composition. Their physical and chemical properties depend on cation exchange capacity and shrinking capacity, etc. [31].

5.1 Physical Properties

The physical properties of silicate minerals are necessarily linear when these are plotted against composition. Generally, there is no any accepted theory that can explain this type of behavior, although it is not unreasonable to consider that, these properties of the minerals in which there are stoichiometric relations are indefinitely variable can exhibit such a simple relationship [32]. However, these properties can be enhanced by some bioactive materials. Some of the important physical properties of silicates are as follows.

5.1.1 Solubility of Silicates

Sodium silicates are white powders, or crystalline solids, or colorless glass. Silicates are readily soluble in water except for those which are rich in silicon. Sodium silicates are very stable in alkaline and neutral solutions [33].

5.1.2 Refractive Index

The medium refractive index glasses are "**fused silica**" and "**fused quartz**", both have silica in amorphous form. Fused quartz refers to the glass which is formed out of the manufacturing process in which quartz crystals are heated to the temperature of almost 2000 °C, opposite to natural mineral or crystal [34]. The ultraviolet transmission of fused quartz is better than other common glasses, and this condition makes it ideal for the applications in the spectral region of sub-400 nm. Fused silica/ fused quartz is generally resistant to water, acid, and scratches and has very low thermal expansion of coefficient. Strong absorption bands in the infrared region are due to normal varieties of quartz having water, although water-free varieties are available commercially. The refractive index of fused quartz varies from 1.40 to 1.55. As it is consistent and optically stable, many use it as a standard for spectroscopic measurements and as a reference [35]. Typical sample of silica at extinction coefficient and refractive index at 632.8 nm are 0 and 1.45704. Figure 3 shows different values of refractive index of fused silica [36].

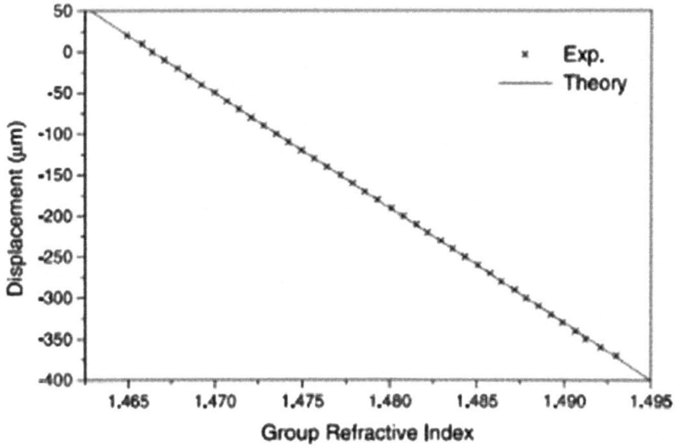

Fig. 3 Refractive index and thermos expansion variation of SiC with temperature. Reprinted from Hlubina [37], with permission from Elsevier. Copyright (2001)

5.2 Chemical Properties

Chemical properties of silicate minerals are those which become evident during or after a chemical reaction. Chemical properties of silicates include reactions, detection, and zeolite formation.

5.2.1 Chemical Reactions

Silicate anions are conjugate bases of silicic acids. These are the compounds with Si–O–H functional group. The fourfold deprotonated orthosilicic acid (Si(OH)) is viewed as orthosilicate. These are generally weak acids and so, can be isolated. They generally exist in aqueous solution as partially protonated anions and mixtures of condensed structures in a dynamic equilibrium. The processes in equilibrium are **condensation/hydrolysis** and **deprotonation/ protonation** [38].

$$-\mathrm{Si}-\mathrm{O}-\mathrm{Si}- + \mathrm{H_2O} \leftrightarrow \mathrm{Si}-\mathrm{OH} + \mathrm{HO}-\mathrm{Si}- \quad (1)$$

$$-\mathrm{Si}-\mathrm{OH} \leftrightarrow \mathrm{Si}-\mathrm{O}^- + \mathrm{H}^+ \quad (2)$$

By increasing the acidity of medium or concentration of silicates, this dynamic equilibrium can be shifted to larger anions. For example, orthosilicate anion is considered as predominant form of the silica which is dissolved in seawater naturally, the concentration is less than hundred parts per million. In low pH or higher concentration, polymeric anions are predominating [39].

5.2.2 Detection

The yellow silicomolybdate complexes are generated by chemical reaction between silicate anions and molybdate anions. In the typical preparation, only in 75 s monomeric orthosilicate reacts completely. The reaction time of dimeric pyrosilicate is only 10 min while higher oligomers have considerably long reaction time. This reaction in general is not observed with the suspensions of the colloidal silica [38].

5.2.3 Zeolite Formation

Zeolites are the main mineral components of modified volcanic-elastic rocks, differing in age and composition. They are formed by the change of predominantly volcanic glass in different geological environments, under changing geochemical and temperature conditions. Proposed genetic models for zeolite deposits include weathering, open, or closed hydrologic systems, low temperature hydrothermal systems, primary igneous environments, and impact craters. The most common types of zeolites that can be found in decayed deposits are clinoptilolite-heulandite, mordenite, chabazite, analcime, and phillipsite. Maini zeolite deposits are widespread in many countries of the world. The annual world production of natural zeolites has remained almost constant for the past 10 years at about 3 million tons. While several large deposits of high-quality zeolites are currently exploited and numerous studies have been conducted on the suitability of zeolite materials for various applications, the majority of the annual production of zeolites is devoted to massive low-value applications such as pozzolan cement additives and lightweight aggregates.

Nature of soluble silicates is applicable to the understanding of aluminosilicates and biomineralization, like industrially important catalysts called as zeolites [38] (Table 3).

Table 3 Physical and chemical characteristics of silicones

Physical characteristics	Chemical characteristics
Soluble in water	Weak acids
Refractive index is 1.45	Protonated anions formation
Electrical insulator	Condensation products formation
High gas permeability	Formation of polymeric anions at low pH
Resistance to oxygen, ozone, and UV-light	Zeolite formation
Thermal stability	Weathering
Low toxicity	Superior solvent for organic compounds
Low thermal conductivity	Low chemical reactivity

6 Applications of Silicates

Silicate-based compounds and silicates have variety of applications in different industries. The primary structure of silicates is pyramid or tetrahedral. It consists of 4 oxygen atoms, 1 located at each corner of the regular tetrahedron. At the center, there is a single silicon atom in the structure of silicate [40] (Scheme 5).

The aspiration to get the combined properties of the different materials in one while this same product generated composite materials on the macroscopic scale. Therefore, examples may be laminates like automotive screens, filled polymers, fiber-reinforced epoxides, and even paint coatings of all types of materials. The other possibility is to combine the properties of different parts on a molecular scale, like it is appeared on large scale with the organic copolymers [41]. On a microscopic scale, the idea to unite organic with inorganic components demands an inorganic network formation process which is similar to thermal stability of organic components [31].

One of the major applications of silicates is their use as **copolymers, fillers in filling materials like glass-ionomer cements, adhesives,** and **composite systems.** In such materials, there is a reaction between filler material and acid during a setting process. During this process they enhance their mechanical properties by increasing thermal expansion coefficient, radio capacity in acrylic filling materials, and physical resistance. They increase esthetics and handling properties. These materials also reduce the polymerization shrinkage [42].

Minerals having silica combined with metal oxides, constitute a major part of the rocks of earth's crust. Almost 40% of common minerals are composed of silicates. These are the most abundant class of the minerals on the earth. Quartz crystals and sand are silicates, and their central part is silicon. Silicates have many amazing technological uses because of their unique properties [1].

6.1 Quartz Crystals

Silica or Silicon dioxide, having SiO_2 chemical formula, is a complex chemical compound of silicon. Silicon is commonly present in the nature as glass or quartz and almost more than 90% of minerals of the world are made up of silicates. The crystalline quartz which is enriched and reinforced with phase diagram along with many polymorphs is the most widely studied material in various fields. Research on quartz has been managed and conducted in different fields such as **micro-fabrication**, **biophysics**, **geology,** and **mechanical engineering** [43].

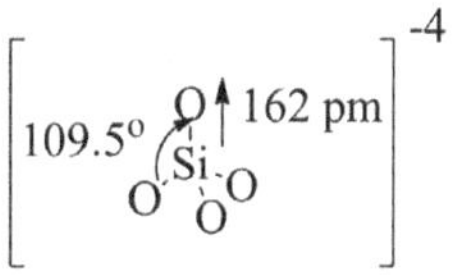

Scheme 5 Tetrahedral structure of silicate

Quartz is widespread in ceramic and glass industries with an enormous potential to be used in **microelectronics**, **optical fibers**, **catalysis**, **filtration,** and **nanocomposites**. The surface properties of silica and the simulation of bulk became essential in the 80s when the numerical computer mechanisms were discovered and so many developments were made in the coming years. The structure of silica is very complex. There are different forms of crystalline silica; **coesite, stishovite, α- and β-quartz, keatite, tridymite,** and **α- and β-cristobalite**. The large part of these structures consists of tetrahedral SiO_2 units having the two-fold coordinated corners of oxygen atoms. Each of these structures has variable bond angle and bond length.

α-quartz refers to an ordinary quartz. This is also known as low quartz while the low term represents the α-quartz is polymorphic silica is stable at low temperature, for the left-handed form which has space group of P3121. **β-quartz** is a polymorphic silica which is known as high quartz due to its stability at high temperature. In crystalline quartz, α-β transition occurs at the temperature of 846 K. Figure 4 describes the supercell and unit cell of α-quartz in 3D. Quartz exhibits highly anisotropic properties, with very direction-related physical properties [44]. Oxide ceramics has significant thermal conductivity in both technological applications and dielectric material physics where these materials are used as substrates for **refractory material processes, thermal barrier coatings,** and **microelectronic devices**. In addition to this, in the geothermal investigations, thermal diffusivity, specific heat capacity, and thermal conductivity are considered as main thermal properties of the fundamental interest. If we talk about geothermal investigations then these include seismicity, volcanism to drilling, applied mining, geothermal techniques, and geothermal resources used in geophysics and environmental exploration [45]. Among the rock forming minerals, **crystalline quartz exhibits a high thermal conductivity**. Regarding thermal conductivity, minerals like nepheline, polyhalite, and mica have lower values. This shows that thermal conductivity depends on mineral components [46].

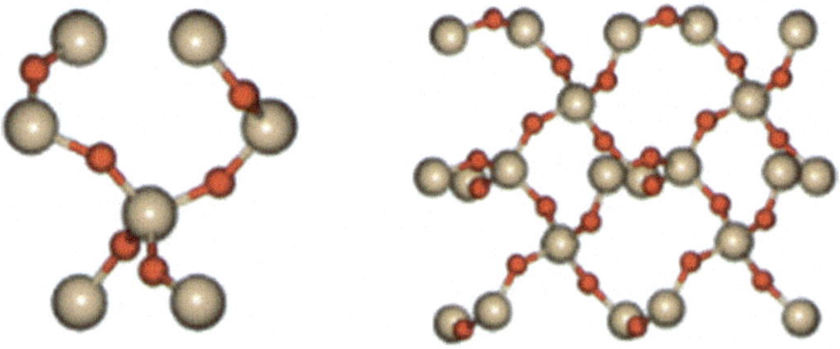

Fig. 4 A unit cell of α-quartz (left), and a super cell of α-quartz (right). Reprinted from Molaei and Siavoshi [46] with permission from Elsevier. Copyright (2020)

6.2 *Mica*

A group of minerals that are chemically and physically similar is called "**Mica**". All of these are silicate minerals. These are generally known as sheet silicates as they are formed in distinct layers. Micas are relatively soft, and somewhat tight. The flakes and sheets of mica are very flexible. Micas do not conduct electricity and they are heat resistant. There are almost 37 different kinds of mica minerals. The most common mica minerals are **black biotite, clear muscovite, brown phlogopite,** and **purple lepidolite.**

White mica has been broadly utilized to date tectonic events and microstructures in folds, shear-zone, and faults because of having suitability for the radiogenic dating. However, complicated properties include (i) individual chemical evolution of K-bearing phases, (ii) microstructural evolution, (iii) combining the "detrital grains" with newly recrystallized or formed as well as chemically reset grain (iv) diffusion in volume which can cause apparent K–Ar ages. Here, the specimens from the prograde sediment sequent of an exhumed fossil European Alpine accretionary wedge were utilized to look into the settling process of white mica with the help of intensity of deformation and type, and from the peak metamorphic conditions. K–Ar system is combined with mineral and mass quantities from fractions of grain size to find out the actual amount of white mica, recrystallized in each fraction of grain size along with metamorphic gradient. This is important because the increase in recrystallization with the increase in grated metamorphic is connected to dissolution-precipitation creep, and activated pressure solution, which is seen by the creation of recrystallized phyllosilicates with spaced cleavage that is documented through scanning electron microscopic (SEM) imaging, and synchrotron X-ray fluorescence microscopy techniques [47].

Addition of inorganic fillers to polymer electrolyte gives rise to composite polymer electrolytes that can enhance the electrochemical performance of the polymer-based batteries. The utilization of poly nanocomposites instead of using the dimensions of filler particles can increase the performance of electrolytes [48].

Now let's have a look at some common applications of mica minerals.

6.2.1 Electric Heating Appliances

Mica minerals are used in rice cookers, cigarette lighters, flat irons, curling irons, permanent wave machines, space heaters, toasters, immersion heaters, hair dryers, hot plates, percolators, soldering irons, steam iron presses, 55 waffle iron, and portable dryers.

6.2.2 Electrical Apparatus

These are used in many electrical apparatuses such as heat sinks, rectifiers, transformers, rotating field coils, armature coils, and commentators.

6.2.3 Electrical Control

In this category, mica minerals are used in grid resistance boxes, motor control panel, industrial mortar starters, electrical and thermal regulators, pyrometers, rheostat rods and terminals, relays, and starting boxes of the trolley cars [49].

6.2.4 Industrial Electrical Heating Appliances

Heating appliances in which mica minerals are used include glue pots, lead pots, branding irons, laundry irons, and thermostats.

6.2.5 Mechanical Uses

Mechanical uses of mica minerals include use in diaphragms for hearing aids, non-breakable googles, petromax chimney, shields for the liquid level gauges of a high-pressure steam boiler, compass cards, guided missiles, Mommsen lungs, oxygen breathing apparatus, quarter wave plates for optical instruments, furnace and stove inspection windows and in the synthetic optical crystals.

6.3 Asbestos

The word **asbestos** is the derivation of a Greek word which means "inconsumable, unquenchable or inextinguishable". This is a generic name given to a class of fibrous silicate minerals. So, a group of naturally occurring minerals that are collectively known as "asbestos" has been utilized for many years because of their distinctive fibrous properties of **high tensile strength, electrical resistance, chemical degradation, resistance to heat, flexibility,** and **large surface area to mass ratio**. The commercial or industrial world that processes and mineralizes those minerals makes the use of the term asbestos to represent a group of 6 fibrous silicate minerals. These six minerals are naturally occurring having the above listed distinctive properties [50]. These fibers are extracted from certain kinds of hydrated alkaline silicates that consist of two families' amphiboles and serpentines. The chrysotile consists of five fibrous forms while serpentine consists of a single form of amphiboles. The five forms of amphiboles are amosite, actinolite, crocidolite, anthophyllite, and tremolite. The most common fibrous silicate minerals are **amosite**, **crocidolite,** and **chrysotile**.

Asbestos is extremely flexible and durable. The most common amphibole species is tremolite in the fibrous form among the 6 asbestos minerals [51]. It is characterized by particles/crystals which consist of fibers with aspect ratio >3, width <3 μm and length >5 μm.

The Romans and Greeks later in the history have found more applications and uses for asbestos as the lamp wicks and have also acknowledged that a woven cloth of soiled asbestos could be cleaned very quickly by throwing them into fireplace. Although the applications of asbestos continued and during reign of industrial revolution, in the modern times only an asbestos industry was evolved. Although all the different asbestos minerals are hydrated fibrous silicates, all of these are different in crystalline structure, chemical properties, fiber dimensions, and chemical composition. Thus, all of these minerals share certain unique properties to the varying degrees, but each of them has unique properties. During 1935 to mid 1970s in United States, asbestos minerals, mainly chrysotile, and amosite in blend or individually were used widely in ceilings, walls, thermal systems and to build thermal insulation. All of this is to provide the buildings with fireproofing and thermal insulation [50, 52].

If we talk about the applications of asbestos, then it covers many different areas. It has so many wonderful uses that have made our lives even more easier. Things made from asbestos have been used since time immemorial. The asbestos was given a nickname as "**the black mineral**" due to its unique and amazing physical and chemical properties that have made this silicate mineral suitable for use in many products from the sewage pipes to the insulating mattresses, from the floor tiles to the road signs [53]. History shows that this silicate mineral has been used by man for more than 4,000 years; during this century asbestos was used in more than 3,000 products which includes insulating mattresses, cement building material, insulation board which are resistant to fire, fire-proofing sprayed products, pipework lagging, sewage and water pipes, floor tiles and coverings, gas masks, friction materials for vehicle brakes and clutches, machinery, and lifts. Pipework and boilers were loaded with asbestos in power stations, hospitals, and throughout the heavy industry. Products with asbestos insulation were very popular in railway industries, in dockyards and in shipbuilding, etc. The products made with asbestos have been utilized on a very large scale in the British buildings for acoustic and thermal insulation, for fireproofing, reinforcement, and condensation protection in the asbestos-cement products [52].

6.4 Cement

Binding materials and cement are very essential for the construction of structures. These are very important for durable and good structures, and they play a very important role to get good bond and good strength between two materials. We can determine the strength and properties of binding materials by using different tests [54].

For all the types of constructions in this world, cement is one of the most important materials. Cement is used for all kinds of works which includes large, medium, and small works. Currently there are many kinds of cements and binders in the market. Many researchers are working on binding materials particularly GGBS, fly ash, and cement. All these materials are used with cement to get better results. Manufacturing of cement is a lengthy process which includes many steps. Nowadays global warming and emission of greenhouse gases play an important role in damaging the structures of buildings. So, using appropriate binding materials and cement we can protect our buildings from the harmful external agent.

Cement is manufactured first by crushing and then milling and proportioning the following chemicals and materials.

- Silica, SiO_2: from old bottles, sand, argillaceous rock, or clay
- Calcium oxide or line, CaO: from chalk, limestone, shale, calcareous rock, or shell
- Iron, Fe_2O_3: from iron ore, clay, fly ash and scrap iron
- Alumina, Al_2O_3: from clay, recycled aluminum, bauxite
- Gypsum, $CaSO_4.2H_2O$: found along with $CaCO_3$ (limestone)

Without gypsum, the materials, are proportioned to make a mixture with a desired chemical composition. The mixture is then grounded and blended by any of the two processes, i.e. **wet process** or **dry process**. After that materials are fed through the kiln at almost 2600 °F to make grayish black pellets which are known as **clinker**. The iron and alumina act as the fluxing agents which then lower the M.P of silica from almost 3000 to 2600 °F. After this step, the clinker produced is cooled down, grounded and then gypsum was added to regulate the setting time. It is then pulverized very finely to produce cement [31].

Nowadays fiber reinforced mortars (FRM) are growing rapidly in different building technology (e.g., roofing, masonry structures, raised floors, facade panels) as building materials. Textile waste from cloth wastes is one of the most promising types of material that can be used by these composite materials. These are used as cement reinforcement materials to increase the sustainability of cement [55].

6.5 Ceramics

Ceramics are defined as the solid compounds created by application of heat. Sometimes they are formed by applications of heat and pressure, and consist of at least 2 elements, provided that one of them is a metalloid or a nonmetal. The other elements/element can be metal/metals or other metalloid/metalloids [56]. A simpler definition is given by Kingery, according to which "*the science and art of forming and using the solid articles, which have their important component, and are made in large portion of non-metallic, inorganic material*" [57].

Examples are **magnesia**, or MgO is a solid compound of metal and is a ceramic. Here Mg is bonded to nonmetal oxygen. In **Silica**, metalloid silicon is bonded to a nonmetal and it is also a ceramic. **SiC** is a ceramic as it is a combination of two metalloids. Ceramic compounds are not restricted to the binary compounds:

$YBa_2Cu_3O_3$, Ti_3SiC_2, and $BaTiO_3$ all of these are ideally respectable class members. The nitrides, carbides, silicides, oxides, and borides of all metalloids and metals are ceramics which leads to a relatively large number of the compounds. **Silicates by definition are also ceramics**. Because of abundance of silicon and oxygen in nature, these silicates are everywhere; mud, sand, mountains, dust, rocks, clay, in short, we can say that major portion of earth's crust is made up of the **silicate-based minerals**. As we know that concrete, bricks, and cement are silicates then a case could be formed that we are living in a ceramic world [58].

Ceramics are very common, from the sanitary ware to porcelains and fine Chinas to the glass products. Recently, these compounds are being considered for their uses that some decades ago were unbelievable; their applications ranging from the electro optic applications to the laser materials, from the substrates used in electronic circuits to the electrodes used in photochemical devices and from the ceramic engines to the optical communications [57]. Historically, these compounds are widely famous for having electrical insulating properties, out of which alumina and porcelains are primary examples. Nowadays, the so-called electronic and electrical ceramics still have an important role in any of modern technological society. For example, the insulated properties all together with amazing environmental and thermal stability and low loss factors make these compounds a good choice for the substrate materials in the electronic packages [56]. Table 4 lists the properties and applications of ceramics.

6.6 *Zeolites*

Nature of soluble silicates is applicable to the understanding of aluminosilicates and biomineralization, such industrially important catalysts are called as zeolites. **Zeolites** are the microporous crystalline aluminosilicates formed of tetrahedral $TO_{4/2}$, where T is for Al, Si. Zeolites' crystallization is done under the hydrothermal conditions by binding these Al (Si) tetrahedral into the corner sharing network. The primary building blocks (tetrahedral) are linked to secondary building blocks (corner sharing network) through the bridge of oxygen atoms that are formed by prisms and rings of different sizes. The formation of aluminosilicate zeolites requires incorporating Al and Si species, organic molecules, metal cations, and water. After that, this mixture is treated hydrothermally and converted into microporous crystalline aluminosilicate. This method can also be designated as **zeolitization** [59].

Zeolite was first synthesized in 1848 when Wholer first recrystallized the apophyllite. This was executed by heating the apophyllite in water solution at around 180–190 °C under 10–12 atm. After this many attempts were carried out during the 1860s. Claireveille became the very first person who synthesized levynite through hydrothermal process in laboratory. In 1940s, the synthesis of zeolite gained a huge attention due to the great work of Milton, Barrer and their coworkers. In 2001 Yoshimura and Byrappa proposed the three processes that are generally used to form zeolite. These are as follow:

Table 4 Properties and applications of ceramics [57]

Property	Applications
Electrical and dielectric	
Ferroelectricity	Capacitors
Conductivity	Heating elements for use in furnace (ZrO_2, SiC, $MoSi_2$)
Low voltage insulators	Ceramic insulation (Steatite, porcelain, forsterite)
Ion conducting	Fuel cells and sensors (ZrO_2, Al_2O_3 etc.)
Gas-sensitive conductors	Gas sensor (ZnO, SnO_2)
Optical	
Transparency	Cables for an optical communication and windows
IR transparency	Infrared laser window (SrF_2, NaCl, CaF_2)
Chemical inertness and translucency	Corrosion and heat resistance materials generally Na lamps
Nonlinearity	Switching devices for the optical computing ($LiNbO_3$)
Thermal	
Thermal conductivity	Heat sinks for the electronic packages (AlN)
Insulation	High-temperature furnace linings for the insulation
Refractoriness	High-temperature furnace linings for the containment of slags, molten metals, and insulation
Chemical	
Anticorrosion	Chemical equipment in the corrosive environment, heat exchangers
Catalysis	
Biocompatibility	Fitters; the purification of exhaust gases Artificial joint prostheses
Nuclear applications	
Fusion	The fusion reactor lining (Si_3N_4, SiC, C), Tritium breeder materials silicates of Li_2O, Li and zirconates)
Fission	Fuel cladding, neutron moderators and nuclear fuel

1. Synthesis of the molecular sieve zeolites as crystalline powders of high quality.
2. Transformation of the clay minerals into zeolites.
3. Processes in which other naturally occurring raw materials are used.

More than forty zeolites that are naturally occurring are known in the form of unique 229 zeolite frameworks. In the framework of zeolite, Al is trivalent while Si is tetravalent. $SiO_{4/2}$ is the electro-neutral species while $AlO_{4/2}$ is negative species. Extra framework of cations compensates this charge. Hydrated zeolite has the following chemical composition [60].

$$|M_{x/n.}(H_2O)_z|\left[(AlO_2)_y\right] \tag{3}$$

Here M is an extra framework of cation with the valence x, y, and n in the subscript are the pertinent values of the molar concentrations of Si and Al in framework of zeolite and z in subscripts represents the molar concentration of H_2O. Loewenstein's rule is obeyed in the distribution of Al in framework of zeolites [61]. Zeolites are very important nanoporous aluminosilicates that are broadly used in industrial applications for their separation and catalytic properties. These are typically formed from the aqueous solution gels that contain different heteroatomic compounds, with organic or/and inorganic cations. These cations act as the directing agents of mobilizing agents (hydroxyl or fluoride anions).

Since, the systematic studies on synthesis of zeolites have been revealed in 1940s. In the late 1940s, a zeolite technology on a very large scale was initiated by a group of Milton and Barrer. They have prepared zeolite through hydrothermal process using a reactive alkali-metal gel of aluminosilicate at low pressure (autogenous) and temperature (~100 °C). In this process the starting materials which have Al–O and Si–O bonds are transformed into crystalline product of Si–O–Al bonds by mineralizing medium (F^- and/or OH^-), the microsporosity of which is determined by crystal structure. In the framework of zeolite, Al is substituted isomorphously by some of the other elements like Zr, Ti, and B. Some of the zeolites can be purely siliceous [60]. In 1982, it was reported by Barrer that the first zeolite material without any natural counterpart has been prepared (Fig. 5).

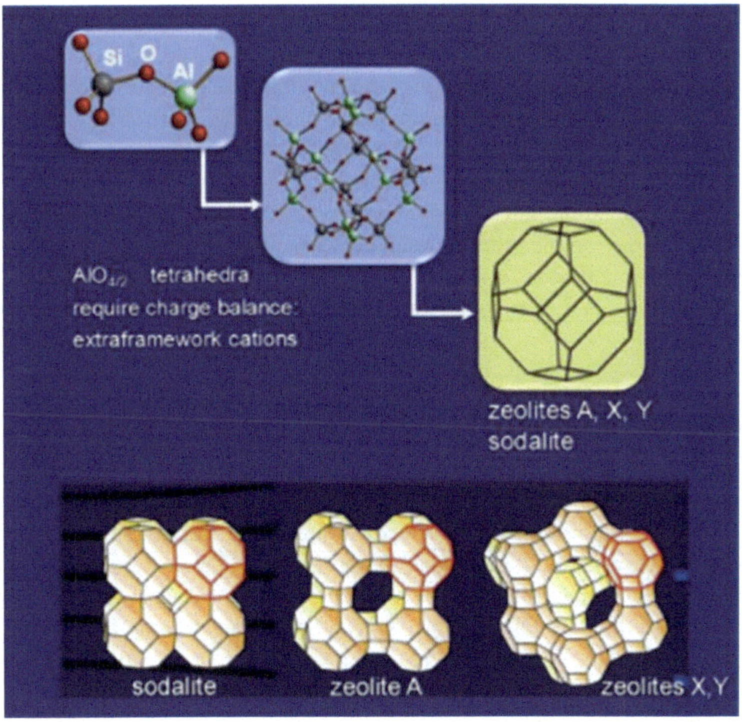

Fig. 5 Scheme of structure of sodalite, zeolites X, Y, and zeolite A. Reprinted from Bacakova et al. [60], with permission from Biomaterials Science. Copyright (2018)

Table 5 Applicability and significance of various zeolites in medicine and biotechnology [60]

Code	Zeolite	Applications
FER	Ferrierite	Antioxidant
ERI	Erionite	Highly carcinogenic
ANA	Analcime	Antioxidant effects in the human beings
ABW	Zeolite A	Antimicrobial coating of the bone implants, Antimicrobial wound-healing dressing, inhibition of the osteoclast
LTL	Zeolite L	Detection of cancerous cells, cell separation, DNA delivery
PHI	Phillipsite	Antioxidant effects in the human beings

Zeolites have wonderful applications in medicine and biotechnology (Table 5).

7 Conclusion

Silicones are a potential class of minerals, having practical applications in almost every field from construction, and industrial to medical units. The inorganic framework of silicone backbone works for different purposes in conjugation with organic groups of methyl or ethyl. Siloxane unit can be cross-linked easily at both room temperature and elevated temperature for the formation of elastomers, gels, and adhesives. Such polymers are formed by the arrangement of tetrahedral siloxane in different arrangements. All these silicones have high gas permeability, low surface tension, thermal stability, and hydrophobicity. These properties make silicones a biocompatible and biodurable material used in various daily life purposes. Silicones have a vast range of applications as quartz crystals used in thermal barrier coatings and microelectronic devices, mica, electrical heating appliances, electrical heaters, and mechanical strengthening products like asbestos, cement, and zeolites.

8 Future Perspectives

The whole study shows the importance of silicones and its applications. A variety of studies point to the development of new techniques and materials. Advances silicone materials are very important in number of applications like electronic applications.

As a class of materials, they can offer proven benefits including resistance to many chemicals, hydrothermal stability, viscosity, and mechanical strength that will curtail stress on the electronic components and will provide them a good protection against vibration and shock, non-toxicity, long pot life and many more. Because of these properties, electronic industrialists are thinking of replacing many costly materials with silicones. There are number of ways by which these high-performance silicones can reduce the cost of manufacturing by increasing productivity and replacing costly materials. In future, we will have many high-performance electronic devices with minimum costs. All this will be because of silicones.

References

1. M. Imtiaz, M.S. Rizwan, M.A. Mushtaq, M. Ashraf, S.M. Shahzad, J. Environ. Manag. **183**, 521 (2016)
2. A. Malashevich, S. Ismail-Beigi, E.I. Altman, D.A. Saeed, J. Phys. Chem. C **120**, 26770 (2016)
3. X. Zhao, W. Shunqing, L. Xiaobao, M.C. Nguyen, C.-Z. Wang, Sci. Rep. **5**, 1 (2015)
4. N.R. Leonardi, J.M. Compoginis, E.A. Luce, Ann. Plast. Surg. **77**, 485 (2016)
5. F.C. Hawthorne, Am. Miner. **70**, 455 (1985)
6. S. Zheng, M. Liao, Y. Chen, M.A. Brook, Green Chem. **22**, 94 (2020)
7. A. Victor, J. Ribeiro, F.F. Araújo, J. Mech. Eng. Biomech. **4**, 1 (2019)
8. G. Lorenz, A. Kandelbauer, Elsevier **3**, 555 (2014)
9. K. Mojsiewicz-Pieńkowska, M. Jamrógiewicz, K. Szymkowska, D. Krenczkowska, Front. Pharmacol. **7**, 132 (2016)
10. P.C. Hess, Struct. Dyn. Prop. Silic. Melts **32**, 145 (2018)
11. M. Geng, H. Jónsson, J. Phys. Chem. C **123**, 464 (2018)
12. C. Zhang, R. Zuo, J. Zhang, Y. Wang, J. Am. Ceram. Soc. **98**, 702 (2015)
13. F. Tavangarian, A. Fahami, Li. Guoqiang, M. Kazemi, A. Forghani, J. Mater Sci. Technol. **34**, 2263 (2018)
14. A. Kottmann, E. Mejía, T. Hémery, J. Klein, U. Kragl, Chem. Asian J. **12**, 1168 (2017)
15. F. Liebau, De Gruyter **5**, 1 (2018)
16. J. Curtis, S.D. Steichen, Biomater. Sci. Introd. Mater. Med. **67**, 109 (2020)
17. P. Mazurek, S. Vudayagiri, A.L. Skov, Chem. Soc. Rev. **48**, 1448 (2019)
18. M.-A. Campeau, A. Lortie, P. Tremblay, M.-O. Béliveau, D. Dube, E. Langelier, L. Rouleau, Biomed. Eng. Online **16**, 1 (2017)
19. J. Schweitzer, S. Merad, G. Schrodj, F. Bally-Le Gall, L. Vonna, J. Chem. Educ. **96**, 1472 (2019)
20. Li. Xiuwen, W. Qiang, M. Zheng, Li. Qian, S. Wang, C. Zhang, Cellulose **25**, 473 (2018)
21. A.L.R. Pires, Â.M. Moraes, J. Appl. Polym. Sci. **1**, 132 (2015)
22. B. Nedelec, A. Carter, L. Forbes, S.-C.C. Hsu, M. McMahon, I. Parry, J. Burn Care Res. **36**, 345 (2015)
23. S.B. Lin, Int. J. Adhes. Adhes. **14**, 185 (1994)
24. J. Curtis, S.D. Steichen, Biomat. Sci. Introd. Mater. Med. **5**, 109 (2020)
25. L. O'Brien, D.J. Jones, Cochrane Database Syst. Rev. **3**, 456 (2013)
26. S. Zolkiewski, Trans. Tech. Publ. **837**, 290
27. I. Low, M. McGrath, D. Lawrence, P. Schmidt, J. Lane, B. Latella, K. Sim, Compos. A Appl. Sci. Manuf. **38**, 963 (2007)
28. G. Lorenz, A. Kandelbauer, Elsevier **43**, 555 (2014)
29. M.-A. Einarsrud, E. Nilsen, J. Non-Cryst. Solids **226**, 122 (1998)
30. A.F. White, S.L. Brantley, Chem. Weathering Rates Silic. Miner. **1**, 53 (2018)
31. P.M. Kumar, G. Ramesh, D.A.R. Prakash, Int. J. Mod. Trends Sci. Technol. **7**, 119 (2021)
32. R.M. Morsi, M.A. Basha, M.M. Morsi, J. Non-Cryst. Solids **439**, 57 (2016)
33. K. Urban, J. Neuhaus, D. Donnermeyer, E. Schäfer, T. Dammaschke, J. Endod. **44**, 1736 (2018)
34. O. Wada, D. Ramachari, C.-S. Yang, T. Uchino, C.-L. Pan, Opt. Mater. Express **10**, 607 (2020)
35. C. Racles, V. Cozan, A. Bele, M. Dascalu, Des. Monomers Polym. **19**, 496 (2016)
36. An. Guowen, L. Shuguang, X. Yan, X. Zhang, Z. Yuan, Y. Zhang, JOSA B **33**, 1330 (2016)
37. P. Hlubina, Opt. Commun. **193**, 1 (2001)
38. C.T. Knight, R.J. Balec, S.D. Kinrade, Angew. Chem. **119**, 8296 (2007)
39. L. Picard, P. Phalip, E. Fleury, F. Ganachaud, Prog. Org. Coat. **80**, 120 (2015)

40. U. Eduok, O. Faye, J. Szpunar, Prog. Org. Coat. **111**, 124 (2017)
41. T. Aziz, H. Fan, F.U. Khan, M. Haroon, L. Cheng, Polym. Bull. **76**, 2129 (2019)
42. L. Namitha, M. Sebastian, Ceram. Int. **43**, 2994 (2017)
43. X. Huang, B. Qingsong, H. Jianguo, H. Dong, Sensors **17**, 1785 (2017)
44. K.I. Vatalis, G. Charalambides, N.P. Benetis, Procedia Econ. Finance **24**, 734 (2015)
45. F. Dirri, E. Palomba, A. Longobardo, E. Zampetti, B. Saggin, D. Scaccabarozzi, Sens. Actuators A **287**, 48 (2019)
46. F. Molaei, H. Siavoshi, Solid State Commun. **320**, 114020 (2020)
47. I.V. Akker, A. Berger, H. Zwingmann, A. Todd, C.E. Schrank, M. Herwegh, Tectonophysics **800**, 228708 (2021)
48. H. Nuranian, A. Keshavarz, M.A. Kabiri, J. Adv. Mater. **2**, 42 (2016)
49. N. Andraschek, A.J. Wanner, C. Ebner, G. Riess, Polymers **8**, 201 (2016)
50. B.R. Strohmeier, J.C. Huntington, K.L. Bunker, M.S. Sanchez, K. Allison, R.J. Lee, Int. Geol. Rev. **52**, 801 (2010)
51. E. Solbes, R.W. Harper, J. Investig. Med. **66**, 721 (2018)
52. D. Di Giuseppe, N. Perchiazzi, D. Brunelli, T. Giovanardi, L. Nodari, Sci. Rep. **11**, 1 (2021)
53. V. Paolini, L. Tomassetti, M. Segreto, D. Borin, F. Liotta, M. Torre, F. Petracchini, J. Mater. Cycles Waste Manag. **21**, 205 (2019)
54. J.J. Assaad, Constr. Build. Mater. **163**, 139 (2018)
55. P. Sadrolodabaee, J. Claramunt, M. Ardanuy, A. de la Fuente, Case Stud. Constr. Mater. **14**, e00492 (2021)
56. G. Corman, R. Upadhyay, S. Sinha, S. Sweeney, S. Wang, S. Biller, K. Luthra, Springer **59** (2016)
57. M.W. Barsoum, CRC Press **32**, 633 (2019)
58. V.K. Yadav, K.K. Yadav, V. Tirth, A. Jangid, G. Gnanamoorthy, N. Choudhary, S. Islam, N. Gupta, C.T. Son, B.-H. Jeon, Crystals **11**, 1067 (2021)
59. Y. Li, L. Li, Yu. Jihong, Elsevier **3**, 928 (2017)
60. L. Bacakova, M. Vandrovcova, I. Kopova, I. Jirka, Biomater. Sci. **6**, 974 (2018)
61. M. Moliner, C. Martínez, A. Corma, Angew. Chem. Int. Ed. **54**, 3560 (2015)

Overview of Some Production Routes for Hydroxyapatite and Its Applications

Medhat Ibrahim, Rasha A. Youness, and Mohammed A. Taha

Abstract It is well-known that bone is one of the most commonly replaced organs worldwide after a blood transfusion and therefore, the development of different strategies is necessary for treating bone defects. Based on this concept, bone surgical problems have been solved. Thanks to biomaterials, which involve calcium phosphates/silicates, porous metals, bioactive glasses, glass–ceramics, and synthetic polymers. Notably, hydroxyapatite (HA) is one of the calcium phosphate family, which possesses marvelous ability to integrate with hard tissues without triggering foreign body reactions and consequently, save patients from the possible undesirable complications. It is meaningful to mention that in spite of the attractive applications of HA in medicine, it has other important applications in many fields like agriculture, chemistry, and the environment. This chapter briefly discusses the different bone grafting materials, and their defects, 3D bioprinting of scaffolds, the definition and types of biomaterials, host responses to biomaterials, different methods for producing HA, and the different properties and applications of HA. Noteworthy, one can say that the preparation of HA with certain characteristics is very complicated due to the potential production of toxic intermediary phases that may occur during the synthesis process. Meanwhile, further investigations and descriptions of the HA structure and fictionalization could be achieved with molecular modeling at different levels.

Keywords Calcium phosphate · Hydroxyapatite · Production routes · 3D bioprinting · Host responses to biomaterials · Applications · Molecular modeling

M. Ibrahim (✉) · R. A. Youness
Spectroscopy Department, National Research Centre, El Buhouth St., 12622 Dokki, Giza, Egypt
e-mail: methat.ibrahim@gmail.com

M. A. Taha
Solid State Physics Department, National Research Centre, El Buhouth St., 12622 Dokki, Giza, Egypt

S. J. Ikhmayies (ed.), *Advances in Minerals Research*, Advances in Material Research and Technology, https://doi.org/10.1007/978-3-031-49175-7_6

1 Overview of Bone Grafting Materials

Generally, the ability of bone to self-regenerate is well established. However, some pathological reasons such as trauma, neoplasm, infection, or inherent genetic disorders leave large bone defects that cannot be healed without medical help. According to this concept, bone is considered as the second most widely transplanted tissue and there are four million surgical operations carried out worldwide annually [1]. In tight correlation with medical needs, the field of bone regeneration has been highly improved to treat bone damage. Unfortunately, bone regeneration is considered a complex task because it requires many important factors to ensure this process's success. The factors include appropriate scaffold materials, tissue-inducing substances (growth factors) along with the existing tissue. Scaffolds are 3D porous solid materials fabricated to fulfill one or more of the following functions:

- Stimulate cell-material interactions, deposition of extracellular matrix (ECM), and cell adhesion.
- Facilitate the passage of gases, nutrients, and other necessary factors for cell survival, proliferation, and differentiation.
- Biodegrade with an adjusted rate corresponding to the rate of tissue regeneration.
- Exhibit a low degree of toxicity [2].

As a rule, the bioactivity of scaffold material is greatly affected by the interaction of its constituent molecules and osteoprogenitor cells at the interface. Based on this principle, manipulating the implant's composition is a critical point for raising the efficiency of treatment. This concept opens the way for incorporating protein growth factors like insulin-like growth factors and bone morphogenetic proteins to activate cellular signals in the stem cells surrounding a bone defect to promote active repair. Regardless of great importance of growth factors in the bone repair mechanism, they have many drawbacks like instability, immunogenicity, and high cost. To face this challenge, small molecules commonly derived from natural products can be used to take therapeutic osteoinductive action taking into account that they show less prevalence of the aforementioned drawbacks of growth factors. For a more comprehensive view, researchers paid great attention to studying specific ions' roles in the bone regeneration process. They discovered that some ions are able to induce osteoprogenitor cell differentiation via growth factor signaling pathways and promote processes for supporting bone growth. These ions involve boron, calcium, cobalt, copper, fluoride, lithium, magnesium, niobium, phosphate, silicate, silver, strontium, vanadium, and zinc. Unlike protein growth factors, these ions exhibit great simplicity, better stability, and cost-effectiveness [3].

2 3D Bioprinting of Scaffolds

3D printing of scaffolds is the most attractive method which is gaining interest in the materials science community. The reason for this great interest is the ability of this approach to adopt the needed structure layer by layer or pixel by pixel and thus, improves the formation of inherent vascular structures. In traditional methods, milling process is used to cut the raw materials, thus obtaining the desired shape. Unfortunately, the 3D placement of seeded cells onto the pre-fabricated scaffold is not precise which consequently, restricting capacity to form complex hierarchical tissue constructs [1, 4, 5]. It is interesting to note that the terminology "3D printing" was first used in 1993 to modify a standard inkjet printer into custom processing equipment. Since then, various technologies have been developed [6]. Here, we will briefly mention these technologies and their importance in the advancement of bone tissue engineering (BTE) application.

3 The Different Techniques Used for 3D Bioprinting

There are four technologies used for the purposes of 3D bioprinting process. These are inkjet (drop-on-drop-demand), laser assisted bioprinting (LAB), microvalve, and extrusion bioprinting [1].

- **Inkjet (drop-on-drop-demand) bioprinting**

The first relies on the use of thermal force/sound waves to expel the droplets from a pint head nozzle. In thermal inkjet printers, a droplet of bioink is ejected due to its use of thermal energy to generate a pressure pulse inside the print head. On the other hand, piezoelectric crystals are the basis of other systems where the application of voltage leads to mechanical stress and thus, changes its shape. In this case, acoustic waves are generated and consequently, form pressure suitable to remove the droplets from the nozzle. The main benefits of this technology are cost-effectiveness, high print speed, and availability. On the flip side, its disadvantages include frequent blockage, and undesirable exposure of cells and materials to thermal/mechanical stress [7–9].

- **Laser-assisted bioprinting (LAB) systems**

These systems take advantage of a pulsed laser beam to create a high pressure bubble. In turn, the materials containing the cells are pushed toward the collecting substrate from an initial print material "ribbon". Keriquel et al. [10] aided in the regeneration of bone of a mouse skull defect by using LAB to print mesenchymal stromal cells combined with collagen and nano-HA. The exploitation of this method has some limitations such as the risk of cell destruction due to heat, difficulty in producing a 3D structure, the high cost, and long time consumption; especially in the case of creating ribbons with a large number of cells with high concentrations of a biomaterial [11].

- **Microvalve bioprinting**

This method depends on opening and closing a small valve to dispense fluids a constant pneumatic pressure from cartridge tips. It should be noted that this valve can be controlled mechanically, electrically, or magnetically. This method is suitable for printing mesenchymal stem cells (MSCs) with excellent viability and functionality along with the deposition of collagen and bone morphogenic protein [1].

- **Extrusion bioprinters**

Unlike other techniques, they are capable of depositing filaments of materials. This technique relies on applying pneumatic/mechanical pressure to a syringe to make a bioink extrusion through a nozzle. The great benefit of this method is its ability to deposit a large number of cells [12]. Indeed, diverse bioinks were successfully produced by this method including tissue spheroids, tissue filaments; cell pellets, etc. Despite this great success, this production faces many challenges like the difficulty of obtaining high print resolution, advances in printable bioinks, and the effect of shear stress on cells inside the print nozzles [1].

Generally, bone grafting (scaffold) materials can be classified into two main groups: biological and synthetic ones [3]. The first type includes autografts, allografts, corals, natural polymers-like collagen type I, and demineralized bone matrix (DBM). Autografts are the so-called "gold standard" as they are characterized by their ability to directly bond to bone (osteoconduction) and encourage the differentiation of local stem cells to bone ones (osteoinduction) without detectable immunological response [13]. These graftings depend upon the transplantation of fresh cortical/trabecular bone from one body location to another within the same patient. Unfortunately, this transplantation is restricted by the limited bone supply and donor site morbidity. Accordingly, these limitations can be avoided by using allograft which is dependent on the transplantation of cortical/trabecular bone or demineralized bone matrix from the donor to the patient. Notably, these grafts exhibit osteoconductivity and even, in some cases, show limited osteoinductivity. However, this type of bone grafting may lead to disease transmission, bacterial infection, and immune response. Therefore, one of the most common approaches to overcome these disadvantages is to use the second type of bone grafting materials; namely biomaterials [3].

4 Biomaterials

Biomaterial science deals with studying the interactions between materials and tissues to identify the mechanisms that contribute to the material integration in biological systems to be valid for biomedical applications. Importantly, this science is a multidisciplinary approach where it integrates chemistry, biology, medicine, and material science. However, researchers have a great challenge to combine the mechanical, chemical, and tribological properties of materials with their biocompatibility that have not yet been fulfilled [14]. This is why these implants rapidly suffer from failure, which in turn leads to the subsequent replacement of prosthetic devices for young patients. The most commonly used materials include bioactive glasses, glass ceramics, porous metals, and calcium phosphates/silicates [15–20]. These materials can be briefly discussed as follows:

- **Bioactive glasses (BGs)**: These materials possess a unique set of properties regardless of their brittleness. They are known to be biodegradable taking into account that their degradation can be controlled. Importantly, this controlled degradation has a great role in promoting osteogenesis and angiogenesis processes. Furthermore, the released ions can be effectively converted to a hydroxyapatite/carbonated hydroxyapatite (HA/CHA) layer that binds to hard and soft tissues even in some cases. There are two routes for the production of BGs, i.e. melt-quench and sol–gel. The prepared BGs by sol–gel technique are characterized by better bioactivity behavior and cellular response than those prepared by melt-quench method. The reason behind this enhancement in their biological activity is closely related to a high volume fraction of mesopores along with high specific surface area. Depending on the preparation conditions, the sol–gel method allows the control of the phase composition process, structure, texture, and microstructure of the prepared glasses. For example, altering the process' pH value is beneficial for obtaining mesoporous structure with micro/nanoparticles. Notably, the doping of BGs with some useful elements may help in improving the cellular response, surface reactivity, and solubility in physiological fluids [21, 22].
- **Glass–ceramics (GCs)**: They represent an important family of materials which can be used as bone substitute materials due to its great ability to bond with living hard tissues (bone) through creating a HA/CHA layer. They are obtained through proper crystallization of glasses which contributes to encouraging the in-situ precipitation of crystalline phases along with residual amorphous matrix. Due to the crystallization process, there is a noticeable enhancement in the mechanical properties of the resultant GCs. Unfortunately; there is a strong relationship between the mechanical strength of a material and its bioactive behavior. Based upon this fact, glass–ceramics exhibit better mechanical properties and less bioactivity than those of glasses.
- **Porous metals**: They are considered as ideal candidates for load-bearing orthopedic applications because they have high mechanical properties. Additionally, another advantage of porous metals is that they facilitate bone ingrowth leading to a rapid fixation for the implants. It is worth to note that these metals morphologically and mechanically resemble trabecular bone. Although these aforementioned promising properties of porous metals, they possess several disadvantages such as a lack of bioactivity, which is responsible for implant loosening, and a relatively high level of toxicity due to the undesirable accumulation of metal ions due to their corrosion [3].

- **Calcium silicate (Ca-Si) compounds**: Depending on the ratio of Ca/Si, there are three Ca-Si compounds; namely monocalcium ($CaSiO_3$), dicalcium silicate (Ca_2SiO_4), and tricalcium silicate (Ca_3SiO_5), keeping in mind that Portland cement is mainly composed of Ca_2SiO_4 and Ca_3SiO_5. Notably, these three compounds have good bioactivity and biocompatibility [19].
- **Calcium phosphate compounds**: It is one of the most promising materials for different biomedical applications due to its marvelous properties including their high chemical similarity with natural bones, bioactivity, solubility, biocompatibility, etc. [23]. Therefore, this chapter focuses on giving readers interested in biomaterials an overview about such precious materials, especially, hydroxyapatite, and its various properties and applications in different fields, particularly biomedical ones.

5 Host Responses to Biomaterials

Knowing the host's response to biomaterials is absolutely essential to trying to find strategies to overcome the potential complications that arise due to the presence of implants within the human body [24]. In this regard, before talking about the calcium phosphate group in some detail, we must first talk about the interactions of biological materials with cells as follows:

- **Inflammation, wound healing, and the foreign body response**

In general, any injury can lead to the occurrence of inflammation which serves as the starting point for a protective biomechanism. Based on this biomechanism, wound healing is stimulated to restore the damaged part of the organism. After the signal occurs, the affected blood vessels along with the local blood cells release some biochemical/chemical signals to call the white blood corpuscles (WBCs) to the site of the injury. Notably, inflammation is the first step toward wound healing. If the integrity of the vessel is cut due to this injury, thrombocytes (platelets) adhere to the edges of the wound to form a clot plug. In response to the action of these platelets, the liberated molecules increase the permeability of this vessel which leads to the transfusion of water from it to the affected tissue, causing edema (tissue swelling). Alongside with these successive steps, the released biomolecules help relax the muscles surrounding the vessels with the goal of controlling blood flow through the vasodilation process. Therefore, one can expect the site to become reddish. To warn the nervous system (NS) to draw attention to our wound care, some of these liberating molecules act as pain modulators. When the WBCs reach the site of the injury, they start to fight against the cause of this injury and restore the vascular lumen by resorbing the blood clot. Additionally, new connective tissue is produced by the fibroblasts considering that the newly formed tissue will be replaced by a clot until tissue regeneration takes place. It is possible to notice that epithelial cells grow and multiply forming a new covering around the wound area [24, 25].

If a biomaterial remains inside the human body, the body's response is closely related to the biocompatibility of that material. In other words, if this material is biocompatible, then the initial acute response occurs followed by a great ability to survive within the organ without problems. On the other hand, if this material lacks the required biocompatibility behavior, the rejection mechanism is promoted by the immune system to withdraw this implant (foreign body). This description can be better understood in the knowledge that both the inflammatory reaction and the fibroblasts continue their work to isolate this unwanted foreign body to avoid any potential threats to the host body. Accordingly, edema, pain, and a reddish appearance continue until the withdrawal process ends. It should be noted that this process is so-called "foreign body reaction" and represents a chronic development [26].

- **Immunology**

It is well-established that the body's immune system is the defense system against the different dangerous factors surrounding us, having in mind that this system is also very sensitive to risks that develop internally such as tumors. At the outset, we must briefly mention that this important system is basically composed of two components; namely the cellular and the biochemical reactions. The first is carried out by WBCs that form in bone marrow and lymph vessels and can be considered as the battle soldiers. These cells include polymorphonuclear neutrophils, polymorphonuclear eosinophils, polymorphonuclear basophils, monocytes, lymphocytes, and plasma cells.

Interestingly, the major function of these cells is ingestion (phagocytosis) of the invading bodies. However, the biochemical defense is in turn divided into innate and acquired to eliminate/inactivate the biologic and chemical agents. It is worth to note that some naturally produced secretions such as the acid produced during digestion process can destroy bacterial cells entering our digestive system. Apart from that, there are about twenty proteins that include serum proteins, serosal proteins, and cell membrane receptors produced by the liver and found in blood as inactive precursors. These proteins make up the complement system. If stimulated by a suitable trigger, they can be activated to kill bacterial cells. In the presence of a target, the antibodies (immunoglobulins) produced by the triggered system along with the sensitized lymphocytes are able to kill it [24].

- **Systemic toxicity**

It is acceptable to note that the introduction of natural/synthetized materials (biomaterial, drug, etc.) into the human body may damage living tissues. The degree of this harm is closely related to the dose, efficacy, method of administration, and metabolism. Under the influence of this damage to living structures, internal/external cell functions are adversely influenced. In the case of chronic toxicity, the adverse effects may extend for a long period (possibly for life). Based on these facts, implant corrosion or biodegradation must be carefully controlled to avoid toxicity risks. It is worth noting that once toxicity takes place, the expected complications for the patient will vary from irritation, at the implant site, to severe and even life threatening conditions unless the cause of this toxicity is removed or neutralized. The process of neutralizing (inactivating) the threat (antigen) is done by the immune system and is also called "allergic reaction" [26].

- **Tumorogenesis**

Deoxyribonucleic acid (DNA) is known to control the mitosis process. Moreover, it carries a specific code to induce programmed cell death (apoptosis). Unfortunately, some external factors (chemical substances, gamma/X-ray radiation, some bacterial/viral infections, etc.) can deteriorate this property or even change its structure [24, 25]. Therefore, its inhibitory effect on the mitosis process is not only lost, but also encourages a continuous cell division while suppressing the apoptosis process. As a result, it causes an unstoppable cell number also called "tumorogenesis". Notably, some materials are able to promote the mitosis process by inducing DNA mutation. It is important for researchers interested in the field of biomaterials to exercise caution when designing implants as the survival of these materials in the human body may lead to chronic irritation, which in turn stimulates the mitosis process to repair the injured area [27].

- **Implant-associated infection**

The success of implants is estimated through their ability to remain in the human body which depends on many internal/external factors [27]. One of these factors is implant failure due to infection by regarding the implant is a foreign material. Therefore, it can be rejected. One promising solution to overcoming this problem is to produce implants from biocompatible materials and sterilize them before introducing into the body.

Mucosal barrier and skin protect our bodies against diverse types of microbes by blocking their entry. The injury spoils the integrity of the skin-mucosa barrier and thus, microbes can invade the body. Then, they replicate themselves and form colonies. Additionally, they are able to secrete toxins to weaken the defense system and make the surrounding environment more conductive to their survival. The point of entry of these microbes into the implant may be direct or indirect contact through blood. In this perspective, infection occurs which increases the likelihood that the body will reject this implant. It should be noted that although the roughness of the implant surface is of great importance in increase the contact area between the implants and bone tissues, it is considered as an ideal environment for bacterial colonization [28].

- **Blood-biomaterial interactions**

These reactions include protein adsorption, platelet reactions, intrinsic coagulation, fibrinolytic activity, erythrocytes, leucocytes, and complement activation. Factors affecting these interactions are the structure of the implant, the existence of an antithrombotic agent, the condition of the patient, and the nature of the application. Therefore, it is important to assess the anticipated interactions of biomaterials with blood to advance their development. Possible options for this include clinical, in vivo, in vitro, and ex vivo [29]. As described former, any injury causes a corrupt in the integrity of blood vessels and consequently, bleeding takes place. One of the

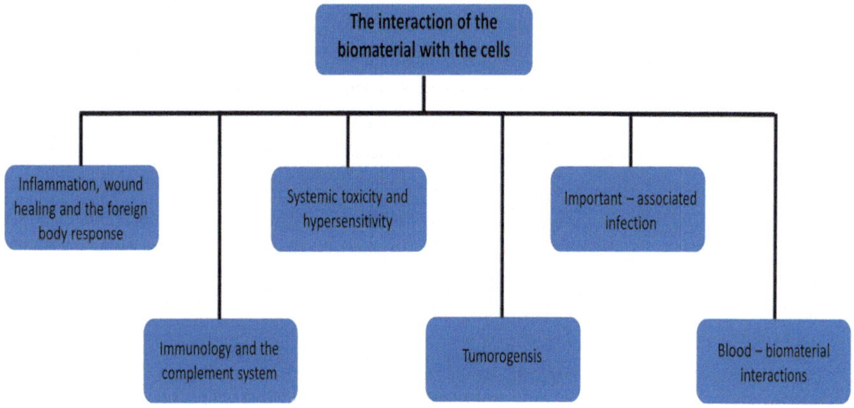

Fig. 1 A summary of the biomaterial's interaction with cells

protective mechanisms that happen to stop this bleeding is coagulation (hemostasis) mechanism. Also, blood when in close contact with biomaterials tends to initiate a thrombotic response. According to this information, one can expect that this response is a serious problem when a biomaterial is introduced into the body to treat vascular diseases, which leads to the failure of this device and the formation of emboli that prevent the flow of the bloodstream causing a stroke [30, 31]. A summary of the biomaterial's interaction with cells is represented in Fig. 1.

6 Calcium Phosphate (Ca-P) Materials

Calcium phosphate (Ca-P) compounds are a group belonging to bioceramics and have gained extensive interest for use as alternatives to hard tissues. There are many reasons for this interest including their high chemical similarity with natural bones, bioactivity, solubility, biocompatibility, etc. Therefore, using biomaterials based on Ca-P for fracture treatment has been begun in 1920 [32] and since then, they were extensively used for different purposes such as thin coatings on metallic implants, while sintered Ca-P can be used as a bone graft substitute. Noteworthy, these materials achieved great clinical success [33]. Owing to the abundance of phosphate compounds, Ca-P family is considered as the most complex material families. It is important to highlight that many factors are affecting the chemical durability of calcium phosphate compounds like composition, pH, solvent type, temperature, etc. [34]. In accordance with this concept, Ca-P compounds can be divided, according to calcium-to-phosphorous (Ca/P), into the following classes [35]:

a. **Hydroxyapatite (HA; $Ca_{10}(PO_4)_6(OH)_2$)**: Since this chapter is mainly concerned with HA, the following sections will introduce its properties, synthesis processes, and different applications in detail.

b. **Calcium deficient hydroxyapatite (CDHA)**: Like HA, it is found in natural bone and has a lower Ca/P ratio than that of stoichiometric one, i.e. 1.67 which facilitates the doping of HA with different traces and controls the solubility. With the help of these facts, CDHA can also be beneficial to many fields like solving substantial environmental problems such as the purification of water and soil through removing heavy metals [36]. Besides this amazing use in environmental applications, it can be widely used as alternative material for bones due to its close similarity to human bones, and therefore, it shows the best bioactivity among other types of Ca-P compounds. Unlike the stoichiometric HA, previous studies revealed its rapid precipitation of HA when it was soaked in a simulated body fluid (SBF) solution [37].
c. **Dicalcium phosphate dihydrate (DCPD)**: It is also known as "brushite and has the chemical formula ($CaHPO_4 \cdot 2H_2O$) with a Ca/P ratio of 1. More to the point, it is naturally found in bone and kidney stones. One of the most important advantages of such phase is its use as a raw material for HA preparation. Under physiological conditions, DCPD shows a higher solubility than HA. Thus, its stability after storage on the shelf is challenging for researchers due to its tendency to lose bonded water molecules. It is important to emphasize that this substance is favorable for use in various applications including gene delivery and bone cement.
d. **Tricalcium phosphates (α, α′ or β-TCP)**: It can also be termed as "whitlockite" and its chemical formula, $Ca_3(PO_4)_2$ with Ca/P ratio of 1.5. It is well-recognized that it has three allotropic forms, namely α-, α′-, β-TCP. The latter is formed at relatively low temperatures and can easily convert to the high temperature form of α-TCP at 1125 °C. It is extremely important to emphasize that β-TCP cannot be precipitated under normal aqueous circumstances because it is considered to be unstable in water. Alternatively, it can be obtained by the heat treatment of calcium-deficient hydroxyapatite above 700 °C. Interestingly, β-TCP is highly resorbable material. However, its dissolution is followed by an osteoclastic cellular phagocytotic response. On the other hand, α′-TCP has no clinical benefits since it only appears at temperatures above 1430 °C and easily converts to α-TCP when cooling below the transition temperature. Conversely, both α- and β-TCP are commonly used in many clinical applications such as dentistry, maxilla-facial surgery, and orthopedics bearing in mind that β-TCP is considered as the main component of commercial mono- or biphasic bioceramics and composites. However, α-TCP is majorly utilized in various hydraulic bone cement [38, 39].
e. **Biphasic calcium, triphasic, and multi-phasic calcium phosphates**: In 1985, scientists found during the Preparation of TCP that the resulting product contains 80% of TCP, while the rest contains HA and thus this product is called "biphasic calcium phosphate; BCP. Thereafter, this accident discovery encouraged many researchers to prepare BCP containing α-TCP and β-TCP phases or even three-phase calcium phosphate containing α-TCP, β-TCP and HA mixture. Actually, preparing high-purity Ca-P compounds is a very expensive and complex process, and therefore the final product has one or two major phases. Generally, their physicochemical properties are highly based on the relative amounts of the

constituents. Accordingly, the manipulation of phases' ratios greatly influences the in vivo resorption and bioactivity. For instance, the biphasic composition shows a higher bioresorption rate than that of monophasic one as different Ca-P phases possess a different solubility rate in the order of HA < β-TCP < α-TCP and thus, BCP principle achieves a good balance of the most soluble phase and the most stable one through the manipulation of phase composition. Moreover, the osteoinductive behavior of BCP (HA + β-TCP) is better than that of monophasic HA or β-TCP [33]. The solubility and biodegradation behaviors will be better explained later in this chapter.

f. **Tetracalcium phosphate (TTCP)**: It is also termed as "hilgenstockite and its chemical formula is $Ca_4(PO_4)_2O$. It should be noted that TTCP can be produced through the heat treatment of $CaO–P_2O_5$ system at high temperatures, i.e. 1300 °C having in mind that it is the only Ca-P compound with a higher Ca/P ratio than stoichiometric HA. Owing to the high solubility of TCP, TTCP, and Ca_2SiO_4, they are more suitable for use as fertilizers than HA. Some caution should be taken into account when preparing a pure TTCP phase due to that it is a metastable phase, which requires fast quenching or absence of humidity to avoid its potential decomposition into HA.
g. **Octacalcium phosphate (OCP)**: The chemical formula of OCP is $Ca_8H_2(PO_4)_6{\cdot}5H_2O$ and is not only a precursor to bones, enamel, and dentine but also encourages bone growth. Based on this concept, one may expect it to be used also as a precursor to HA. In this sense, bone defects can be reconstructed using OCP granules considering that they exhibit a higher osteoconductivity behavior than HA and β-TCP. According to its superior properties, it has gained great interest over the last few decades. On the other hand, in spite of these amazing properties, OCP is an unsuccessful material for the purpose of bone replacement since it cannot be prepared in block form. The differences in physico-chemical and biological properties, advantages, and disadvantages and applications of different types of Ca-P compounds are listed in Table 1.

Ca-P compounds are always used as sintered ones or granules keeping in mind that if these compounds were prepared as porous blocks (pre-form), they achieve excellent infilling for many defects like burr holes provided that these defects are uniform in their shapes to maintain good contact between the implant and bone. However, this limitation can be overcome by forming Ca-P compounds as granules regardless of the difficulty in handing them. Moreover, it is very difficult to fill a narrow defect with granules and in the event that the granules fall outside the boundaries of the defect, the removal of these granules is very disturbing. Consequently, matching granular size to bone defect is necessary to achieve a cleaner application. In this regard, the granules used in the dental application should be smaller than 1 mm, while those used in the orthopedic field must be ranged from 1 to 6 mm. With the aim of providing injectable and/or moldable pastes, extensive efforts have been made for developing Ca-P materials. Injectable Ca-P cements are made from TTCP and DCPD and an aqueous solution. Within less than an hour, solidification occurs through a combination of reactants dissolution and crystallization. By applying such

Table 1 A comparison table of the physico-chemical, and biological characteristics and applications of the different Ca-P systems

	HA	CDHA	DCPD	TCP	TTCP	OCP
1. Chemical formula	$Ca_{10}(PO_4)_6(OH)_2$		$CaHPO_4 \cdot 2H_2O$	$Ca_3(PO_4)_2$	$Ca_4(PO_4)_2O$	$Ca_8H_2(PO_4)_6 \cdot 5H_2O$
2. Mineral name	Hydroxylapatite	–	Brushite	Whitlockite	Hilgenstockite	–
3. Natural presence	Found in bone and teeth	Found in bone and teeth	Found in bone and kidney stones	–	–	A precursor to bone, enamel, dentine and HA
4. (Ca/P) ratio	1.67	Lower than 1.67	(Ca/P) = 1	(Ca/P) = 1.5	(Ca/P) = 2	(Ca/P) = 1.33

(continued)

Table 1 (continued)

	HA	CDHA	DCPD	TCP	TTCP	OCP
5. Advantages	• Excellent biocompatibility • Good bioactivity • Less soluble than TCP and CDHA	• Doping with different traces and controlling its solubility are easy • It shows the best bioactivity among other types of Ca-P compounds	• It can be used as a raw material for HA preparation • Under physiological conditions, it shows a higher solubility than HA	• More soluble than HA	• Exhibit high solubility	• It encourages bone growth • Exhibits a higher osteoconductive behavior than HA and TCP
6. Disadvantages 7. Applications	• Has low mechanical properties • Bone and dental implants • Coating of metal alloys • Drug delivery system • Remove water pollutants • Removal of soil pollutants • Catalysis of some reactions • Fertilizer	• Has low mechanical properties • Non-load bearing orthopaedic and dental applications	• Has poor mechanical properties • Its stability after storage on the shelf is a critical issue • Gene delivery system • Bone cement	• Has poor mechanical properties • Dental applications • Maxilla-facial surgery and orthopedics • Different hydraulic bone cement	• Has poor mechanical properties A metastable phase • Suitable for use as a fertilizer	• Has low mechanical properties • It cannot be used as a bone substitute material due to that it cannot be produced in block form • The reconstruction of bone defects

cement in the biomedical field, the clinicians are able to treat damaged bones and stabilize their fractures. Most importantly, fixing these fractures may help eliminate ordinary metallic internal fixators such as plates and screws if their low mechanical properties are improved. However, one of the main problems associated with the injection system of pastes is the migration of liquid between solid Ca-P particles during injecting the paste, leading to a wet densified powder block at the plunger part, unamendable to injection. This phenomenon is so-called "filter-pressing". Therefore, increasing the viscosity of this liquid is a key factor for overcoming this serious problem. Unfortunately, this strategy has a set of drawbacks including high cost, requires manual operations prior to injection and less time to inject cement after preparation [32].

7 Hydroxyapatite (HA)

Generally, "apatite" term was derived from the Greek word "Apatow" and was used for the first time by Werner in 1786 [40]. It was found that the X-ray diffractograms of enamel, dentin, and bone are nearly similar to those of mineral apatites; namely HA and fluorapatite (FA). Additionally, the chemical analyses confirmed that both calcium and phosphorous constitute these minerals. This finding strongly supports the belief that the inorganic part of bones and teeth is basically composed of HA which is represented by the chemical formula, $Ca_{10}(PO_4)_6(OH)_2$ as will be discussed later. On the opposite side, in spite of these mentioned findings, the exact structure of biological apatites is still unclear. Myriad forms of their morphology and variations in non-stoichiometry are probably the most important reasons for the unclearness of their exact structure [41].

It is well-known that bone is a natural nanocomposite where it mainly consists of biological hydroxyapatite (BHA; $Ca_{10}(PO_4)_6(OH)_2$), which occupies about 70% of its weight, while type I collagen fibrils constitute 20% only. It is interesting to note that hydroxyapatite, according to the molar ratio of Ca/P, can be divided into stoichiometric and non-stoichiometric. The Ca/P ratio of the former type is 1.67, while that of the non-stoichiometric one is slightly higher than 1.67. Importantly, BHA is non-stoichiometric and considered as Ca- or P-efficient one. These deficiencies in Ca and P are that there are some substitutions in Ca positions by Na^+, Mg^{2+} and Al^{3+}. On the other hand, other elements can substitute phosphate (PO_4) positions and carbonate (CO_3) groups. Furthermore, CO_3 and fluorine (F) can also substitute OH^- positions [42–48]. It has been reported in the literature that the presence of CO_3^{2-} ions, in HA lattice, is an important constituent in biological apatite and they occupy 5–8 wt% in bone minerals. In fact, this incorporation of CO_3^{2-} ions, can greatly affect a set of physicochemical properties such as mineralization, demineralization, and remineralization processes. Notably, this variation in the Ca/P ratio relies upon many factors such as the species, age, sex, and bone type [45]. Actually, HA possesses amazing properties such as bioactivity and osteoconductivity which are the major advantages of use in repairing and reconstructing bone tissue defects.

However, although these wonderful biological properties, HA has a relatively slow bone-bonding ability and cannot prohibit microbial growth onto the implant surface. Therefore, the addition of some beneficial ions can solve this problem [49].

8 Synthesis Methods

To date, several approaches including wet, dry, high-temperature processes and synthesis based on biogenic sources have been adopted for the preparation of HA. Therefore, this chapter is intended to list the most commonly successful ways of preparing HA.

8.1 Wet Methods

As expected from their names, chemical reactions are carried out in the presence of solution. Noteworthy, they can be performed using various solvents, precursors, and devices and at different calcination temperatures. As a rule, these methods are the most popular ones for producing HA because they offer interesting advantages like the precise control for the morphology and the powders' sizes. Accordingly, the resulting HA lies in nano-scale range with a homogenous morphology. Unfortunately, they have serious drawbacks such as the production of other undesirable Ca-P phases along with HA, the product having low crystallinity and the incorporation of trace impurities into the crystal structure of HA due to the presence of some ions in the aqueous solution [50]. It is important to stress that HA's properties, produced by these methods, are strongly related to the preparation method, the calcination temperature, pH, and Ca/P ratio [45].

8.1.1 Sol–Gel Method

It is well-accepted that tdhe sol–gel method is highly relied on the mineralization of precursors in a solution keeping in mind that the most preferred compounds are the organometallic ones. Briefly, this route begins with the hydrolysis of the used precursors and giving micelles around templates in an aqueous or an organic phase and subsequently, they are polycondensed through the formation of a 3D network [50]. Notably, the gel formation has strongly relied on the solvent, reagents, and the conditions of reactions like pH and temperature. Moreover, imprecise control of specific factors during HA growth may lead to the formation of undesired phases such as calcium oxide (CaO), calcium pyrophosphate ($Ca_2P_2O_7$), calcium phosphate ($Ca_3(PO_4)_2$) and calcium carbonate ($CaCO_3$). It is substantial to note that this method is recommended for synthesizing porous HA, at soft-chemistry circumstances, where it does not require high temperature unlike many other conventional methods. On the other hand, this method is not suitable for preparing the desired products for the industry due to the following reasons:

a. High cost along with the relative shortage of commonly used alkoxide-based precursors.
b. Requires precise control, especially for time-consuming processes [51].

8.1.2 Co-precipitation Method

This method is the most straightforward route for the production of HA. Simply, this method is dependent on the fact that HA becomes less soluble in aqueous solution when pH equals to 4.2 at room temperature. Based on this principle, this reaction takes place at higher pH values and temperature ranges from room temperature to 100 °C [50]. In this current method, calcium and phosphate sources are used as precursors in the existence of acid/base additives [52]. In particular, the most commonly used reagents are calcium hydroxide ($Ca(OH)_2$), calcium nitrate ($Ca(NO_3)_2$), orthophosphoric acid (H_3PO_4), and diammonium hydrogen phosphate ($(NH_4)_2HPO_4$). Briefly, under continuous stirring, one precursor is gently added to the other one keeping the molar ratio of Ca/P at 1.67, i.e. the stoichiometry of HA. Finally, aging may be performed for the suspension under atmospheric pressure or it is washed, filtered, dried, and crushed into powders [49]. Additionally, it can also be carried out in the presence of templates [53]. Importantly, the final characteristics of HA are strongly related to the conditions of reaction and ripening time where the slow addition of phosphate precursor leads to a lower nucleation rate and faster crystal growth. On the opposite side, the resulting HA particles' sizes increase linearly with increasing of temperature [50, 54, 55].

8.1.3 Hydrolysis Methods

Generally, there are two necessary steps for taking place in the aqueous hydrolysis of Ca-P to form HA; namely dissolution and precipitation. Many Ca-P precursors are highly recommended for this method including dicalcium phosphate dihydrate ($CaHPO_4$), tricalcium phosphate ($Ca_3(PO_4)_2$), and octacalcium phosphate ($Ca_8H_2(PO_4)_6 \cdot 5H_2O$). Over the last few decades, little attention was paid to studying the preparation through the hydrolysis of the latter source. Later, some studies revealed that this precursor has a slow hydrolysis rate along with its high ability to incorporate unfavorable species which leads to the belief that this precursor is not appropriate for this method. Under the effect of pH and temperature, hydrolysis process is successfully applied to these chosen precursors. Moreover, other calcium and phosphate precursors may be added to obtain stoichiometric HA bearing in mind that the resultant product's porosity is highly dependent upon the aforementioned parameters [50].

8.1.4 Hydrothermal Methods

Several techniques are included in hydrothermal processes for carrying out the crystallization of the prepared materials in the aqueous medium. This route is highly dependent on applying high temperature along with pressure in the presence of calcium and phosphate sources beside water or organic solvent. Therefore, it requires autoclave or a pressure vessel to help in the formation of chemical bonds and the creation of nuclei to form the crystalline HA. Instead, applying high temperature and vapor pressure can be replaced by the presence of microwave irradiation. The resultant HA powders have brilliant characteristics such as high crystallinity with a Ca/P ratio close to that of stoichiometric HA. Indeed, its size distribution is usually broad. Interestingly, the existence of high pressure is the major reason for controlling the particle sizes and the porosity of HA. It should be noted that hydrothermal processes can be classified into two fundamental categories; namely hydrothermal conversion and hydrothermal growth techniques. Noteworthy, this method can be combined with other ones like co-precipitation and sol–gel [56].

8.1.5 Emulsion Method

This method is deemed as efficient, simple, and appropriate for preparing nanostructured HA powders [57]. Additionally, it permits a good control for the morphology and the distribution of grains size. Generally, it is frequently utilized for obtaining porous materials and the elimination of particle agglomeration problem. Many sources of calcium and phosphate can be used, however the most commonly used ones are $Ca(NO_3)_2$ and H_3PO_4 [58]. It is important to underline that using a surfactant, like dioctyl sodium sulfosuccinate salt, dodecyl phosphate, etc., in this method is essential [59]. Lately, one can say that the final product's main characteristics are mainly related to the synthesis factors such as surfactant type, pH, temperature, and ratio of aqueous and organic phases [56].

8.1.6 Sonochemical Method

Another interesting approach for preparing nanostructured HA is sonochemical method that relies heavily on the activation of chemical reactions between liquid and solid reactants by powerful ultrasonic waves keeping in mind that the rate of HA crystal growth is progressively increased through sonication up to 5.5 times. This method's benefits are that it is responsible for producing HA with homogenous, small, and pure crystals, and minimum agglomeration [50]. It is possible to notice that nanostructured HA is characterized by better sintering kinetics and consequently, good mechanical properties resulting from higher surface area to volume ratio.

8.2 Dry Methods

At first glance, we can expect that these methods do not need the usage of solvents in the preparation. By these ways, we can also prepare the desired products with large amounts without careful control for the processing parameters where they do not significantly affect the final properties of the obtained powders [55].

8.2.1 Solid-State Synthesis

Overall, solid-state synthesis method is very appealing for producing HA because it is a simple and cost-effective one. In this way, calcium and phosphate precursors are first mixed, followed by calcination process at elevated temperature, i.e. about 1000 °C for a specific period [60]. However, despite the aforementioned amazing characteristics, the obtained powders are inhomogeneous and unwanted phases are formed beside the final product which is considered a serious drawback to this method. The most likely explanation for the formation of such undesirable phases is that they need a long calcination time to achieve the high diffusion necessary for homogenizing the system composition. Based on this fact, this method is not effective for producing biomaterials [55].

8.2.2 Mechanochemical Synthesis

This method is also called mechanical alloying. Noteworthy, it is valid for preparing various advanced materials like nanocrystalline and ceramics [50]. This method depends on promoting a chemical reaction giving a product lies in nano-scale range. According to Heinincke [61], mechanochemical method is a branch of mechanochemistry that is concerned with studying the promotion of chemical and physicochemical transformations by mechanical energy. In this sense, nano-sized CHA powders' preparation has been successfully achieved in our recent work through this method after specific milling conditions like time and ball-to-powder ratio (BPR). This chemical reaction is achieved because mill balls apply a strong impaction on the reactant powders toward the internal wall of the vial leading to the generation of energy and a significant reduction in powder particles [56]. It is important to note that this method has many advantages because it is simple and cost-effective; and the resulting product is produced in nano-scale range [62]. Moreover, it does not require melting which is very beneficial for saving energy [46, 47]. Unlike solid-state method, the produced powders usually have a well-defined structure. Youness et al. [45] reported that the mechanochemical method was efficient for producing nano-sized carbonated hydroxyapatite (CHA) after 8 h only of milling. This result is strongly evidenced by X-ray diffraction (XRD) technique and Fourier transform infrared (FTIR) spectroscopy as seen from Figs. 2 and 3, respectively. It is interesting to see from Fig. 2 that the characteristic XRD pattern of CHA according to

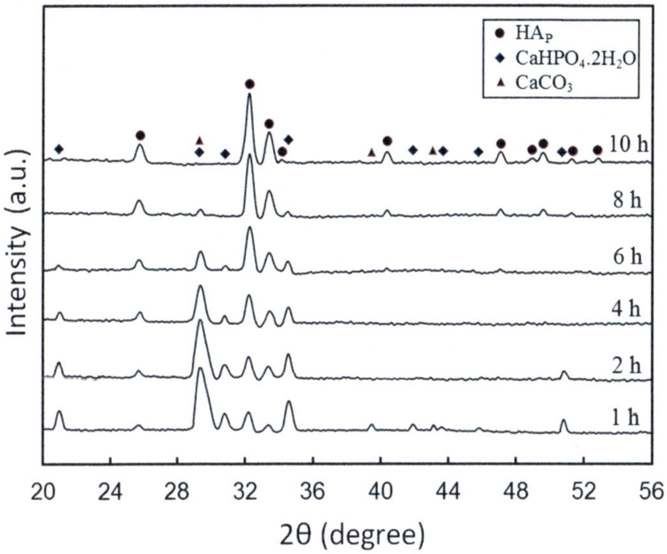

Fig. 2 XRD patterns of the formed CHA after the milling of reactants at different times. Reprinted from Youness et al. [45], with permission from Elsevier. Copyright (2017)

(JCPDS No. 19-0271) has clearly appeared in terms of the position of lines at $2\theta = 32.19°$, 33.43°, 25.74°, and 34.19° which correspond to (1 1 2), (3 0 0), (0 0 2), and (2 1 0), respectively. On the other hand, the vibrational modes of OH^-, CO_3^{2-}, and PO_4^{3-} groups were obviously distinguished at 1630, 1470, 1420, 1100, 1055, 975, 605, 560, and 470 cm^{-1}. Moreover, as discussed formerly in our recent article [63], the scanning electron microscopy (SEM) images of CHA powder, at two different magnifications as shown in Fig. 4, revealed that mechanochemical synthesis is a major reason for the appearance of CHA spherical particles with relatively high agglomeration.

It is well-established that nanomaterials have great biological significance over conventional ones. These outstanding benefits are improving the adhesion and the proliferation of osteoblasts and giving a good environment for protein adsorption and cellular activity. Additionally, they are also responsible for better mechanical properties and particular surface properties [63].

8.3 High-Temperature Processes

As expected, these processes are dependent on high temperature. Based on this principle, these processes can be carried out by two techniques; combustion and pyrolysis [50].

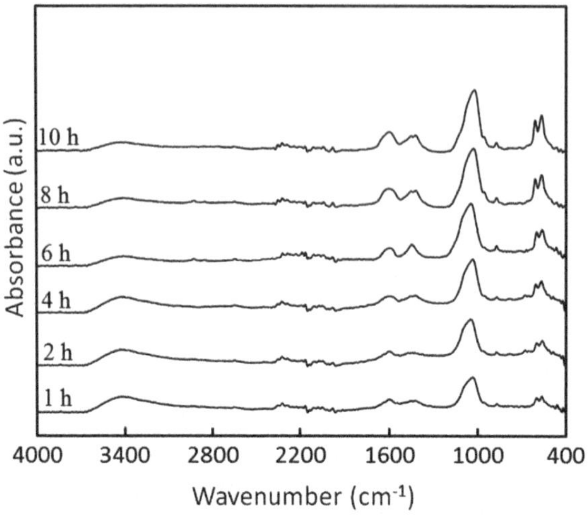

Fig. 3 FTIR absorption spectra of the formed CHA after the milling of reactants for different times Reprinted from Youness et al. [45], with permission from Elsevier. Copyright (2017)

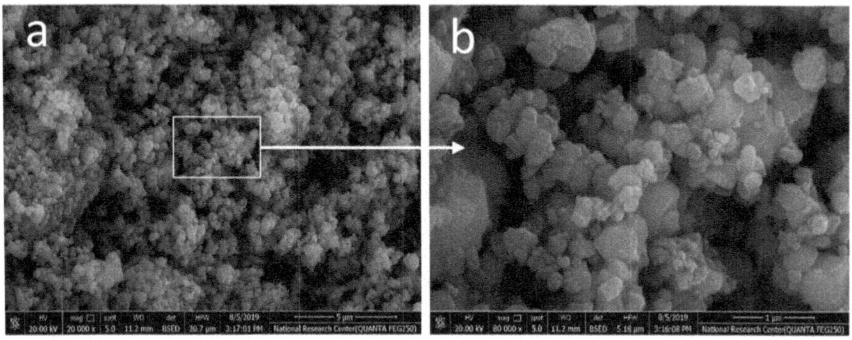

Fig. 4 SEM micrographs of the prepared CHA nanopowders at two different magnifications. **a** 20,000× and **b** 80,000×. Reprinted from Youness et al. [63], with permission from Elsevier. Copyright (2019)

8.3.1 Combustion Method

It is generally considered that this method is an effective way to prepare different ceramics, and in particular, to prepare HA nanopowders. This method is based on thermochemical concepts that are usually used in the chemistry of fuels and explosive materials. It is meaningful to note that this process is rapid and gives the desired product a high degree of purity with a good homogeneity in a one-step operation. Besides these brilliant advantages, this process also offers economic viability. In this way, rapid exothermic and self-sustaining redox reaction between the reactants, i.e. $Ca(NO_3)_2$ and nitric acid (HNO_3) has happened in the presence of appropriate organic fuel such as glycine, urea, sucrose, citric acid, and succinic acid. Firstly, the aqueous solutions of $Ca(NO_3)_2$ and diammonium phosphate ($(NH_4)_2HPO_4$) are well-mixed and subsequently, the concentrated $Ca(NO_3)_2$ solution is added with the aim of dissolving the resultant precipitate. It should be mentioned that one or more of organic fuels are also incorporated into the resultant solution. Moreover, this reaction can be promoted by heating the mixture at 300 °C and later on, the temperature is suddenly increased. Finally, rapid quenching is done to achieve the maximum nucleation and prohibit further particle growth [50].

8.3.2 Pyrolysis

Obviously, it is possible to obtain a stoichiometric single-crystal HA using the above methods if the conditions of chemical reactions are carefully controlled and some heat treatments are also required. By contrast, pyrolysis method achieves this goal easily and quickly. In typical pyrolysis, evaporation of the liquid reactants can form the desired product particles in the gas phase. Compared to the previous method, i.e. combustion, there is no need for fuel. It is important to emphasize that this method belongs to a broad category called "aerosol methods. Indeed, such method also can be sometimes called spray pyrolysis [50].

8.4 Synthesis-Based on Biogenic Sources

It cannot be denied that the extraction of HA from natural sources is a good step toward obtaining such valuable product with low cost and a great possibility for controlling particle sizes. Additionally, this method can also be considered a golden chance to eliminate unuseful waste [64]. Moreover, the obtained HA has the native architecture of animal osseous tissue [65]. Notably, BHA can be extracted from several natural sources such as bovine, pig, camel, horse, dog, corals, eggshells, seashells, and fish. Despite these diverse sources for extracting HA, bovine bones, especially the cortical part of the femoral bone, are the most preferable sources for such purpose due to that they closely resemble human bones [66]. It should be noticed that, as discussed in the literature, the pretreatment of the excised bone is necessary for removing the fats, adhered impurities, and bone marrow. Some articles discussed the importance of boiling bones for 8 h or more to achieve this goal. On the contrary, others reported the importance of boiling and chemical treatment by acetone, chloroform, or alkali solutions. However, some cautions should be taken into account when treating bones with alkali solutions as they leave some traces which cannot be completely eliminated causing a noticeable cytotoxic effect. Moreover, they lead to noticeable increases in the method cost [65, 66].

Briefly, this preparation method begins with cutting bone into small pieces and then, boiling is done for several hours or treating with chemicals to remove undesirable adherents. Then, they should be put in an electric furnace at specific heat temperatures range from 600 to 1400 °C to get rid of organic part of bones and kill the pathogens that may be present [42]. According to many published articles [67–70], extraction of HA from biogenic sources is more beneficial than synthetic methods for the following reasons:

- It has economic and environmental benefits.
- HA obtained from natural animal sources is less likely to be rejected by the human body.
- The extracted HA contains many trace elements which are of biological significance for biomedical applications. Unfortunately, doping HA with these traces by synthetic preparation methods is very difficult.
- Unlike HA which is derived from natural sources, synthetic HA offers low osteoinduction ability.

Finally we would like to mention that, as discussed in the literature, it is so difficult to obtain nano-sized hydroxyapatite particles with crystal structure, orientations, and morphology similar to that found in human bone tissue by conventional chemical methods. The reasons behind this difficulty are that the aforementioned characteristics are heavily dependent on many factors like the pH value of solution, the temperature, the ratio of reactants, etc. The different methods for producing HA are summarized in Fig. 5.

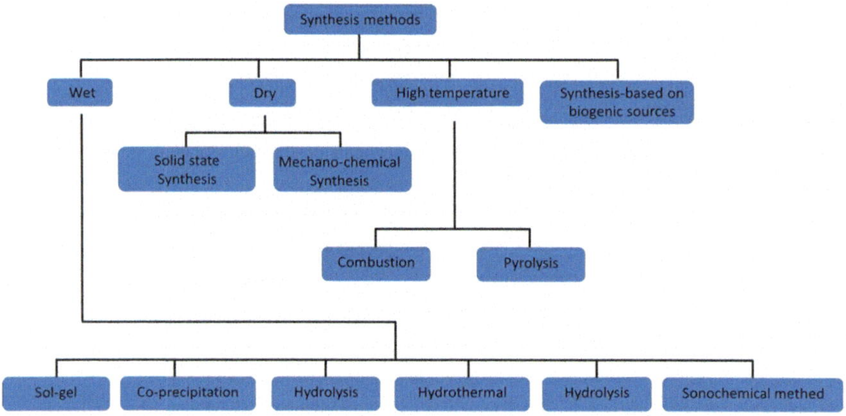

Fig. 5 Scheme for illustrating the different synthesis methods for hydroxyapatite

9 Properties of HA

This section provides an overview of the different characteristics of Ca-P compounds, particularly HA; to give non-specialist readers the knowledge needed to improve the benefits and limitations of these valuable compounds in different purposes, especially biomedical applications.

9.1 Physical Properties

9.1.1 Mechanical Properties

Generally, the load applied on human bones is a combination of bending, torsion, tension, and compression. Moreover, by knowing that fresh bones gradually replace the grafts during the healing process, one can expect that the loss of mechanical support is undesirable during this substantial process. Indeed, implant material's mechanical properties must be comparable to those of surrounding bone tissue to avoid the possible failure of such implant due to stress forces in the implant-tissue interface. Nevertheless, due to their high strength ionic bonds, Ca-P materials are characterized by poor mechanical properties limiting their uses in non-load-bearing implants.

Dense HA possesses Young's modulus ranges from 35 to 120 GPa which is in a good agreement with that of the hardest calcified tissues. On the contrary, its mechanical resistance is about 100 MPa, while that of natural bones is around 300 MPa [41, 71, 72]. Since materials' mechanical properties are greatly affected by porosity and grain size, they are highly improved by reducing grain size and porosity. It has been also reported the dependence of strength on the Ca/P ratio, i.e. decreases when Ca/P tends to be 1.67 which distinguishes a stoichiometric HA. Additionally, the prepared HA's high crystallinity contributes to enhancing other mechanical properties such as the stiffness, fracture toughness, compressive, and tensile strength [41]. In accordance with these facts, various attempts to improve HA's mechanical properties will be discussed later in this chapter.

9.1.2 Porosity

Depending upon the interconnectivity, there are two types of porosity; namely closed (isolated) and open (interconnected). The last type, i.e. interconnected porosity is very favorable to biomedical applications due to its ability to increase the surface area, allows cell adhesion to the implant, increases the ability of bone bonding through the suitable fixation of biomaterial, and improves bone ingrowths and vascularization. It is worth stressing that porosity can also be divided into three categories based on the pore size; namely macro (>100 μm), micro (<10 μm) and nano (<100 nm) given that each type of porosity has its own advantage, so each type has specific uses. For example, macroporosity in the chosen bioceramics is quite useful for cell colonization and vascularization. On the opposite side, microporosity is helpful for facilitating the impregnation of the implants by fluids, improves the solubility, and gives a higher surface area for protein adsorption. Also, nanoporosity acts to enhance cell adhesion, proliferation, and differentiation [41].

As previously discussed in the literature, the synthetic HA's physicochemical properties and microstructure are strongly based on the preparation method. Following this concept, we find that the synthetic HA possesses low crystallinity, high surface area along with high porosity. By contrast, HA obtained from animal sources has excellent crystallinity after its calcination at 800 °C. Amazingly, this material easily forms chemical bonds with surrounding bone tissues. Noteworthy, it is characterized by excellent biocompatibility behavior due to that it resembles the structure and chemistry of natural bone. Moreover, the addition of barium titanate ($BaTiO_3$) to HA has been reported to increase permittivity and AC conductivity making these compounds suitable for polarized bone substitutes [42, 73, 74].

It cannot be denied that HA, derived from bovine source, is characterized by porosity and pore structure comparable to those of native bone. However, there is no correlation between size of the obtained HA and the extraction method irrespective to the additional milling is very useful for reducing the size of HA particles and therefore, it is very similar to the size of a human HA. In comparison with micron-sized HA, nanosized HA particles exhibit several superior characteristics such as high-surface activity, ultrafine structure, excellent bioactivity, and better resorbability. In addition, nano-sized HA displays better sinterability and densification behavior which helps in reducing microcrack formation and programmed cell death [42]. Moreover, as will be discussed later one, drug delivery is one of the most promising applications for HA. The reason behind the validity of HA for such purpose is that it possesses a good wettability with water and organic solvents which allow ceramic loading with different drugs. In the case of whiskers, morphology and structural characteristics of HA are strongly correlated to the initial Ca/P ratio, pH, and calcium concentrations. Additionally, it has been reported that the selected sinter temperature is a key parameter for many critical factors such as density, microstructure, and thermal stability of the HA phase. Otherwise, when the sintering temperature exceeds a certain threshold, it can lead to a partial /complete degeneration of HA giving other Ca-P phases such as β- or α-TCP which possess different solubility, bioactivity, etc., than those of HA. It is worth to note that the macroporous HA network has biological significance, where it facilitates the attachment, spreading, and proliferation of osteoblastic cells [75].

9.2 Biological Properties

9.2.1 Biodegradability

As a rule, there are two different mechanisms for biodegradation process; namely active and passive resorption. The cellular activity of macrophages and osteoclasts is responsible for the occurrence of the first mechanism. Conversely, the second has happened due to the action of either dissolution or chemical hydrolysis. It is worth to note that the dissolution is strongly affected by many physico-chemical factors like basicity/acidity, solubility, and surface area of the chosen biomaterial, local acidity, fluid convection, and temperature of the medium. Indeed, the solubility products such as TCP, HA, and FA contribute to the supersaturation of the surrounding fluids bearing in mind that biodegradation is expedited by cells.

Osteoclasts (a type of bone cells responsible for breaking down bone tissue) play a strong role in dissolving Ca-P materials due to their ability to secrete the carbonic anhydrase enzyme or lactic acid which dramatically reduces pH to 4 or 5. On the other hand, osteoblasts (a type of bone cells responsible for manufacturing bone tissue) help to increase pH to higher than 8.5 through the excretion of ammonia.

Nanoparticles of Ca-P materials can be exposed to phagocytosis process and when they are introduced into the cytoplasm, they are dissolved as a result of acid attack and/or enzymatic processes taking into account that these processes compete with each other and happen simultaneously during biodegradation of Ca-P. Surprisingly, the dissolution rate of the monophasic Ca-P compound reduces in the order α-TCP > β-TCP > CDHA > HA when pH value is 7.3. However, as previously discussed, the biodegradation of bi-, tri-, and multi-phasic Ca-P is greatly dependent on the phase ratios. Noteworthy, the doping of Ca-P phase with some ions could alter their solubility and biodegradation behaviors. According to this information, one can expect the positive effect of CO_3^{2-} ions on the biodegradation of CDHA and HA. On the contrary, Mg^{2+} and Zn^{2+} ions negatively affect the biodegradability of β-TCP [40].

9.2.2 Bioactivity

It is well-known that bioactive materials such as HA, when inserted into the human body, are able to interact with physiological fluids and surrounding bone tissue to form a CDHA layer at the bone-biomaterial interface which aids the ability of bone to bond. The bioactive behavior of materials is significantly influenced by crystal phases, surface roughness, and porosity. Given the great significance of bone bonding ability, researchers have paid more attention to the mechanism of CDHA layer formation. One way to find out this mechanism is to examine various media that are similar to body fluids. According to these attempts, simulated body fluid (SBF), ringer solution or Tris-HCl have been prepared to be used for this purpose.

9.2.3 Cellular Response

The immediate adsorption of proteins and bioorganic compounds onto the surface after implantation influences the different cell responses including cell proliferation and differentiation to bone cells are affected. Also, cell responses are affected by compositional and morphological factors. Importantly, the nanostructured Ca-P compound exhibits better mineralization improvement and enhanced osteoblast functions compared to the microstructured one. These findings may be attributed to the close similarity between these compounds and BHA in natural bones. It should be mentioned surface roughness of the implant may effectively facilitate cell seeding and fixation.

9.2.4 Osteoinductivity

Extensive research has been dedicated for demonstrating the osteoinductive properties of Ca-P materials. However, an osteoinductive mechanism is still unknown at this writing. Some hypotheses revealed the importance of some physicochemical factors like microporosity, surface area, surface topography, and geometry in possessing this unique property, while others attributed the osteoinduction property of the material to its ability to concentrate growth factors in physiological fluid or adsorption of osteoinductive substances on their surface [41].

10 Molecular Modeling Approaches

Molecular modeling is a class of computational calculations based on the fundamental laws of physics to describe and understand the different properties and interactions of chemical structure. Correlating these tools with basic principle of Physics one can classify modeling into two main classes. The first one is molecular mechanics in which classical mechanics is used for modeling molecular systems. Simply this class of computation assumed that the Born–Oppenheimer approximation is valid, and then the potential energy of the molecular system is calculated as a function of the nuclear coordinates using force fields. This assumption allows this computation valid for molecular systems ranging in size and complexity from small to large molecules. The application extended to cover biological systems or material assemblies with many thousands to millions of atoms [76]. This is because such class of modeling performs computations based upon the interaction among the nuclei, the electronic effects are approximated, this makes the computations quite inexpensive—consume less time—, also it makes it suitable for computing very large systems. Molecular mechanics simply depend on the classical Physics, as one moves to modern Physics, quantum mechanics will be of concern. The class of computational work depends on quantum mechanics is termed as electronic structure method, which applies the Schroedinger wave equation. Practically the exact solution of Schroedinger equation is not enough so, electronic structure method has three famous approximations to its solution namely:

Semi-empirical methods: this approximation uses parameters derived from experimental data in order to simplify the computation. It is inexpensive and provides to some extent considerable qualitative descriptions of molecular system, but its quantitative predictions of energies and structures for systems are considered accurate as compared with other approximations, where good parameter sets exist. Accordingly, it can be applied to large systems.

Ab initio: it is the second approximation of the electronic structure method, and it is based only on quantum mechanics. This means that no need for experimental data rather than semi-empirical methods. It provides high-quality quantitative predictions for a broad range of systems. Concerning the accuracy it is more accurate than semi-empirical methods

Density functional methods (DFT): the third approximation of the electronic structure methods termed as DFT. This level of computation is similar to ab initio methods in many ways, it requires about the same amount of computation resources as Hartree–Fock theory, the least expensive ab initio method. DFT is considered attractive because it includes the effect of electron correlation. It provides more accurate results than ab initio [76].

The main task in molecular modeling is to build the model molecule in order to simulate the molecular phenomena. The model molecules must be simple, short, descriptive, and representing the phenomena of concern.

Molecular modeling employs computer-generated images for a given chemical structure to simulate the structure, which helps in understanding physicochemical properties of such chemical structure. Such prediction and understanding of physicochemical properties dedicates this class of computational work for many applications in Physics, Chemistry, Biology, and Environment. Moreover, these applications covered both basic and technological sciences whereas molecules and molecular phenomena are of concern. These computational tools are now widely applied for biological molecules.

It is stated that, density functional theory (DFT) and ab initio calculations are providing biological molecules such as hydroxyapatite with different descriptors and parameters reflecting the reactivity of the structure.

For biological molecules, it is important to have some descriptors to describe the reactivity of a given molecule in terms of its surrounding molecules.

Some parameters such as total dipole moment, highest occupied molecular orbital HOMO, lowest unoccupied molecular orbital LUMO, and molecular electrostatic potential MESP could be good indicators for the interactions of molecules [77–79].

It is stated that, total dipole moment increased as the reactivity of the studied molecule. Moreover, the molecule is considered active as its HOMO/LUMO energy is low. For MESP it could map the electrostatic potential of the studied molecular surface. MESP map by color the sites could accept or denote electron. This could be considered as an indicator for the active sites of a given molecular system. In biology this parameter beside total dipole moment and HOMO/LUMO are important factors. From this point of view molecular modeling specially those depending on quantum mechanical calculations are important tools for elucidating molecular structure of given biological molecules in terms of the surrounding structures and medium [80–82].

Molecular modeling at DFT: B3LYP level was used to study the possible interaction between chitosan and hydroxyapatite HA. Eight schemes were proposed to simulate this interaction and further supported with experimental work [83]. Each scheme indicates possible structure for the HA. The proposed schemes are indicated as shown in Fig. 6.

The unit cell of hydroxyapatite crystal was optimized by generalized gradient approximation of DFT in order to determine and optimize the positions of hydroxyl ion inside the cell [84]. For bone tissue engineering applications molecular modeling is also supported this class of applications. As shown in Fig. 6 there are 8 schemes to indicate the possible interaction between chitosan and HA. Collecting these together by generalized model for chitosan/hydroxyapatite is indicated as shown in Fig. 7a. If we are going to describe the model, chitosan is indicated as a model of three chitosan units. Ongoing from Fig. 6 to Fig. 7 one can describe that, the interaction between chitosan and HA is tried in many schemes through OH, NH, and O. Three separate structures representing hydroxyapatite are proposed forming the 8 schemes. In order to verify the interaction of OH, Ca, and PO_4 a new chitosan model structure must be built with 3 units to verify the proposed interactions indicated in Fig. 6. From the studied 8 schemes, the most likely structures were chosen depending on their binding energies. The optimized structure of chitosan/hydroxyapatite is indicated such that, the Ca^{3+} is attached with the two units through N-atom of (NH_2) of chitosan unit 2, oxygen of unit 1, and to the oxygen of the linkage between the two units. This leads to adding a third chitosan unit instead of chitosan 2 units as in part 4 of Fig. 6. PO_4 is chosen to interact with chitosan unit 3 through its oxygen with N-atom of (NH_2). The interaction lengths are indicated as 2.25, 2.28, 2.950, and 1.080 Å respectively. Based on this knowledge of hydroxyapatite, a model molecule is further constructed for carbonated hydroxyapatite with the formula $Ca_{10}(PO_4)_6(OH)_2 \cdot 14H_2O$ is presented as indicated in Fig. 7b, It is worth to mention that, this model is described earlier [45].

Simulation methods including quantum mechanical, Monte Carlo, molecular dynamics, mesoscopic, and macroscopic scale are used to calculate and investigate physical properties of polymer nanocomposites [85]. Multiscale modeling was employed for prediction of the compressive mechanical values of the prepared composites and comparing these values with those of experimental method [86]. Digimat was also chosen as the major modeling software, because it is usually considered as a nonlinear multiscale material and structure modeling platform that can be used to predict the behavior of multi-phase and heterogeneous materials such as multi-component composite and nanocomposites [87].

As an overall conclusion for the research of molecular modeling in the biological interactions including hydroxyapatite (HA) research one can conclude that the applications of different levels of modeling leads to a continuous descriptions of electronic

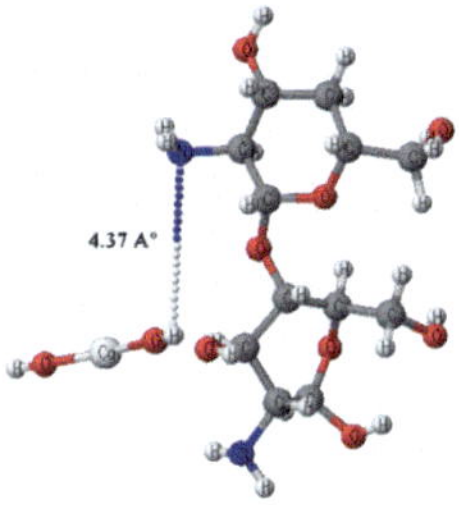
4.37 A°

1

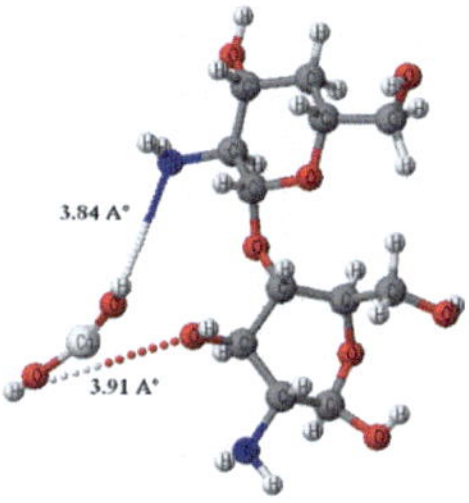
3.84 A°
3.91 A°

2

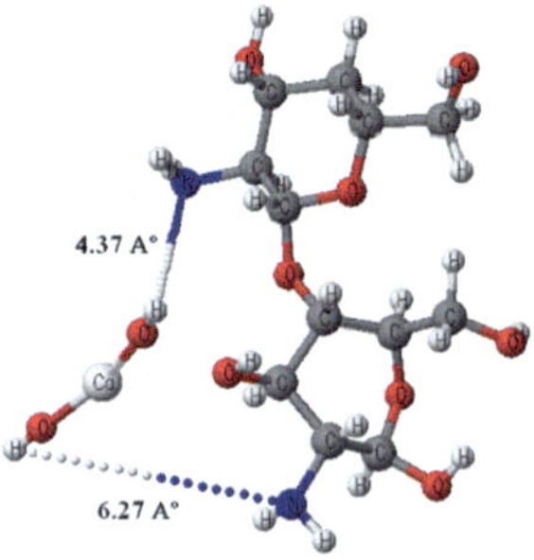
4.37 A°
6.27 A°

3

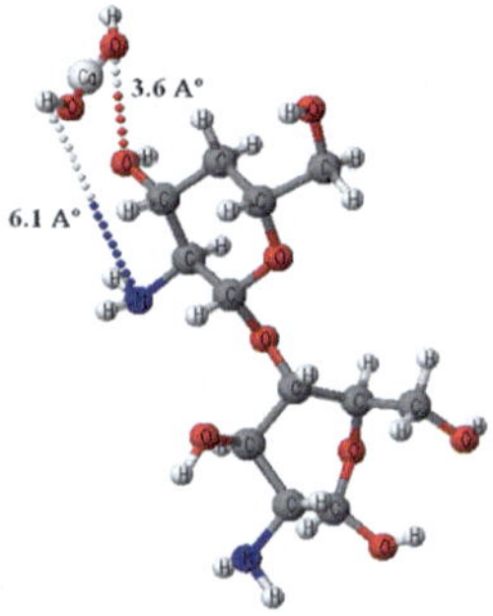
3.6 A°
6.1 A°

4

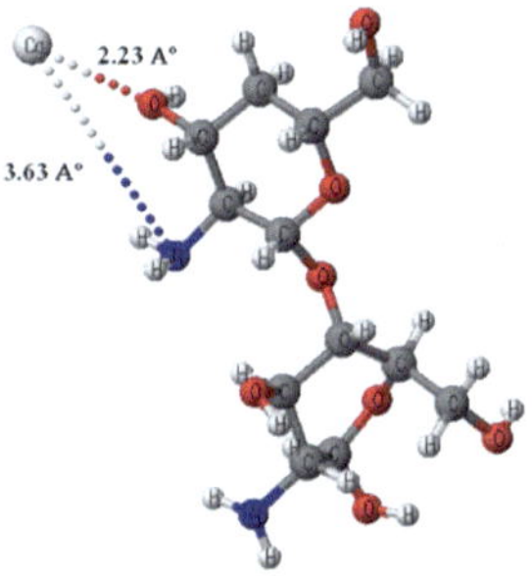
2.23 A°
3.63 A°

5

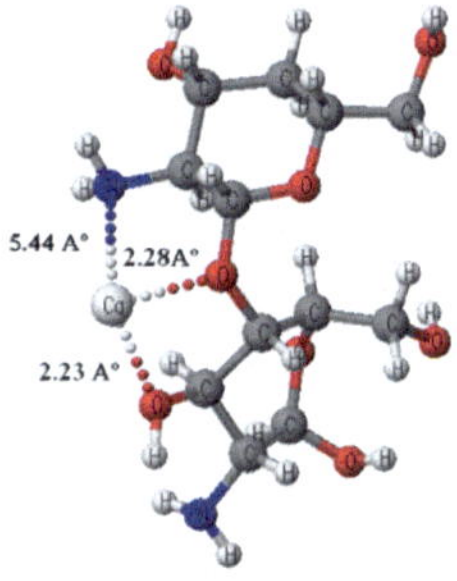
5.44 A°
2.28A°
2.23 A°

6

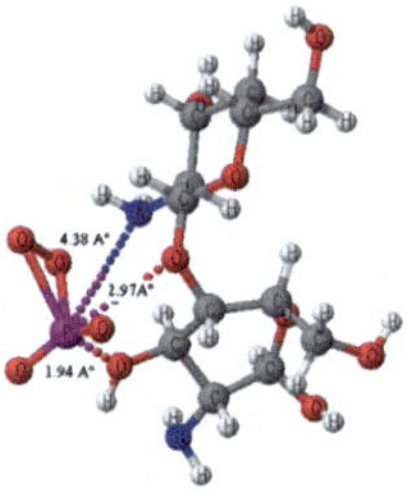
4.38 A°
2.97A°
1.94 A°

7

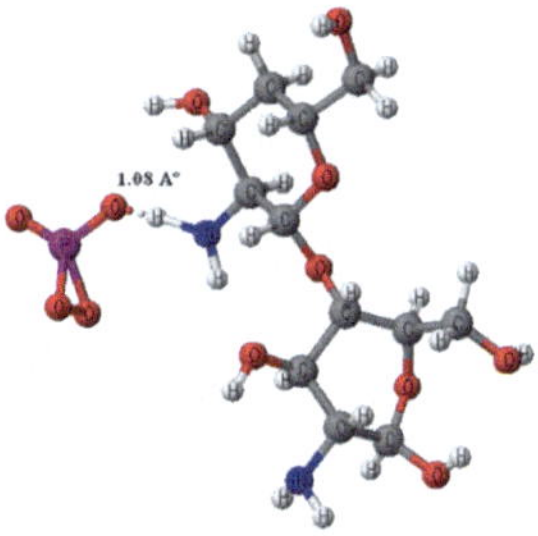
1.08 A°

8

◂**Fig. 6** Chitosan/hydroxyapatite 8 schemes of interactions [83]. Each scheme could be described as in the following, (1) the OH of $Ca(OH)_2$ interacted with the N–H of chitosan unit 2, (2) the first OH of $Ca(OH)_2$ is coordinated with N–H of chitosan unit 2; second one is coordinated with OH of chitosan unit 1, (3) OH of $Ca(OH)_2$ is coordinated with N–H of both chitosan units, (4) OH of $Ca(OH)_2$ is coordinated with N–H of chitosan unit 2 while, the second one coordinates with oxygen (OH) of the same chitosan unit, (5) Ca is freely coordinated with N–H and OH of chitosan unit 2, (6) Ca is freely coordinated with N–H of chitosan unit 2 also to OH of chitosan unit 1 and to the oxygen of the linkage, (7) P of PO_4 is coordinated with N–H and of chitosan unit 2, to OH of chitosan unit 1, and to the oxygen of the linkage, and (8) O of PO_4 is coordinated with N–H and of chitosan unit 2

properties of HA in different interaction which is an important step toward the functionality and applications of HA in different fields of applications. This is in a good agreement with the recent modeling findings and reports dealing with hydroxyapatite [88, 89].

11 Applications of HA

It is very well-known that HA is an attractive material for several important fields like agriculture, environment, chemistry, biology, and medicine [41]. On the other hand, its applications in medicine are still the major ones for such material. Based upon this importance, we will also briefly mention these applications as follows.

11.1 Biomedical Applications

11.1.1 Bone Fillers and Tissue Engineering Scaffolds

As previously discussed, HA possesses various attractive properties that make it very suitable for BTE applications. These properties are biocompatibility, controlled degradation, and excellent bone osteoconductivity and osteoinductivity [90, 91]. Also, it is able to promote the expression of bone cell growth factors such as bone morphogenetic protein (BMP) and enhance alkaline phosphatase (ALP) activity in mesenchymal stem cells (MSCs) having in mind that ALP activity plays an important role in the early mineralization process [92–94]. Despite these surprising characteristics, its poor mechanical properties are the major drawbacks to its uses in load-bearing sites applications. Based on this fact, several approaches to enhance these low properties received a great attention from a handful of researchers in worldwide. One promising strategy to improve the mechanical properties is to prepare organic/inorganic composites which can greatly mimic living bones. Noteworthy, the major benefit of such composites is a combination of good compressive strength of inorganic material and the elasticity of organic one. Moreover, they also have a porous structure which in turn, serves as a temporary extracellular matrix and promoting

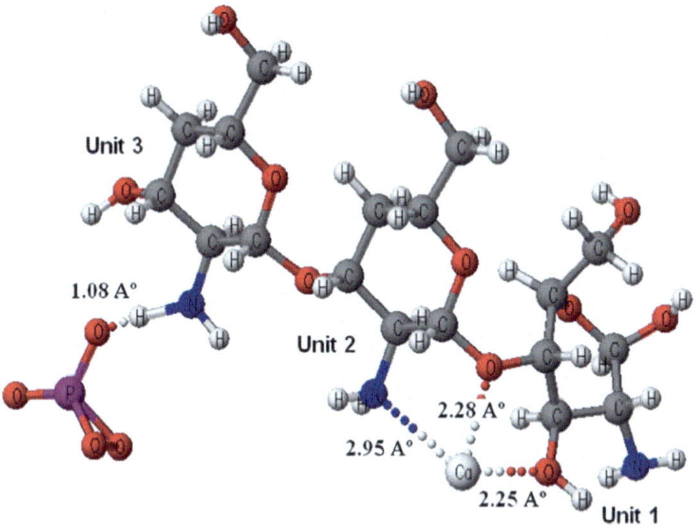

a

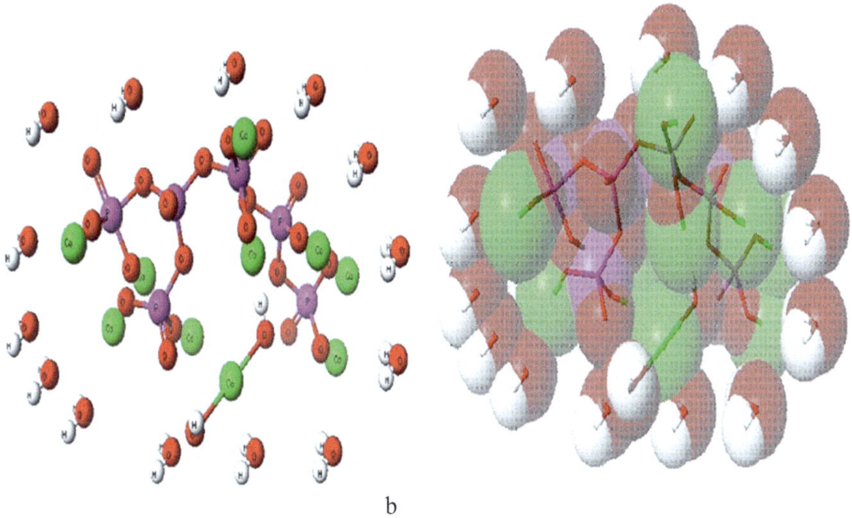

b

Fig. 7 Collecting the different schemes in Fig. 6 the above scheme is a more general scheme for chitosan/hydroxyapatite as indicated in **a**, in this model, Ca is freely coordinated with N–H and of chitosan unit 2, OH of unit 1 and to the oxygen of the linkage, O of PO_4 is coordinated with N–H and of chitosan unit 3. The interaction lengths are indicated as 2.25, 2.28, 2.950 and 1.080 Å respectively. A model molecule for carbonated hydroxyapatite $Ca_{10}(PO_4)_6(OH)_2{\cdot}14H_2O$ is indicated in **b**, the model structure is surrounded with 14 water molecules as described earlier [45] tissue regeneration process. It should be noted that there are two types of polymers (organic phase) those commonly used for such purpose; namely natural and synthetic. The former one, i.e. natural polymers display outstanding biological properties like appropriate cellular affinity, high bioactivity, and excellent biocompatibility on the condition that care must be taken to avoid the possible denaturation during scaffold preparation. On the other hand, synthetic one exhibits better mechanical strength along with a controlled degradation rate. Natural polymers involve chitosan, gelatin, alginate, collagen, silk fibroin, and cellulose. On the opposite side, synthetic ones include poly(vinyl)alcohol, poly-2 hydroxyethyl methacrylate, polycaprolactone, polylactic-co-glycolic acid (PLGA), and polylactic acid (PLA) [64].

We are also very interested in the improvement of the mechanical properties of HA through preparing nanocomposites containing two materials. The first is HA, while the second is metal or ceramic with the aim of obtaining nanocomposites characterized by good mechanical properties and excellent biological significance. Youness et al. [95] reported the successive addition of HA to TiO_2 helps in improving the bioactivity of the resultant nanocomposites without sacrificing their mechanical properties, provided that the chosen sintering temperature should not exceed 900 °C. Otherwise, HA is decomposed to form other phases that contribute in reducing the abovementioned properties. Youness et al. [63] studied the influence of adding different weight percentages of HA to ZrO_2 and the results pointed out that the obtained nanocomposites can be valid for orthopedic and dental applications due to their excellent properties. Youness et al. [96] performed a novel study through enhancing the bioactivity behavior of borosilicate inert glass via successive addition of HA and selenium dioxide (SeO_2) to this glass system and the results clarified that the presence of a noticeable improvement in bioactivity performance of the prepared nanocomposites as a result of the formation of $CaSiO_3$ phase which is characterized by a good bioactivity character. Furthermore, biological and mechanical properties of HA have been significantly improved by adding different contents, up to 40 wt%, of SiO_2 as reported in our recent work [97].

11.1.2 Coatings

Titanium, stainless steel, and shape memory (Ni–Ti) alloys are known to be widely used as implants due to their suitable mechanical properties, low density, and endurance. Apart from considering that, pure Ti is an inert metal, it can be subjected to electrochemical corrosion after a long implantation period, thus resulting in ions being released into the blood stream and the formation of inflammatory reactions causing a serious effect to the patient. In this respect, these severe consequences require a second surgery to remove the implant. In turn, this surgery raises the chances of infection and increases the financial burden on the patient and/or healthcare system. In other cases, a second surgery is responsible for the harmful aesthetic effects due to the formation of scars and therefore, bad psychological problems for the patient. Thus, these consequences are greatly avoided through using HA as a coating material for such alloys [44, 98, 99]. Interestingly, the metal implant coating with HA must have a suitable thickness for optimum performance. On the other hand, although the lower thickness of the HA coating layer achieves better mechanical properties, about 10–15 μm of such layer dissolves during the first few months of implantation. On the opposite side, fatigue failure under tensile load can take place if the coating layer is above 100–150 μm [55].

11.1.3 Drug Delivery System

At present, wide attention is paid for using phosphate ceramics; especially HA as drug delivery carriers. The secret lies in the physical/chemical incorporation of drug molecules into HA keeping in mind that they are retained until they reach the desired site. Then after, the drugs degrade with a controlled rate. On the contrary to the conventional oral/intravenous drug delivery routes, this approach achieves the delivery of medications only to the diseased organs without healthy ones. In this sense, nanorods/nanoparticles of HA have been utilized for the controlled release of different proteins and drugs. Accordingly, it has been successfully used for delivering anticancer substances, anti-inflammatory drugs, anti-osteoporotic substances, gene delivery agents, osteoinductive growth factors, osteogenic cell population, vitamins, hormones, and proteins. Noteworthy, the major benefit of HA is that it can be manufactured as microspheres which consequently, gives HA the ability to load a high capacity of drugs [100].

11.2 Other Applications

11.2.1 Environmental Applications

Water pollution has been known to have a negative role in human health, as a result of the harmful effect of both industrial effluents and toxic elements such as heavy metals. Scientists found that lead (Pb) and cadmium (Cd) are expected not only to be carcinogenic heavy metals but also extremely toxic at low concentrations. Their presence can be attributed to that they are poured into water/groundwater from mining and petrochemical industries. Due to the serious hazards of such metals, plentiful approaches have been developed to enhance and purify the quality of water by decreasing the concentrations of such metals. According to the World Health Organization (WHO), the permissible limits for Pb and Cd are 0.01 and 0.003 mg/L, respectively. Moreover, Algeria has determined precisely the marginal values for such metals in drinking water that should not exceed 0.01 mg/L (for Pb) and 0.003 mg/L (for Cd). On the other hand, discharge in sewage should not exceed 0.5 mg/L (for Pb) and 0.2 mg/L (for Cd). Unfortunately, the treatment of polluted water is a serious problem facing governments around the world because of its high cost. Therefore, the best solution to overcome this trouble is to find cost-effective materials for this purpose. To meet this challenge, considerable attempts were made to remove heavy metals using mesoporous and nanoporous materials. Among nanoporous materials, HA has shown great efficacy to remove such undesirable metals. Moreover, it is also able to remove both nickel (Ni) and zinc (Zn) from the soil [101, 102]. In continuation of the substantial role of HA in the environmental applications, it can be extensively used for the treatment of soil.

One can expect that the soil tends to be the main sink for metals and serves to prohibit their entry into food. However, mining and metals melting is a major reason for the gradual transfer of massive toxic metals from the Earth's crust to the soil causing more pollution due to the spreading of heavy metals. Unfortunately, this problem has become very serious in recent decades around the world. For example, 13.33% of all soil specimens in China collected from 6.3 million km^2 of land are contaminated with elevated levels of metals. Indeed, Cd was detected in all agricultural lands. Interestingly, Cd is chemically durable and therefore, it does not expose to chemical/microbial degradation. Furthermore, it is an unnecessary element for living organisms. Instead, it is the most toxic heavy metal for living organisms causing deleterious consequences. In the soil, Cd hurts plants through changing metabolic pathways leading to a severe damage to different organelles like chloroplasts and mitochondria. Consequently, lipids and proteins are damaged. Moreover, it can also interfere with the uptake and transport of vital elements such as phosphorous (P) and potassium (K). Additionally, excessive Cd amounts negatively affect photosynthesis process by reducing iron (Fe) uptake by plants. Regrettably, it also accumulates in the bodies of plants/animals and therefore, enters the human body via food chain. These serious consequences shed light onto thinking of promising approaches for remediation of heavy metal-contaminated soils using physical, chemical, and biological techniques. In respond to this principle, one of the precious efforts has been done recently [103] using some specific microorganisms which are highly tolerant to different types of heavy metals along with HA to remediate polluted soils and they obtained good results.

11.2.2 Chemical Applications

One of the important applications of HA is its usage as catalyst for many chemical reactions like the production of ribose from simple carbon sources in one-pot process [104], one-pot Biginelli reaction for the synthesis of 3,4-dihydropyrimidin-2-(1H)-one [105], titanium dioxide catalyst support [106], methane oxidation reaction [107], aqueous reduction of nitroaromatics to the corresponding amines [108] and formaldehyde combustion at room temperature [109].

11.2.3 Agricultural Applications

The high cost of fertilizers, in developing countries, is the major reason for limiting food supply. Therefore, reducing this cost may help to overcome this problem. In spite of the importance of mined phosphate (P) rock as a phosphate source in agricultural sector, its supply is limited. Therefore, there is a great need for other phosphate sources in the future. It is important to emphasize that the complex chemical reactions in soils are responsible for the reduced use of commercial phosphorous fertilizers by plants. Depending on soil properties, the efficiency of commercial P fertilizers for plants is approximated by 20% only. Hence, the development of other effective strategies is highly required having in mind that as the world's population grows, the depletion rate of P fertilizers is progressively increased. Based on these facts, synthetic HA is believed to be one of the best candidates to solve this issue. It can be argued that the mobility of soluble phosphates in the soil is high and therefore, a large amount of these phosphates ends up in surface waters through runoff which leads to eutrophication. On the opposite side, direct use of solid HA as P source has many obstacles such as large particle size along with the high pH value of the soil restricts P mobility in the soil which prohibits phosphate availability to plants. Since BHA is not soluble in water, the chemical reactions in the soil are limited and thus, its precipitation and/or adsorption on soil colloids is significantly reduced. In response to this information, the availability to plants is remarkably increased. With regard to the tremendous significance of nanomaterials in different applications, BHA nanoparticles are highly applicable. Interestingly, the uptake of these nanoparticles by the cell wall of plants is also dependent also on the pore diameter of cell wall (5–20 nm). Hence, if the diameters of nanoparticles are less than this range, they easily enter and reach the plasma membrane [110]. Based on this concept, HA is extensively utilized as a nano-fertilizer for some crops.

Despite the great success of the slow release of the chemicals used as delivery systems, the effort to apply this principle in the agricultural field is much less. Urea is a commonly used fertilizer in agriculture since it is a rich source of nitrogen, but its high solubility may be considered as an important drawback. In this sense, it can be combined with HA to get nanoparticles that have the required nutrients with the controllable release [111]. The aforementioned amazing applications of HA in different fields are illustrated in Fig. 8.

12 Conclusions and Outlook

Numerous researchers around the world pay great efforts to develop modern strategies to treat serious bone defects like trauma, tumors, infection, etc., which cannot be cured without medical assistance, which opens the way for bone regeneration field to be considered as one of the most promising strategies to solve this problem. However, this task is so sophisticated where enormous factors are essential. These factors include the optimum choice of scaffold material, growth factors, etc. Notably, these scaffold materials include natural ones such as autografts, allografts, and natural polymers. On the other hand, synthetic ones involve bioactive glasses, glass–ceramics, porous metals, calcium phosphates, and calcium silicates. Owing to the good chemical similarity with natural bones, bioactivity, solubility, biocompatibility, calcium phosphates; hydroxyapatite (HA) is an excellent scaffold material.

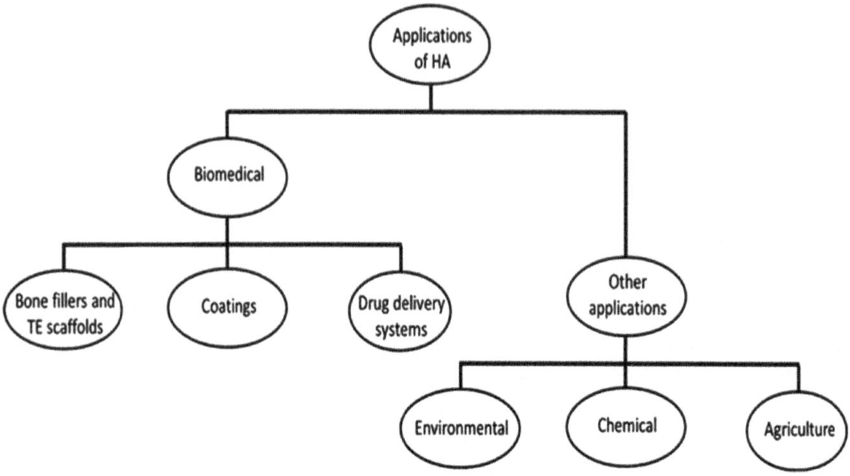

Fig. 8 Different applications

As described in this chapter, the preparation methods of HA can be classified into four main categories, i.e. wet, dry, high-temperature processes, and synthesis based on biogenic sources. Of course, each method has its own privileges and drawbacks where some methods need expensive chemicals, consume time, energy, etc. On the contrary, others are cost-effective ones but need carefulness and good experience while performing the preparation. Since HA exhibits a unique set of properties, it can be used in different applications including tissue engineering, drug delivery systems, coatings for metals and their alloys, treatment of wastewater, the catalysis of some chemical reactions, controlling the release of some fertilizers, and work also as a fertilizer for some crops. It is worth to mention that this study may be the first to discuss, in a comprehensive manner, the experimental and theoretical issues related to hydroxyapatite, its physico-chemical and biological properties and its applications in different fields, not only in the biomedical fields. Moreover, this chapter briefly discussed the host responses to biomaterials while stating the 3D bioprinting techniques used. It also briefly mentioned other calcium phosphate compounds, their properties, and applications. In light of this chapter, we invite readers to broaden their interests toward this valuable material for use in other applications that could be beneficial to the industrial sector.

References

1. G. Turnbull, J. Clarke, F. Picard, P. Riches, L. Jia, F. Han, B. Li, W. Shu, 3D bioactive composite scaffolds for bone tissue engineering. Bioact Mater. **3**, 278–314 (2018). https://doi.org/10.1016/j.bioactmat.2017.10.001
2. B. Dhandayuthapani, Y. Yoshida, T. Maekawa, D.S. Kumar, Polymeric scaffolds in tissue engineering application: a review. Int. J. Polym. Sci. **2011**, 1–19 (2011). https://doi.org/10.1155/2011/290602
3. E. García-Gareta, M.J. Coathup, G.W. Blunn, Osteoinduction of bone grafting materials for bone repair and regeneration. Bone **81**, 112–121 (2015). https://doi.org/10.1016/j.bone.2015.07.007
4. S.V. Murphy, A. Atala, 3D bioprinting of tissues and organs. Nat. Biotechnol. **32**(8), 773–785 (2014). https://doi.org/10.1038/nbt.2958
5. A. Shafiee, A. Atala, Printing technologies for medical applications. Trends Mol. Med. **22**(3), 254–265 (2016). https://doi.org/10.1016/j.molmed.2016.01.003
6. J. Lu, C. Yan, 3D printing of scaffolds for tissue engineering. IntechOpen (2018)
7. J. Cui et al., Polydopamine-functionalized polymer particles as templates for mineralization of hydroxyapatite: biomimetic and in vitro bioactivity. RSC Adv. **6**(8), 6747–6755 (2016). https://doi.org/10.1039/C5RA24821C
8. R.E. Saunders, B. Derby, Inkjet printing biomaterials for tissue engineering: bioprinting. Int. Mater. Rev. **59**(8), 430–448 (2014). https://doi.org/10.1179/1743280414Y.0000000040
9. S. Bose, S. Vahabzadeh, A. Bandyopadhyay, Bone tissue engineering using 3D printing. Mater. Today **16**(12), 496–504 (2013). https://doi.org/10.1016/j.mattod.2013.11.017
10. V. Keriquel et al., In situ printing of mesenchymal stromal cells, by laser-assisted bioprinting, for in vivo bone regeneration applications. Sci. Rep. **7**(1), 1778 (2017). https://doi.org/10.1038/s41598-017-01914-x
11. L. Koch, M. Gruene, C. Unger, B. Chichkov, Laser assisted cell printing. Curr. Pharm. Biotechnol. **14**(1), 91–97 (2013). https://doi.org/10.2174/1389201113804805368
12. Y. Yu et al., Three-dimensional bioprinting using self-assembling scalable scaffold-free "tissue strands" as a new bioink. Sci. Rep. **6**, 28714 (2016). https://doi.org/10.1038/srep28714
13. J. Zhang, W. Liu, V. Schnitzler, F. Tancret, J.M. Bouler, Calcium phosphate cements for bone substitution: chemistry, handling and mechanical properties. Acta Biomater. **10**, 1035–1049 (2014). https://doi.org/10.1016/j.actbio.2013.11.001
14. M. Catauro, F. Bollino, Advanced glass-ceramic materials for biomedical applications. J. Bone Rep. Recommend. **3**, 1–3 (2017). https://doi.org/10.4172/2469-6684.100035
15. R.A. Youness, M.A. Taha, A.A. El-Kheshen, N. El-Faramawy, M. Ibrahim, In vitro bioactivity evaluation, antimicrobial behavior and mechanical properties of cerium-containing phosphate glasses. Mater. Res. Expr. **6**, 1–13 (2019). https://doi.org/10.4172/2469-6684.100035
16. E.M.A. Khalil, R.A. Youness, M.S. Amer, M.A. Taha, Mechanical properties, in vitro and in vivo bioactivity assessment of Na_2O-CaO-P_2O_5-B_2O_3-SiO_2 glass-ceramics. Ceram. Int. **44**, 7867–7876 (2018). https://doi.org/10.1016/j.ceramint.2018.01.222
17. R.A. Youness, M.A. Taha, M. Ibrahim, A. El-Kheshen, FTIR spectral characterization, mechanical properties and antimicrobial properties of La-doped phosphate-based bioactive glasses. Silicon **10**, 1151–1159 (2018). https://doi.org/10.1007/s12633-017-9587-0
18. S.M. Abo-Naf, E.M. Khalil, E.M. El-Sayed, H.A. Zayed, R.A. Youness, In vitro bioactivity evaluation, mechanical properties and microstructural characterization of Na_2O-CaO-B_2O_3-P_2O_5 glasses. Spectrochim. Acta A **144**, 88–98 (2015). https://doi.org/10.1016/j.saa.2015.02.076
19. M.A. Taha, R.A. Youness, M.F. Zawrah, Phase composition, sinterability and bioactivity of amorphous nano-CaO-SiO_2-CuO powder synthesized by sol-gel technique. Ceram. Int. **46**, 24462–24471 (2020). https://doi.org/10.1016/j.ceramint.2020.06.231
20. W.S. AbuShanab, E.B. Moustafa, M.A. Taha, R.A. Youness, Synthesis and structural properties characterization of titnia/zirconia/calcium silicate nanocomposites for biomedical applications. Appl. Phys. A **126**, 1–12 (2020). https://doi.org/10.1007/s00339-020-03975-8

21. G. Kaur, G. Pickrell, G. Kimsawatde, D. Homa, H.A. Allbee, N. Sriranganathan, Synthesis, cytotoxicity, and hydroxyapatite formation in 27-Tris-SBF for sol-gel based CaO-P_2O_5-SiO_2-B_2O_3-ZnO bioactive glasses. Sci. Rep. **4**, 1–14 (2013). https://doi.org/10.1038/srep04392
22. R. Ravarian, F. Moztarzadeh, M.S. Hashjin, S.M. Rabiee, P. Khoshakhlagh, M. Tahriri, Synthesis, characterization and bioactivity investigation of bioglass/hydroxyapatite composite. Ceram. Int. **36**, 291–297 (2010). https://doi.org/10.1016/j.ceramint.2009.09.016
23. K. Labgairi, A. Borji, M. Kaddami, A. Jourani, Kinetic study of calcium phosphate precipitation in the system H_3PO_4–$Ca(OH)_2$–H_2O at 30°C. Int. J. Chem. Eng. **2020**, 1–9 (2020). https://doi.org/10.1155/2020/2893298
24. E. Mariani, G. Lisignoli, R.M. Borzì, L. Pulsatelli, Biomaterials: foreign bodies or tuners for the immune response? Int. J. Mol. Sci. **20**, 1–42 (2019). https://doi.org/10.3390/ijms20030636
25. R.M. Berne, M.N. Levy, B.M. Koeppen, B.A. Stanton, *Berne and Levy Physiology*, 6th edn. (Mosby, St. Louis, 2009)
26. A.C. Guyton, J.E. Hall, *Textbook of Medical Physiology*, 12th edn. (Elsevier Saunders, Philadelphia, 2010)
27. A.B.H. Yoruç, B.C. Şener, Biomaterials, a roadmap of biomedical engineers and milestones, ed. by Sadik Kara (InTech, 2012). ISBN: 978-953-51-0609-8
28. F.O. Costa, S. Takenaka-Martinez, L.O. Cota, S.D. Ferreira, G.L. Silva, J.E. Costa, Peri-implant disease in subjects with and without preventive maintenance: a 5-year follow-up. J. Clin. Periodontol. **39**(2), 173–181 (2012). https://doi.org/10.1111/j.1600-051X.2011.01819.x
29. M. Esposito, M.G. Grusovin, V. Loli, P. Coulthard, H.V. Worthington, Does antibiotic prophylaxis at implant placement decrease early implant failures? A Cochrane systematic review. Eur. J. Oral Implantol. **3**(2), 101–110 (2010)
30. A. Tathe, M. Ghodke, A.P. Nikalje, A brief review: biomaterials and their application. Int. J. Pharm. Pharm. Sci. **2**(4), 19–23 (2010)
31. S.R. Hanson, E.I. Tucker, R.A. Latour, Blood coagulation and blood–material interactions, in *An Introduction to Materials in Medicine*, Biomaterials Science, 4th edn. (Academic Press, 2020), pp. 801–812. https://doi.org/10.1016/B978-0-12-816137-1.00052-0
32. F. Albee, H. Morrison, Studies in bone growth: triple calcium phosphate as a stimulus to osteogenesis. Ann. Surg. **71**, 32–38 (1920). https://doi.org/10.1097/00000658-192001000-00006
33. W. Habraken, P. Habibovic, M. Epple, M. Bohner, Calcium phosphates in biomedical applications: materials for the future? Mater. Today **19**(2), 69–87 (2016). https://doi.org/10.1016/j.mattod.2015.10.008
34. P.N. Kumta, C. Sfeir, D.H. Lee, D. Olton, D. Choi, Nanostructured calcium phosphates for biomedical applications: novel synthesis and characterization. Acta Biomater. **1**, 65–83 (2005). https://doi.org/10.1016/j.actbio.2004.09.008
35. M.N. Hassan, M.M. Mahmoud, A. Abd El-Fattah, S. Kandil, Microwave-assisted preparation of nano-hydroxyapatite for bone substitutes. Ceram. Int. **42**, 3725–3744 (2016). https://doi.org/10.1016/j.ceramint.2015.11.044
36. A. Nakahira, K. Nakata, C. Numako, H. Murata, K. Matsunaga, Synthesis and evaluation of calcium-deficient hydroxyapatite with SiO_2. Mater. Sci. Appl. **2**, 1194–1198 (2011). https://doi.org/10.4236/msa.2011.29161
37. X. Li, Y. Deng, M. Wang, X. Chen, Y. Xiao, X. Zhang, Stabilization of Ca-deficient hydroxyapatite in biphasic calcium phosphate ceramics by adding alginate to enhance their biological performances. J. Mater. Chem. B **6**, 84–97 (2018). https://doi.org/10.1039/c7tb02620j
38. M. Bohner, B.G. Santoni, N. Dobelin, β-tricalcium phosphate for bone substitution: synthesis and properties. Acta Biomater. **113**, 23–41 (2020). https://doi.org/10.1016/j.actbio.2020.06.022
39. R.G. Carrodeguas, S. De Aza, α-Tricalcium phosphate: synthesis, properties and biomedical applications. Acta Biomater. **7**, 3536–3546 (2011). https://doi.org/10.1016/j.actbio.2011.06.019

40. J.C. Elliott, Structure and chemistry of the apatites and other calcium orthophosphates, ed. by H.M. Hughes, (Elsevier, Amsterdam, 1994)
41. M. Canillas, P. Pena, A.H. de Aza, M.A. Rodríguez, Calcium phosphates for biomedical applications. Boletín de la Sociedad Española de Cerámica Y vidrio **56**, 91–112 (2017). https://doi.org/10.1016/j.bsecv.2017.05.001
42. N.A.S.M. Pu'ad, P. Koshy, H.Z. Abullah, M.I. Idris, T.C. Lee, Syntheses of hydroxyapatite from natural sources. Heliyon **5**, 1–14 (2019). https://doi.org/10.1016/j.heliyon.2019.e01588
43. M.A. Taha, R.A. Youness, M.F. Zawrah, Review on nanocomposites fabricated by mechanical alloying. Int. J. Miner. Metall. Mater. **26**(9), 1047–1058 (2019). https://doi.org/10.1007/s12613-019-1827-4
44. A. Szcześ, L. Holysz, E. Chibowski, Synthesis of hydroxyapatite for biomedical applications. Adv. Colloid Interface Sci. **249**, 321–330 (2017). https://doi.org/10.1016/j.cis.2017.04.007
45. R.A. Youness, M.A. Taha, H. Elhaes, M. Ibrahim, Molecular modeling, FTIR spectral characterization and mechanical properties of carbonated-hydroxyapatite prepared by mechanochemical synthesis. Mater. Chem. Phys. **190**, 209–218 (2017). https://doi.org/10.1016/j.matchemphys.2017.01.004
46. R.A. Youness, M.A. Taha, H. Elhaes, M. Ibrahim, Preparation, Fourier transform infrared characterization and mechanical properties of hydroxyapatite nanopowders. J. Comput. Theor. Nanosci. **14**, 2409–2415 (2017). https://doi.org/10.1166/jctn.2017.6841
47. R.A. Youness, M.A. Taha, M. Ibrahim, In vitro bioactivity, physical and mechanical properties of carbonated-fluoroapatite during mechanochemical synthesis. Ceram. Int. **44**, 21323–21329 (2018). https://doi.org/10.1016/j.ceramint.2018.08.184
48. R.A. Youness, M.A. Taha, M. Ibrahim, Dense alumina-based carbonated fluorapatite nanobiocomposites for dental applications. Mater. Chem. Phys. **257**, 123264 (2020). https://doi.org/10.1016/j.matchemphys.2020.123264
49. A. Refaat, R.A. Youness, M.A. Taha, M. Ibrahim, Effect of zinc oxide on the electronic properties of carbonated hydroxyapatite. J. Mol. Struct. **1147**(5), 148–154 (2017). https://doi.org/10.1016/j.molstruc.2017.06.091
50. M. Sadat-Shojai, M.T. Khorasani, E. Dinpanah-Khoshdargi, A. Jamshidi, Synthesis methods for nanosized hydroxyapatite with diverse structures. Acta Biomater. **9**(8), 7591–7621 (2013). https://doi.org/10.1016/j.actbio.2013.04.012
51. J. Chen, Y. Wang, X. Chen, L. Ren, C. Lai, W. He, Q. Zhang, A simple sol-gel technique for synthesis of nanostructured hydroxyapatite, tricalcium phosphate and biphasic powders. Mater. Lett. **65**, 1923–1926 (2011). https://doi.org/10.1016/j.matlet.2011.03.076
52. H. Eshtiagh-Hosseini, M.R. Housaindokht, M. Chahkandi, Effects of parameters of sol-gel process on the phase evolution of sol-gel derived hydroxyapatite. Mater. Chem. Phys. **106**, 310–316 (2007). https://doi.org/10.1016/j.matchemphys.2007.06.002
53. A. Fihri, C. Len, R.S. Varma, A. Solhy, Hydroxyapatite: a review of syntheses, structure and applications in heterogeneous catalysis. Coord. Chem. Rev. **347**, 48–76 (2017). https://doi.org/10.1016/j.ccr.2017.06.009
54. S.K. Swain, D. Sarkar, A comparative study: hydroxyapatite spherical nanopowders and elongated nanorods. Ceram. Int. **37**, 2927–2930 (2011). https://doi.org/10.1016/j.ceramint.2011.03.077
55. D.S. Gomes, A.M.C. Santos, G.A. Neves, R.R. Menezes, A brief review on hydroxyapatite production and use in biomedicine. Ceramica **65**, 869–872 (2019). https://doi.org/10.1590/0366-69132019653742706
56. M. Okada, T. Furuzono, Hydroxylapatite nanoparticles: fabrication methods and medical applications. Sci. Technol. Adv. Mater. **13**, 1–14 (2012). https://doi.org/10.1088/1468-6996/13/6/064103
57. H.C. Shum, A. Bandyopadhyay, S. Bose, D.A. Weitz, Double emulsion droplets as microreactors for synthesis of mesoporous hydroxyapatite. Chem. Mater. **21**, 5548–5555 (2009). https://doi.org/10.1021/cm9028935
58. S.K. Saha, A. Banerjee, S. Banerjee, S. Bose, Synthesis of nanocrystalline hydroxyapatite using surfactant template systems: role of templates in controlling morphology. Mater. Sci. Eng. C **29**, 2294–2301 (2009). https://doi.org/10.1016/j.msec.2009.05.019

59. G. Guo, Y. Sun, Z. Wang, H. Guo, Preparation of hydroxyapatite nanoparticles by reverse microemulsion. Ceram. Int. **31**, 869–872 (2005). https://doi.org/10.1016/j.ceramint.2004.10.003
60. M. Jamil, B. Elouatli, H. Khallok, A. Elouahli, E. Gourri, M. Ezzahmouly, F. Abida, Z. Hatim, Silicon substituted hydroxyapatite: preparation with solid-state reaction, characterization and dissolution properties. J. Mater. Environ. Sci. **9**, 2322–2327 (2018)
61. G. Heinicke, Tribochemistry, ed. by Carl Hanser Verlag, (Munchen Publishers, 1984), p. 119
62. S. Adzila, I. Sopyan, M. Hamdi, Mechanochemical synthesis of hydroxyapatite nanopowders: effects of rotation speed and milling time on powder properties. AMM **110–116**, 3639–3644 (2012). https://doi.org/10.4028/www.scientific.net/AMM.110-116.3639
63. R.A. Youness, M.A. Taha, M.A. Ibrahim, In vitro bioactivity, molecular structure and mechanical properties of zirconia-carbonated hydroxyapatite nanobiocomposites sintered at different temperatures. Mater. Chem. Phys. **239**, 122011 (2020). https://doi.org/10.1016/j.matchemphys.2019.122011
64. P. Kamalanthan, R. Singh, L.T. Bang, A. Niakan, C.Y. Tan, J. Purbolaksono, H.C. Thambinayagam, W. Teng, Synthesis and sintering of hydroxyapatite derived from eggshells as a calcium precursor. Ceram. Int. **40**(10B), 16349–16359 (2014). https://doi.org/10.1016/j.ceramint.2014.07.074
65. P.A.F. Sossa, B.S. Giraldo, B.C.G. Garcia, E.R. Parra, P.J.A. Arango, Comparative study between natural and synthetic hydroxyapatite: structural, morphological and bioactivity properties. Revista Materia **23**(4), 12217 (2018). https://doi.org/10.1590/s1517-707620180004.0551
66. M.K. Herliansyah, D.A. Nasution, M. Hamdi, A. Ide-Ektessabi, M.W. Wildan, A.E. Tontowi, Preparation and characterization of natural hydroxyapatite: a comparative study of bovine bone hydroxyapatite and hydroxyapatite from calcite. Mater. Sci. Forum **561–565**, 1441–1444 (2007). https://doi.org/10.4028/www.scientific.net/MSF.561-565.144
67. A. Ressler, K. Gudelj, M. Zadro, M. Antunović, M. Cvetnić, M. Ivanković, H. Ivanković, From bio-waste to bone substitute: synthesis of biomimetic hydroxyapatite and its use in chitosan-based composite scaffold preparation. Chem. Biochem. Eng. Q. **34**(2), 59–71 (2020). https://doi.org/10.15255/CABEQ.2020.183
71. B.N. Alhussary, G.A. Taqa, A.A. Taqa, Preparation and characterization of natural nano hydroxyapatite from egg shell and seashell and its effect on bone healing. JAVS **5**(2), 25–32 (2020). https://doi.org/10.21608/JAVS.2020.85567
68. L. Dou, Y. Zhang, H. Sun, Advances in synthesis and functional modification of nanohydroxyapatite. J. Nanomater. **2018**, 1–7 (2018). https://doi.org/10.1155/2018/3106214
69. T. Laonapakul, Synthesis of hydroxyapatite from biogenic wastes. KKU Eng. J. **42**(3), 269–275 (2015). https://doi.org/10.14456/kkuenj.2015.30
70. A.M. Torgalkar, A resonance frequency technique to determine elastic modulus of hydroxyapatite. J. Biomed. Mater. Res. **13**(6), 907–920 (1979). https://doi.org/10.1002/jbm.820130609
72. P.N. De Aza, A.H. De Aza, S. De Aza, Crystalline bioceramic materials. Bol. Soc. Esp. Ceram. Vidr. **44**(3), 135–145 (2005)
73. C.R. Bowen, J. Gittings, I.G. Turner, F. Baxter, J.B. Chaudhuri, Dielectric and piezoelectric properties of hydroxyapatite-$BaTiO_3$ composites. Appl. Phys. Lett. **89**, 1–3 (2006). https://doi.org/10.1063/1.2355458
74. M. Prakasam, M. Albino, E. Lebraud, M. Maglione, C. Elissalde, A. Largeteau, Hydroxyapatite-barium titanate piezocomposites with enhanced electrical properties. J. Am. Ceram. Soc. **100**(6), 2621–2631 (2017)
75. S. Pokhrel, Hydroxyapatite: preparation, properties and its biomedical applications. Adv. Chem. Eng. Sci. **8**, 225–240 (2018). https://doi.org/10.4236/aces.2018.84016
76. J.B. Foresman, Ab initio techniques in chemistry: interpretation and visualization, Chap. 14 in *What Every Chemist Should Know About Computing*, ed. M.L. Swift, T.J. Zielinski (ACS Books, Washington, D.C., 1996)

77. M. Ibrahim, A.A. Mahmoud, Computational notes on the reactivity of some functional groups. J. Comput. Theor. Nanosci. **6**, 1523–1526 (2009). https://doi.org/10.1166/jctn.2009.1205
78. H.A. Ezzat, M.A. Hegazy, N.A. Nada, M.A. Ibrahim, Effect of nano metal oxides on the electronic properties of cellulose, chitosan and sodium alginate. Biointerface Res. Appl. Chem. **9**(4), 4143–4149 (2019). https://doi.org/10.33263/BRIAC94.979986
79. A. Ibrahim, H. Elhaes, F. Meng, M. Ibrahim, Effect of hydration on the physical properties of glucose. Biointerface Res. Appl. Chem. **8**(4), 4114–4118 (2019)
80. A.M. Bayoumy, H. Elhaes, O. Osman, K.T. Kholmurodov, T. Hussein, M.A. Ibrahim, Effect of nano metal oxides on heme molecule: molecular and bimolecular approaches. Biointerface Res. Appl. Chem. **10**(1), 4837–4845 (2020). https://doi.org/10.33263/BRIAC101.837845
81. A.M. Bayoumy, H. Elhaes, O. Osman, T. Hussein, M.A. Ibrahim, Mapping molecular electrostatic potential for heme interacting with nano metal oxides. Biointerface Res. Appl. Chem. **10**(2), 5091–5095 (2020)
82. G.W. Ali, W.I. Abdel-Fattah, H. Elhaes, M.A. Ibrahim, Spectroscopic and modeling analyses of bimolecular structure of corn silk. Biointerface Res. Appl. Chem. **9**(6), 4481–4485 (2019). https://doi.org/10.33263/BRIAC0102.091095
83. M.M. El-Sayed, A. Omar, M. Ibrahim, W.I. Abdel-Fattah, On the structural analysis and electronic properties of chitosan/hydroxyapatite interaction. J. Comput. Theor. Nanosci. **6**, 1663–1669 (2009). https://doi.org/10.1166/jctn.2010.1363
84. A. Wierzbicki, H.S. Cheung, Molecular modeling of inhibition of hydroxyapatite by phosphocitrate. J. Mol. Struct. THEOCHEM **529**(1–3), 73–82 (2000). https://doi.org/10.1016/S0166-1280(00)00534-0
85. J. Zhao, L. Wu, C. Zhan, Q. Shao, Z. Guo, L. Zhang, Overview of polymer nanocomposites: computer simulation understanding of physical properties. Polymer **133**, 272–2872 (2017). https://doi.org/10.1016/.polymer.2017.10.035
86. N. Zhang, Y. Cheng, X. Hu, J. Yeo, Toward rational algorithmic design of collagen-based biomaterials through multiscale computational modeling. Curr. Opin. Chem. Eng. **24**, 79–87 (2019). https://doi.org/10.1016/j.coche.2019.02.011
87. M. Nouri-Felekori, M. Khakbiz, N. Nezafati, J. Mohammadi, M.B. Eslaminejad, N. Fani, Characterization and multiscale modeling of novel calcium phosphate composites containing hydroxyapatite whiskers and gelatin microspheres. J. Alloys Compd. **832**, 154938 (2020). https://doi.org/10.1016/j.jallcom.2020.154938
88. G.E. Dubinenko, A.L. Zinoviev, E.N. Bolbasov, V.T. Novikov, S.I. Tverdokhlebov, Preparation of poly(L-lactic acid)/hydroxyapatite composite scaffolds by fused deposit modeling 3D printing. Mater. Today Proc. **22**(2), 228–234 (2020). https://doi.org/10.1016/j.matpr.2019.08.092
89. C.M. Garcia, S.A. Toms, A cautionary tale of hydroxyapatite cement use in frontal sinus obliteration. Interdiscip. Neurosurg. **21**, 100702 (2020). https://doi.org/10.1016/j.inat.2020.100702
90. S. Bose, S. Tarafder, Calcium phosphate ceramic systems in growth factor and drug delivery for bone tissue engineering: a review. Acta Biomater. **8**(4), 1401–1421 (2012). https://doi.org/10.1016/j.actbio.2011.11.017
91. U. Gbureck et al., Resorbable dicalcium phosphate bone substitutes prepared by 3D powder printing. Adv. Funct. Mater. **17**(18), 3940–3945 (2007). https://doi.org/10.1002/adfm.200700019
92. T. Tian, C. Wu, J. Chang, Preparation and in vitro osteogenic, angiogenic and antibacterial properties of cuprorivaite ($CaCuSi_4O_{10}$, Cup) bioceramics. RSC Adv. **6**(51), 45840–45849 (2016). https://doi.org/10.1039/C6RA08145B
93. X. Lin, S. Patil, Y.-G. Gao, A. Qian, The bone extracellular matrix in bone formation and regeneration. Front. Pharmacol. **11**, 1–15 (2020). https://doi.org/10.3389/2fphar.2020.00757
94. M.-Y. Shie, S.-J. Ding, H.-C. Chang, The role of silicon in osteoblast-like cell proliferation and apoptosis. Acta Biomater. **7**(6), 2604–2614 (2011). https://doi.org/10.1016/j.actbio.2011.02.023

95. R.A. Youness, M.A. Taha, M.A. Ibrahim, Effect of sintering temperatures on the in vitro bioactivity, molecular structure and mechanical properties of titanium/carbonated hydroxyapatite nanobiocomposites. J. Mol. Struct. **1150**, 188–195 (2017). https://doi.org/10.1016/j.molstruc.2017.08.070
96. R.A. Youness, M.A. Taha, A.A. El-Kheshen, M. Ibrahim, Influence of the addition of carbonated hydroxyapatite and selenium dioxide on mechanical properties and in vitro bioactivity of borosilicate inert glass. Ceram. Int. **44**, 20677–20685 (2018). https://doi.org/10.1016/j.ceramint.2018.08.061
97. M.A. Taha, R.A. Youness, M. Ibrahim, Biocompatibility, physico-chemical and mechanical properties of hydroxyapatite-based silicon dioxide nanocomposites for biomedical applications. Ceram. Int. **46**, 23599–23610 (2020). https://doi.org/10.1016/j.ceramint.2020.06.132
98. D. Arcos, M. Vallet-Regí, Substituted hydroxyapatite coatings of bone implants. J. Mater. Chem. B **8**, 1781–1800 (2020). https://doi.org/10.1039/c9tb02710f
99. R. Chaharmahali, A. Fattah-Alhosseini, H. Esfahani, Increasing the in-vitro corrosion resistance of AZ31B-Mg alloy via coating with hydroxyapatite using plasma electrolytic oxidation. J. Asian Ceram. Soc. **8**, 39–49 (2020). https://doi.org/10.1080/21870764.2019.1698143
100. T.J. Levingstone, S. Herbaj, J. Redmond, H.O. McCarthy, N.J. Dunne, Calcium phosphate nanoparticles-based systems for RNAi delivery: applications in bone tissue engineering. Nanomaterials **146**(10), 1–28 (2020). https://doi.org/10.3390/nano10010146
101. X. Zeng, H. Xu, J. Lu, Q. Chen, W. Li, L. Wu, J. Tang, L. Ma, The immobilization of soil cadmium by the combined amendment of bacteria and hydroxyapatite. Sci. Rep. **10**, 2198 (2020). https://doi.org/10.1038/s41598-020-58259-1
102. J.A.G. del Rio, P.J. Morando, D.S. Cicerone, Natural materials for treatment of industrial effluents: comparative study of the retention of Cd, Zn and Co by calcite and hydroxyapatite. Part I: batch experiments. J. Environ. Manage. **71**, 169–177 (2004). https://doi.org/10.1016/j.jenvman.2004.02.004
103. Ramdani, A. Kadeche, M. Adjdir, Z. Taleb, D. Ikhou, S. Taleb, A. Deratani, Lead and cadmium removal by adsorption process using hydroxyapatite porous materials, Water Pract. Technol. **15**(1), 130–141 (2020). https://doi.org/10.2166/wpt.2020.003
104. K. Usami, A. Okamoto, Hydroxyapatite: catalyst for a one-pot pentose formation. Org. Biomol. Chem. **15**, 8888–8893 (2017). https://doi.org/10.1039/c7ob02051a
105. S. ben Moussa, A. Mehri, B. Badraoui, Magnesium modified calcium hydroxyapatite: an efficient and recyclable catalyst for the one-pot Biginelli condensation. J. Mol. Struct. **1200**, 127111 (2020). https://doi.org/10.1016/j.molstruc.2019.127111
106. D. Milovac, I. Weigand, M. Kovacic, M. Ivankovic, H. Ivankovic, Highly porous hydroxyapatite derived from cuttlefish bone as TiO_2 catalyst support. Process. Appl. Ceram. **12**(2), 136–142 (2018). https://doi.org/10.2298/PAC1802136M
107. S.C. Oh, J. Xu, D.T. Tran, B. Liu, D. Liu, Effects of controlled crystalline surface of hydroxyapatite on methane oxidation reactions. ACS Catal. **8**(5), 4493–4507 (2018). https://doi.org/10.1021/acscatal.7b04011
108. M. Shokouhimehr, S.M.G. Yek, M. Nasrollahzadeh, A. Kim, R.S. Varma, Palladium nanocatalysts on hydroxyapatite: green oxidation of alcohol and reduction of nitroarenes in water. Appl. Sci. **9**, 1–12 (2019). https://doi.org/10.3390/app9194183
109. J. Xu, T. White, P. Li, C. He, Y.F. Han, Hydroxyapatite foam as a catalyst for formaldehyde combustion at room temperature. J. Am. Chem. Soc. **132**(38), 13172–13173 (2010). https://doi.org/10.1021/ja1058923
110. M.B. Taşkın, Ö. Şahin, H. Taskin, O. Atakol, A. Inal, A. Gunes, Effect of synthetic nano-hydroxyapatite as an alternative phosphorus source on growth and phosphorus nutrition of lettuce (Lactuca sativa L.) plant, J. Plant Nutr. **41**(9), 1148–1154 (2018). https://doi.org/10.1080/01904167.2018.1433836
111. N. Kottegoda, C. Sandaruwan, G. Priyadarshana, A. Siriwardhana, U.A. Rathnayake, D.M.B. Arachchige, A.R. Kumarasinghe, D. Dahanayake, V. Karunaratne, G.A. Amaratunga, Urea-hydroxyapatite nanohybrids for slow release of nitrogen. ACS Nano **11**(2), 1214–1221 (2017). https://doi.org/10.1021/acsnano.6b07781

Biomineralization of $CaCO_3$: A Biomimetic Approach Using In Vitro Model Systems

Debojit Paul and Gopal Das

Abstract Our focus is to understand the role of chemistry in understanding the process of biomineralization and eventually helping us strategize the bio inspired synthesis of biominerals. In our pursuit to comprehend the various aspects of calcium carbonate biomineralization, we first need to know the composition, followed by the role of each constituent in contributing to the extraordinary properties of these biogenic materials. This is followed by understanding the crystallography of the compounds, which is of utmost importance in predicting their properties and eventually helps us distribute them into groups based on standard features. Ultimately, we will try to understand the entire process of calcium carbonate biomineralization; starting from the origin to the generation of completely matured crystals and also addressing the various exceptions that we can come across by regulating definite parameters. It is worth mentioning that the polymorphism exhibited by calcium carbonate makes it highly exceptional; therefore, by applying our knowledge and ideas attained from the in-depth analysis of the biological systems, we can easily tune among the different phases to exhibit fascinating morphologies and to fabricate bio-inspired composites with excellent mechanical strength and endurance.

Keywords In vitro biomineralization · Polymorphism · Nucleation · Crystal growth · Additives · Templates

1 Introduction

1.1 Inspiration from Living Organisms

Biomineralization is a fundamental life process that leads to the formation of mass and structure in almost all life forms. The hierarchically structured organic–inorganic hybrid composites, popularly termed biominerals, formed during this biogenic

D. Paul · G. Das (✉)
Department of Chemistry, Indian Institute of Technology Guwahati, Assam 781039, India
e-mail: gdas@iitg.ac.in

S. J. Ikhmayies (ed.), *Advances in Minerals Research*, Advances in Material Research and Technology, https://doi.org/10.1007/978-3-031-49175-7_7

process are widely prevalent among almost all living organisms, ranging from the unicellular marine organisms to shells in tiny mollusks to the highly complex skeleton in higher organisms [1–3]. The properties of these materials, such as mechanical, optical, magnetic, etc., are exploited by the organisms for a variety of purposes [4, 5]. Based on their requirements and to adapt to specific environmental challenges, organisms undergo evolution in the biomineral structures. The ability possessed by the organisms to optimize the shape, size, structure, orientation, texture, and overall morphology of the biominerals is fascinating. Ever since the beginning of the era when scientists involved themselves in understanding this biological phenomenon, they have been thrilled by this special ability and, therefore, have engulfed themselves in understanding the mystery behind this art. Till today, decades since the onset of the preliminary studies followed by an era where full-on investigations by courtesy of the general to the highly-sophisticated scientific tools have been performed, a fairly large number of questions related to biominerals remain unexplored. The scientific efforts in the field of materials science, clubbed together with the understanding of chemistry, have enabled scientists to mimic the design and synthesize various biogenic materials, of which the biominerals commandingly stand out as one of them [6–8]. However, to date, they have not attained success in synthesizing minerals that possess superior properties to their biological counterparts. The quest to obtain the biological standard has in itself been a glory, as every redesigned and recalculated attempt enables the scientists to understand and fabricate these organic–inorganic hybrid materials and in the process, take a step closer to attaining the highest standard.

1.2 How Are Biominerals Different from Minerals?

The biological minerals are composed of two components, one of which is organic and the other inorganic. From the perspective of materials science, the organic molecules though they are soft, are fracture-resistant, whereas the inorganic crystals are hard but brittle [9]. The biominerals are the best-combined output of the two constituents and hence the resulting composites are both hard and fracture resistant. It is this fundamental difference in the composition that makes the biominerals special and gives them an edge over their geological counterparts. The naturally occurring minerals (in geology) have a well-defined chemical composition coupled with a specific crystal structure. These minerals are hard but cannot undergo the same level of wear and tear that is undergone by the biominerals. The lack of hierarchical structure dictated by the organic layer, as in the case of biominerals, makes the natural minerals more vulnerable to cracks and breaking [10]. Hence, it is clear that the biogenic minerals are superior to the non-biological minerals occurring in nature.

1.3 Understanding the Chemistry of $CaCO_3$

After prolonged studies, it is clear now that the organization of the inorganic crystals over the organic matrix is responsible for the exceptional properties of the biominerals [11–14]. Simultaneously, with this organization, the authority over polymorphism and crystallographic arrangement are the additional factors that substantially determine the properties of these materials. As it is evident that the interaction of the organic and inorganic phases is essential in the in vivo generation, hence, both soluble and insoluble directing species have been comprehensively studied; however, a complete understanding of the correlation between these agents and the crystallographic structure and properties of the resulting material is yet to be accomplished [15–18].

In this chapter, we will focus on the biomimetic synthesis of $CaCO_3$ and try to understand the factors that influence the formation, with an insight into the various stages leading to crystallization. $CaCO_3$ being the most abundant biomineral, covers a large part of the biomineral world and is also a material of industrial importance as it finds application in the generation of commercial and personal care products [19, 20]. Among the inorganic materials, $CaCO_3$ has extensive biomedical applications owing to its availability, low cost, safety, biocompatibility, pH sensitivity, and biodegradability [21]. It has been used as a carrier for targeted drug delivery and as a template for biodegradable polymer capsules for applications in nanomedicine [22–24]. $CaCO_3$ has also been used as solid support for active species to be utilized in the field of heterogeneous catalysis leading to environmental applications [25, 26].

The study of biomineralization is a multidisciplinary prospect that engages researchers from biology, chemistry, materials science, geology, and beyond [27]. The key role played by chemistry in biomineralization research can broadly be divided into three main areas:

(i) in studying the chemical composition, biochemistry, and crystallography of the materials;
(ii) in designing in vitro model systems to understand the interactions between the organic and the inorganic phases to understand the actual mechanism in biological systems;
(iii) in developing new synthetic techniques by mimicking the biological systems to generate organic–inorganic composites with enhanced mechanical properties.

The primary source of $CaCO_3$ in nature is the remains from the skeletons of marine organisms that have accumulated over millions of years and make up nearly 4% of Earth's crust. These remains can be seen in the form of chalk, limestone, etc. The minerals of calcium carbonate are also involved in the global carbon cycle [11, 12, 28]. Being the most abundant biomineral, with an alluring science of its formation and characteristic toughness, makes it a very special prospect among chemists and material scientists. But the factor that stands out is the ability of $CaCO_3$ to exhibit polymorphism. For over a century, $CaCO_3$ is known to exist in three main anhydrous crystalline forms; calcite, aragonite, and vaterite, two hydrated crystalline forms; calcium carbonate monohydrate (monohydrocalcite) ($CaCO_3 \cdot 1H_2O$) and calcium carbonate hexahydrate (ikaite) ($CaCO_3 \cdot 6H_2O$), and also in the amorphous state as amorphous calcium carbonate (ACC) [12]. Recently, one more hydrated crystalline form has been reported by Zou et al. as calcium carbonate hemihydrate $CaCO_3 \cdot ½H_2O$ [29]. The anhydrous crystalline forms are the most prevalent and are often the center of discussion among the bio-inspired $CaCO_3$. The polymorph, calcite, is the most stable and has a rhombohedral crystallographic unit cell; aragonite, which is comparatively unstable, is orthorhombic and the metastable vaterite is hexagonal [30]. The Wull's Rule, which predicts the morphology of the crystals, states that the most stable geometry is related to the minimum of the sum of the products between crystal surface area and surface energy for the different faces of the crystal [31]. For calcite, it was observed both experimentally and theoretically that it has a rhombohedral crystal structure with the $\{10\bar{1}4\}$ surface being the dominant surface. This surface consists of layers containing the calcium ions and the carbonate groups and the surface plane is always terminated with oxygen atoms. Aragonite has an orthorhombic crystal structure with {110}, {011}, {010} planes representing the dominant surface. The metastable vaterite has a hexagonal crystal structure and it is their {010} carbonate plane that is the dominant surface. Some other structural properties such as the space group, lattice constants, and density of the anhydrous crystalline phases of $CaCO_3$ are represented in Table 1 [28].

The first member of the hydrated crystal group, calcium carbonate hexahydrate, as was discovered in the nineteenth century, crystallizes with a monoclinic crystal structure [32]. In 1981, it was determined by Effenberger that monohydrocalcite has a crystal structure with trigonal symmetry [33]. Along with the recent discovery of calcium carbonate hemihydrate, Zou et al. reported this newest phase to exist with a monoclinic crystal structure [29].

The formation of matured crystals in nature and also during in vitro studies goes via the nucleation and crystallization mechanisms and $CaCO_3$ is often considered as a reference model system for studying this phenomenon [34, 35]. $CaCO_3$ is also special as its sensitivity toward template effects makes it applicable for elementary studies on template-mineral interactions. When synthesized under ambient conditions without any template or additive, it attains the thermodynamically most-stable calcite phase and appears as rhombohedral crystals, whereas, in the presence of templates or additives, the polymorphs can be tuned from one form to the other based on the synthetic requirements or to understand the dynamic mechanism of formation of $CaCO_3$.

Table 1 Surface properties of the anhydrous crystalline forms of calcium carbonate

Phase	Crystal structure	Space group	Lattice constants	Density (g/cm^3)	Dominant surface planes	Surface energy (J/m^2)
Calcite	Rhombohedral	$R\bar{3}c$	a = b = 4.990 Å, c = 17.061 Å α = β = 90°, γ = 120° [36]	2.71 [12]	{10$\bar{1}$4} [30]	0.59 {10$\bar{1}$4} unhydrated, [30] 0.16 {10$\bar{1}$4} hydrated [30]
Aragonite	Orthorhombic	P_{mcn}	a = 4.959 Å, b = 7.964 Å, c = 3.737 Å, α = β = γ = 90° [37]	2.93 [12]	{110}, {011}, {010} [30]	0.69 {011} CO_3 unhydrated, [30] 0.24 {010} Ca, hydrated [30]
Vaterite	Hexagonal	$P6_3$	a = b = 4.130 Å, c = 8.480 Å α = β = 90°, γ = 120° [36]	2.66 [12]	{010} [30]	0.62 {010} CO_3, unhydrated, [30] 0.22 {010} CO_3, hydrated [30]

2 Stages of Crystal Maturation

2.1 Nucleation

It is the first step in the crystallization of inorganic minerals in which a small number of constituting species such as ions, atoms, or molecules organize themselves in a pattern characteristic of a crystalline solid, following which a new phase (solid) is formed from the previous (liquid) phase [38, 39]. The process begins with the formation of tiny metastable clusters within the liquid phase, where the clusters are in a continuous state of growth and disintegration, which is dictated by a balancing act between the increasing surface energy associated with the increased surface area and the lowering of bulk energy associated with the formation of a crystal lattice, [12, 40] eventually, there arises an instance where the bulk energy neutralizes the surface energy and a critical size is attained. Following this, the subsequent growth causes a reduction in the Gibbs energy of the system as the further decrease of the lattice energy overcompensates the increase in surface energy. These clusters then grow until they reach a critical size and by overcoming an energy barrier which ultimately leads to the generation of a stable nucleus. According to the classical nucleation theory, [41] homogeneous nucleation is directed by elevated supersaturations and heterogeneous nucleation. On the contrary, it is induced by the presence of substrates that behave as active sites and govern the whole process.

2.1.1 Homogeneous Nucleation

Over the period, there has been an evolution in the theories of nucleation. Since the very beginning, the model featuring the nucleation of ions (or pair of ions) has been successfully used to explain many attributes of the nucleation process [42–44]. However, a drawback of this model lies in the fact that it considers only the energy aspects while neglecting the structural information. As a result, this model has been reviewed over time [45, 46] and in this chapter, we shall try to simplify the concept for easy understanding and also provide the bare essentials.

The Gibbs energy associated with nucleus formation is contributed by the Gibbs energy gained by the formation of a nucleus from the supersaturated solution and the energy lost in the formation of a new surface and is represented as;

$$\Delta G = -nkT\ln S + \sigma A \tag{1}$$

where n is the size of the nucleus formed, k *is* the Boltzmann's constant, T *is* the absolute temperature, S is the supersaturation, σ *is* the interfacial Gibbs energy for nucleation, and A is the surface area. The supersaturation S can be determined from the solubility product K_{sp} and the activities a of the species involved, here $a(Ca^{2+})$ and $a(CO_3^{2-})$, via

$$\ln S = \ln[a(Ca^{2+})a(CO_3^{2-})]/K_{sp} \tag{2}$$

As homogeneous nucleation is directed by elevated supersaturations, the chemical parameters such as the Ca^{2+} ion activity $a(Ca^{2+})$ and the CO_3^{2-} ion activity $a(CO_3^{2-})$ determine the nucleation. In any $CaCO_3$ formation reaction, K_{sp} is temperature-dependent and varies depending on the $CaCO_3$ phase. The activity of the constituting ions and hence, the supersaturation ratio is influenced by the conditions, such as temperature, pH, and ionic strength of the system per the Debye–Huckel law. The precipitation kinetics of the reactions are considerably affected by the operating conditions mentioned above, that is why the different crystalline phases, vaterite, aragonite, and calcite, are obtained. The most stable phase at any temperature is calcite, while the other two are metastable phases. According to the study performed by Rodriguez-Blanco et al. [11], the crystallization mechanism of $CaCO_3$ could be divided into two main stages. The initial stage where the transformation of unstable ACC to metastable vaterite phase and in the second stage, the subsequent transformation of metastable vaterite to stable calcite. These observations are in tandem with other reports along the same line; for example, Bots et al. [47] observed the mechanistic features of the transformation of ACC to vaterite and merged their observations with those obtained by Rodriguez-Blanco et al. to understand the complete abiotic transformation mechanism from the amorphous phase to the stable calcite via vaterite. Ogino et al. [48], in their work, pointed out that the rate-determining step of this whole transformation is the dissolution of vaterite, leading to the growth of calcite. In their study performed by Dickinson et al. [49], they illustrated the effect of chemical parameters such as the Ca^{2+} ion activity $a(Ca^{2+})$ and CO_2 partial pressure pCO_2 on nucleation. They varied $a(Ca^{2+})$ and pCO_2 and observed that at low $a(Ca^{2+})$, the process being thermodynamically controlled, the rhombohedral calcite crystals formed were stepped multinucleated. Here, the amount of CO_2 determined the amount of calcite formed. At comparatively higher values of $a(Ca^{2+})$, they obtained perfect rhombohedral crystals of calcite and in this case, the process was a balance between kinetic and thermodynamic factors. At even higher $a(Ca^{2+})$, it becomes a kinetically controlled reaction and a biphasic product, where ~1/3 calcite and ~2/3 vaterite were formed and from then on, with further increase in $a(Ca^{2+})$ and pCO_2, the proportion of the vaterite phase would increase.

2.1.2 Heterogeneous Nucleation

This type of nucleation involves the introduction of a substrate in the reaction medium. The pre-critical nucleus interacts with the substrate and this interaction is said to lower the surface Gibbs energy in most cases which reduces the barrier otherwise set for the development of the critical nucleus. Therefore, the nucleation taking place at the interface is favored over the one occurring in bulk (homogeneous nucleation). This is also the basis for the formation of oriented crystals as the interaction with the surface advocates the nucleation on a specific crystal plane.

For homogeneous nucleation, the relevant Gibbs energy σ_{hom} equals the Gibbs energy γ_{nl} associated with the nucleus (n)-liquid (l) interface, i.e.,

$$\sigma_{hom} = \gamma_{nl} \tag{3}$$

For heterogeneous nucleation, the relevant interfacial Gibbs energy σ_{het} is the sum of the Gibbs energies for the nucleus-liquid and nucleus-substrate interfaces minus that for the liquid-substrate interface, i.e.,

$$\sigma_{het} = (\gamma_{ns} - \gamma_{ls})(1 - \xi) + \gamma_{nl\xi} \tag{4}$$

where, ξ is the ratio of nucleus-liquid area to total area.

On introducing the contact angle θ between the substrate and the pre-nucleation cluster, the relationship between ΔG_{het} and ΔG_{hom} can be expressed as,

$$\Delta G^*_{het} = \kappa f(\theta) \Delta G^*_{hom} \tag{5}$$

where * denotes the critical state.

The contact angle and the shape of the nucleus determine the function $f(\theta)$, while the function κ contains the interface energy γ_{cf} between the growing cluster and the fluid.

It should be noted that some assumptions are to be made before applying the heterogeneous nucleation model [10].

i. The substrate should be rigid.
ii. The nucleating species are all individual ions.
iii. The system is thermodynamical for the nucleus with constant interface energies and solubility products.

However, several drawbacks are associated with the classical heterogeneous nucleation model; for example, it considers the substrate flat and the nucleus to be a spherical cap. This approach makes the model very rigid and it was found to be difficult to correlate with systems associated with different shapes of the substrate and nucleus. Hence, modifications have been made where Liu et al. [43] put forward a new model considering the substrate geometry to be spherical and the concept of a faceted nucleus has been given by Travaille et al. [44]. Therefore, despite the possible shortcomings, the classical heterogeneous nucleation model has been used either directly or post-modification to understand the various aspects of nucleation.

2.2 Crystal Growth

The nucleation process is accompanied by crystal growth which can be classified into two main steps:

i. Mass transfer of the constituting ions/species from the bulk (solution) to the crystal surface.
ii. Surface integration of the crystal unit in the crystal lattice.

In general, the process of crystal growth may or may not be influenced by foreign species such as ions or organic compounds. However, in biomineralization, these species known as additives play a huge part in the crystal growth process, such as the determination of the phase and morphology by influencing the crystallographic parameters of the growing crystal. In the following section, a detailed discussion of the role played by these species on the crystal growth process has been elaborately discussed.

2.2.1 Growth Controlling Factors

This special kind of organization attributed to the hierarchical arrangement is responsible for the outstanding mechanical properties of the biominerals, such as high crack resistance. In biominerals, the organic phase makes up from as low as 0.05 to as high as 40% of the composite [4]. As a result of prolonged research in the field of biomineralization, it is now evident that the organic phase plays a huge role in directing mineral growth [50]. The localization of the organic molecules within the biominerals leads to two subdivisions; intercrystalline and intracrystalline. The biominerals in which

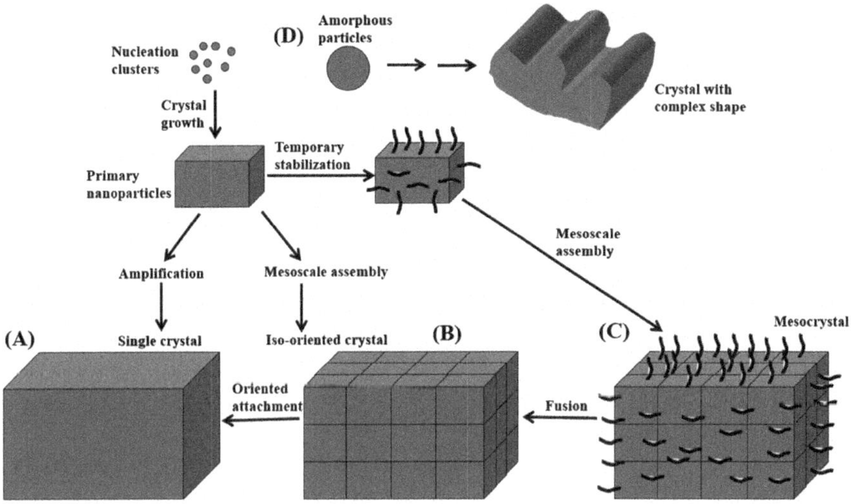

Fig. 1 Schematic representation of different pathways of crystallization. Pre-nucleation clusters grow to produce a critical nucleus which grows further into a primary nanoparticle. (*A*) The primary nanoparticles then mature into a single crystal by the addition of ions; (*B*) The primary nanoparticles mature into a single crystal by oriented attachment of multiple particles; (*C*) The additives attach to the surface of primary nanoparticle to form mesocrystals; (*D*) The pre-nucleation clusters lead to the formation of amorphous particles which crystallize either with or without interaction with an additive to produce different crystal morphologies

the organic phase (molecules) occupies the crystal surface are classified as "inter-crystalline" and the ones in which these molecules are incorporated inside a single crystal are classified as "intracrystalline" [51]. The reports related to intracrystalline molecules are mostly associated with $CaCO_3$. It is of absolute interest to researchers how the macromolecules are successfully incorporated within the dense inorganic host crystals; what may be the advocating criteria for a specific molecule to be incorporated within a specific crystal; and finally, to know the biological function of the resulting intracrystalline composite. In Fig. 1, we represent all the different pathways of crystallization.

2.2.2 Role of Additives

It was in 1965 when a study performed by Norimitsu Watabe reported the presence and also the distribution of intracrystalline molecules in various marine organisms for the first time [52]. He studied the formation of mollusk shells with particular interest to the occluded organic matter within the bio-composite. Since then, taking a cue from the biological specimens, the researchers working on biomimetic materials based on biominerals have moved a long way in their study on the incorporation of species within the composite. It has been demonstrated that not just organic, but also inorganic additives play a crucial role in determining the properties of the composite [53]. As for the intercrystalline composites, the additives adsorbed to the surface of the primary particles prevent the formation of the critical nucleus by facilitating the growth of the different faces. These additives can also promote the growth of crystals by altering the surface properties of crystals post adsorption or can form a bridge between several crystalline units. For the intracrystalline ones, the incorporated additives inside the mineral promote their solubility and also raise the limit in size essential for the development of stable particles. These effects boost the aggregation of smaller units, therefore, forming either randomly organized polycrystalline entities or well-organized, larger aggregates. There are instances where the additive adsorption takes place post the nucleation stage and in this case, they influence the formation of the crystal by hindering the growth at specific crystal faces [54]. As a result, the formation of the equilibrium morphology is disturbed and this will, in turn, affect the structure of the crystallographic unit cell, thereby influencing the overall phase and morphology of the system.

Kim et al. performed an experimental study to understand the role of additives in the early stages of calcite crystal growth [55]. The experiments in this study were based on the fact that the crystallization process advances much more slowly in small volumes as compared to large volumes. They investigated the role of two soluble additives, Mg^{2+} and poly(styrene sulfonate), on the initial stages of calcite crystal growth from an amorphous precursor (ACC). Interestingly, they reported that the first crystals of calcite that were observed in each case were perfectly rhombohedral, and the soluble additives did not influence the morphology till the sizes of the crystals were 100 nm for Mg^{2+} and 1 μm for poly(styrene sulfonate). Based on their findings, they also stated that the concentration, additive binding strength, and supersaturation are the factors that determine the size of the growing calcite crystal at which the additive would start to affect the morphology.

We will now try to observe the role of additives in the bio-inspired $CaCO_3$ crystallization from some recently performed studies. Kim et al. studied the combined effect of Sr(II) and poly(acrylic acid) in the precipitation of $CaCO_3$ by diffusion of CO_2, generated from $(NH_4)_2CO_3$ into an aqueous solution of $CaCl_2$ [56]. They maintained the concentration range of Sr(II) close to its natural abundance (0.08–0.09 mM) in seawater. In their observation, they found that the morphology of the calcite crystals was different from the typical rhombohedral morphology; this was because the incorporation of the additives had generated new types of calcite surfaces. Without any additive, it is the rhombohedral morphology enclosed by {104} faces that prevails as the characteristic of calcite. In this case, with the combined effect of Sr(II) and poly(acrylic acid), the formation of new faces other than {104} was reported leading to the generation of roundness in the shape of the calcite. Another morphological feature that was observed was the appearance of striations on the surface of calcite which is absent in the absence of an additive. In the control experiments, when only one of the two additives was subjected, unlike the combined effect of the additives, here only minute changes were observed in terms of the calcite morphology. When only Sr(II) was used, the rhombohedral morphology of calcite was observed at all concentrations of Sr(II), however, a subtle protrusion on the surfaces was observed with varying concentrations of Sr(II), suggesting the role of the additive in localizing the crystal growth. The use of poly(acrylic acid) also hindered the growth of calcite but in a less regular way as compared to Sr(II) and the overall rhombohedral morphology of calcite remained intact.

Liu et al. performed the precipitation of $CaCO_3$ in the presence of egg white and Cu^{2+} [57]. They reported that copper(II) could induce the formation of metastable vaterite. The influence of different concentrations of copper(II) on the morphology of the composite was observed by the Field Emission Scanning Electron Microscope (FESEM) imaging technique. A micron pearl-like morphology of $CaCO_3$ was obtained at high concentrations of both additives. Ca^{2+} coordinates with functional groups like –OH or –C=O on the egg-white protein and the additive Cu^{2+} also coordinates with peptide bonds on egg white protein; both these interactions are independent of each other. The peptide bonds originating from the protein molecules coordinating with Cu^{2+} cause bending and entanglement of the protein molecule. A higher concentration of Cu^{2+} intensifies the bending and entanglement, leading to the aggregation of protein chains into spheres. As a result, the $CaCO_3$ that is precipitated in the presence of the combination of egg white, and Cu^{2+} assumes a spherical morphology and the crystalline phase is vaterite rather than calcite.

Du et al. successfully showed the influence of many organic and inorganic additives on the crystallization of amorphous $CaCO_3$ when they are exposed to humid environments and elevated pressures [58]. Here, they subjected both low-molecular weight additives such as glycine, citric acid, acrylic acid, and magnesium chloride hexahydrate; and high low-molecular weight additives such as poly(acrylic acid), poly(sodium 4-styrene sulfonate), and poly(allylamine). It is the composition of the additives that influence the kinetic stability of amorphous $CaCO_3$. This is against the crystallization that is induced by humidity and elevated pressures. In humidity-induced crystallization, the water molecules in the surrounding get adsorbed onto the surface of amorphous $CaCO_3$ and partially dissolve them and eventually reinitiate the crystallization process. The mobility of water plays a crucial role here and in the absence of an additive, it is the degree of hydration of amorphous $CaCO_3$ that determines the mobility. In the presence of additives, the mobility of water is influenced by the additives and simultaneously, the kinetics of the crystallization process is changed. It is the interaction of an additive with water that is responsible for the stabilization of the amorphous phase, a stronger interaction results in a smaller amount of mobile water contained in amorphous $CaCO_3$ and, therefore, greater stability of amorphous $CaCO_3$. In this study, they have reported that the interaction of citric acid with water is stronger than the other additives used and hence, the amorphous $CaCO_3$ functionalized with citric acid is highly stable.

Stepic et al. used aspartic acid and its derivatives as models of biomolecules to study their interactions with calcite [59]. They performed kinetic studies to investigate the inhibition of calcite growth using the models and it was found that the inhibitory effect exhibited had increased asymptotically with the length of the peptide chain. They also pointed out that interaction of the aspartates with calcite was initiated by the attachment of the positively charged ammonium group, which was almost always accompanied by carboxylate attachment, but these interactions are always counterbalanced by the solvation of the carboxylates in an aqueous solution. This balance accounted for the asymptotic growth of the binding energies as the chain length aspartate was increased.

Wada et al. [60] demonstrated the effect of electric field combined with the additive poly-lysine on the crystallization of calcite. It was considered that the electric field would direct the reorganization and condensation of clusters at the interface and promote the simultaneous or subsequent adsorption of organic additives. In their experiments, they found that at a small concentration of poly-lysine coupled with polarization treatment, calcite crystals with notches in the edges of the rhombohedra were formed. When the concentration was increased, the calcite crystals formed were observed to be elongated in the direction perpendicular to the substrate. At an even higher concentration of the additive combined with polarization treatment, the calcite crystals displayed a cone-shaped morphology.

Bahrom et al. made use of poly(styrene sulfonate) sodium and dextran sodium sulfate as organic additives in the co-precipitation synthesis of $CaCO_3$ [21]. They performed the syntheses using different concentrations of the additives as well as the ratios of the constituting ions were varied. The morphology of the particles obtained was toroidal-like when the ratio of Ca^{2+}/CO_3^{2-} was maintained at 5:1. Between the two additives, the toroidal-like morphology of the particles was more prominent in the case of poly(styrene sulfonate) as compared to dextran sodium sulphate.

Zhu et al. reported an aggregation-based crystal growth of $CaCO_3$, which is different from the usual mode of solution crystallization [61]. Here, the crystal growth mechanism followed a non-classical particle-by-particle pathway rather than the classical ion-by-ion growth and for this study, they made use of N-stearoyl L-glutamic acid, an amphiphile, as the organic additive. They observed that the crystal growth mechanism is controlled by the concentration of the additive in solution as a higher concentration of the additive could stabilize the precursor nanoparticles to facilitate aggregation-based crystallization by providing building blocks. Therefore, different from the usual growth mechanism, here, the interaction between the additive and the nanoparticles is dominant.

Sun et al. controlled the morphology and surface area of the vaterite particles by tuning the concentration of the additive adipic acid [62]. The morphology of the particles varied between ellipsoidal to spherical and it was shown that the adsorption of adipic acid on the amorphous $CaCO_3$ particles played a pivotal role in determining the morphology. At a low concentration of the additive, the coverage of the molecules on the amorphous precursor was low; in this case, the strength of hydrogen bonds formed between the additive molecules on the surface of the amorphous nanoparticles was weak, resulting in lower shielding on the surface. Hence, the transformation from the amorphous phase to the vaterite phase took place faster and the particles were bigger in size with morphology slightly different from the usual spherical morphology.

Paul et al. successfully synthesized the metastable vaterite phase by subjecting whey protein as an additive [63]. Subsequently, in addition to the role of the concentration of the additive, they also studied the influence of the external parameters such as temperature, pH, and ionic strength on the in vitro mineralization of $CaCO_3$.

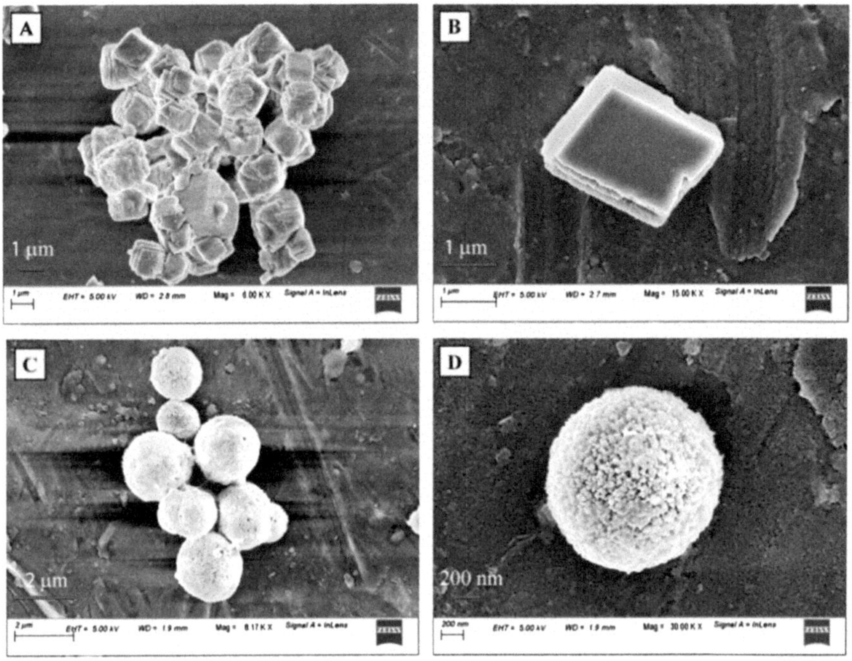

Fig. 2 FESEM images **a** and **b** of $CaCO_3$ under controlled conditions; **c** and **d** in the presence of whey protein [63]

Figure 2 represents the rhombohedral calcite crystals formed without any additive and the vaterite microspheres synthesized under the influence of whey protein.

In another work, they used the organic molecule curcumin and the flavonoid quercetin as the additives in the in vitro mineralization of $CaCO_3$; with curcumin, they were successful in isolating the metastable vaterite, whereas, with quercetin, calcite particles with a dumbbell-shaped morphology was formed [64]. Additionally, they observed the maturation of the particles in solution and it was found that the vaterite particles had completely transformed into rhombohedral calcite after 24 h. Upon changing the parameters such as temperature and ionic strength of the solution, in the presence of the additive quercetin, calcite particles with different morphologies were obtained. At 10 °C, an ellipsoidal shape was observed, at room temperature, 25 °C, a dumbbell shape, and at a high temperature, 80 °C, again a dumbbell shape but with a smoother surface texture was obtained. On increasing the ionic strength of the solution by adding NaCl (100 mM), calcite particles with a dumbbell/peanut shape were obtained. Figure 3 represents the various morphologies of $CaCO_3$ synthesized in the presence of curcumin and quercetin at different temperatures.

In another study, Paul et al. used the tulsi (*Ocimum sanctum*) and green tea (*Camellia sinensis*) leaf extracts in water and methanol as additives in $CaCO_3$ precipitation [65]. With the methanol extract of tulsi and green tea, they isolated the

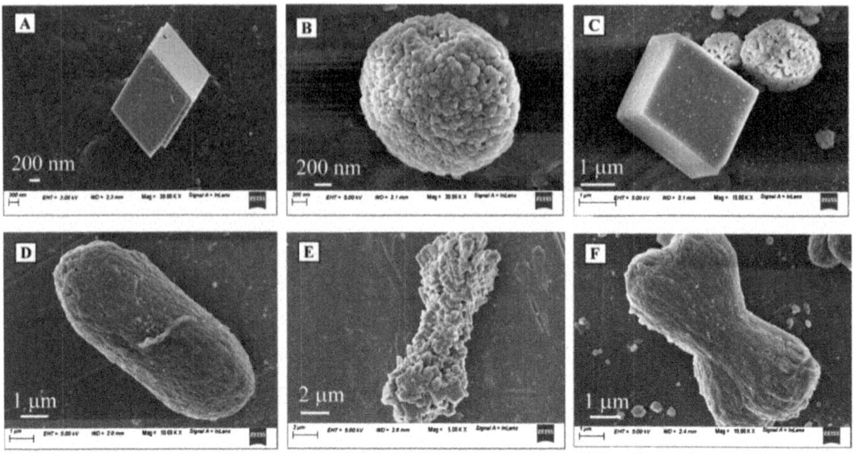

Fig. 3 FESEM Images of $CaCO_3$ **a** at 10 °C in the presence of curcumin; **b** at 25 °C in the presence of curcumin; **c** at 80 °C in the presence of curcumin; **d** at 10 °C in the presence of quercetin; **e** at 25 °C in the presence of quercetin; **f** at 80 °C in the presence of quercetin [64]

metastable vaterite microspheres and calcite with daffodil-like morphology, respectively, Fig. 4. With the water extract of tulsi and green tea, calcite particles with an elegant rose-like morphology and spherical morphology were formed, respectively.

In the second part of the study, they demonstrated how by tuning the sequence of addition and using the methanol extract of tulsi as the additive, the amorphous precursor, ACC phase could be isolated and also by increasing the time of addition,

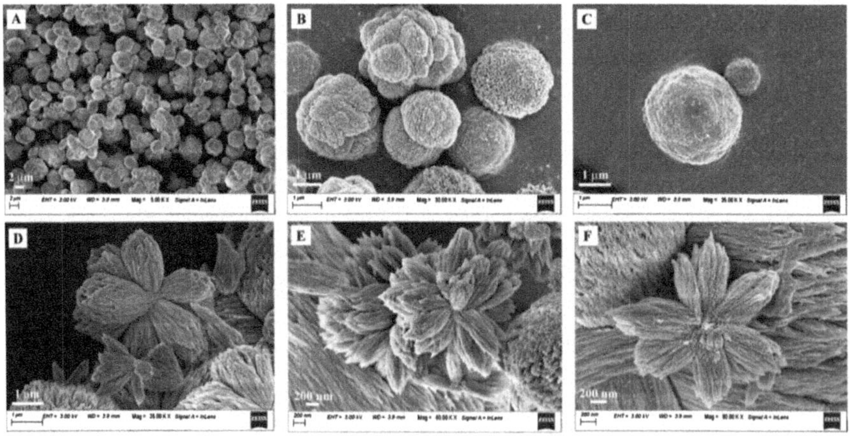

Fig. 4 FESEM images of vaterite **a** microspheres in bulk, **b** vaterite microspheres in cluster, and **c** individual vaterite microsphere precipitated in the presence of tulsi extract in methanol; **d–f** are FESEM images of calcite particles with daffodil-like morphology precipitated in the presence of green tea extract in methanol [65]

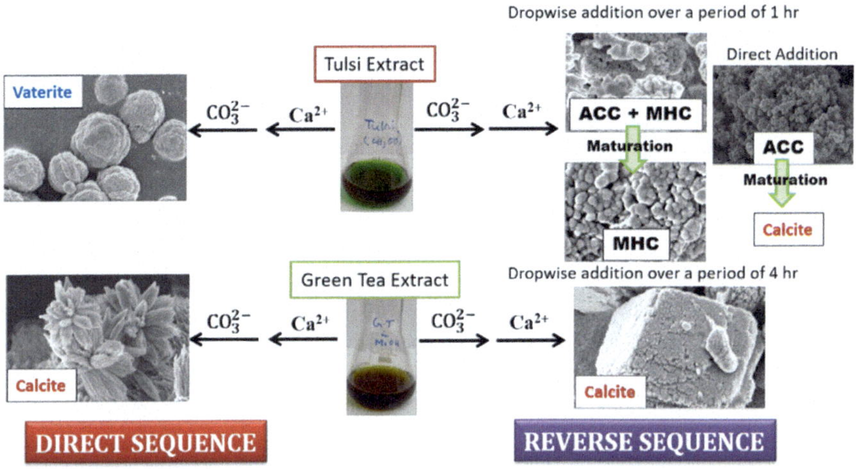

Fig. 5 Schematic representation of the tulsi and green tea-directed precipitation of $CaCO_3$ [65]

one of the hydrated phases of $CaCO_3$, monohydrocalcite was formed. Additionally, they also described, with the help of experimental studies, how the metastable ACC and the monohydrocalcite phases were transforming into the stable calcite phase with time (maturation) and subjecting them to heat. Figure 5 illustrates the different modes of reaction and the various phases obtained using the tulsi and green tea extracts. They described that the Mg ions present in the methanol extract of tulsi played a key role in the formation of both ACC and MHC.

Marzec et al. experimentally showed how in the case of biomineralization, multiple organic additives co-operatively work to control the structure and properties of the composites [66]. They subjected both amino acids and dye molecules as additives during the precipitation of $CaCO_3$; as a result, the dye molecules were successfully occluded within the crystal. Here, the amino acid molecules acted as a chaperone and facilitated the incorporation of the dye molecules. The percentage of incorporation of dye molecules was very high when combined with the amino acid molecules in comparison to when they were used individually. Among the various combinations of amino acid and dye used, the combination of aspartic acid and brilliant blue red showed the best result.

Kim et al. performed the occlusion of block copolymers based on poly(methacrylic acid) (PMAA) and poly(benzyl methacrylate) (PBzMA) during the precipitation of $CaCO_3$ using the ammonia diffusion method [67]. The block copolymers occluded were in the form of worms, $PMAA_{71}$–$PBzMA_{300,}$ and vesicles, $PMAA_{71}$–$PBzMA_{98}$. Calcite crystals of different structures and properties were obtained by varying the concentration of the block copolymers used, where Fig. 6 shows the changes in crystal morphologies with increasing copolymer concentration in solution. In another work, Kim et al., with the help of a series of copolymer nanoparticles subjected as additives during $CaCO_3$ precipitation, studied the crystal habit modification and nanoparticle occlusion within calcite crystals [68]. The copolymer nanoparticles were based on the hydrophobic poly(2-hydroxypropyl methacrylate), which made up the core and was surrounded by hydrophilic tails comprised of mixed carboxylate- and hydroxyl functionalized groups of varying chain length.

Wang et al. subjected alkanolamide 6502 as an additive to study its influence on the morphology of the precipitating $CaCO_3$ [69]. The selective formation of the metastable vaterite with a uniform particle size due to the additive addition was reported. They also explained how the adsorption of the hydroxyl group on the growing crystal surface influences the surface properties as well as inhibits the maturation of the calcite phase. The surfactants reduce the surface tension of water and the organic additives influence the solubility of substrates and products and also the interactions between particles. With the alcoholic additive used here, the hydroxyl group (–OH) is adsorbed onto the surface of $CaCO_3$. Owing to its surface properties, vaterite has a greater affinity for –OH. During the crystallization mechanism,

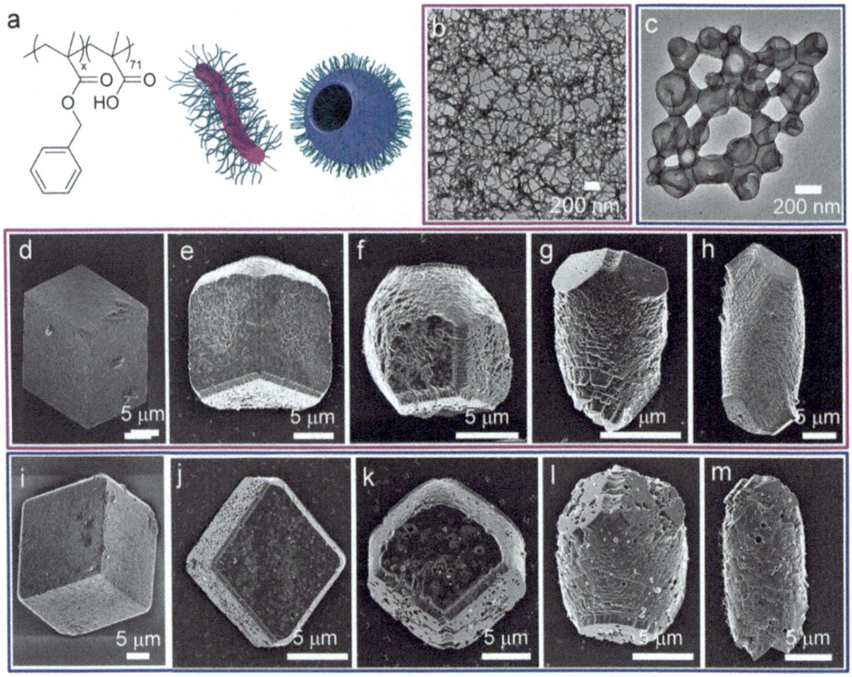

Fig. 6 **a** The chemical structure of the poly(methacrylic acid)-poly(benzyl methacrylate) (PMAA-PBzMA) diblock copolymer particles, where $x = 98$ and $y = 71$ for the worm particles and $x = 300$ and $y = 71$ for the vesicle particles, and schematics of worm and vesicle nano-objects. TEM images of **b** copolymer worms, and **c** copolymer vesicles. SEM images of calcite crystals precipitated in the presence of vesicles (**d–h**), and worms (**i–m**) at fixed $[Ca^{2+}] = 1.25 \times 10^{-3}$ M, and copolymer concentrations of **d** 20 μg mL^{-1}, **e** 50 μg mL^{-1}, **f** 125 μg mL^{-1}, **g** 250 μg mL^{-1}, **h** 500 μg mL – 1, **i** 20 μg mL^{-1}, **j** 50 μg mL^{-1}, **k** 125 μg mL^{-1}, **l** 250 μg mL^{-1}, and **m** 500 μg mL^{-1}. Reprinted from Kim et al. [67], under the terms and conditions of the Creative Commons CC BY license, https://creativecommons.org/about/cclicenses/

the hydrophilic group, –OH, of the additive alkanolamide 6502 was adsorbed on the surface of the intermediate vaterite. The hydrophobic group (long hydrocarbon chain) at the other end prevented the contact of the crystal nucleus with water, in the process inhibiting the dissolution of vaterite and hence, nullifying any chances of crystallization to the calcite phase.

Zou et al. performed experimental studies to recognize the influence of acidic macromolecules during the mineralization of $CaCO_3$ [70]. The experiments were performed at two different concentrations of the organic macromolecule poly(aspartic acid). They reported that at a low concentration of the additive, although the inhibition in nucleation and growth of both the calcite and vaterite phase occurred due to the negative charge on the additive, however, the inhibition was more pronounced for the vaterite phase; hence the calcite phase was more favored. At higher concentrations, the amorphous precursor (ACC) was stabilized for a longer duration which facilitated vaterite formation via the pseudomorphic transformation. Hence, the vaterite phase was favored over calcite at a higher concentration of the additive poly(aspartic acid). Figure 7 represents a schematic illustration of the role of poly(aspartic acid) during the crystallization of ACC in solution. In Table 2, we have summarized the role of all the additives in polymorphic control that have been cited in this section.

It is well known that the organic molecules act as additives and influence the crystallization process of $CaCO_3$. The –O containing groups in these molecules behave as coordination sites for Ca^{2+} attributed to the hard acid-hard base interactions which are responsible for the additive effect. However, differences in the crystal phases of the synthesized $CaCO_3$ and their properties are observed even if we use additives with the same functional group. As for the influence of alcohols on the polymorphism of $CaCO_3$, it is the –OH group that gets adsorbed onto the surface of $CaCO_3$ and dictates

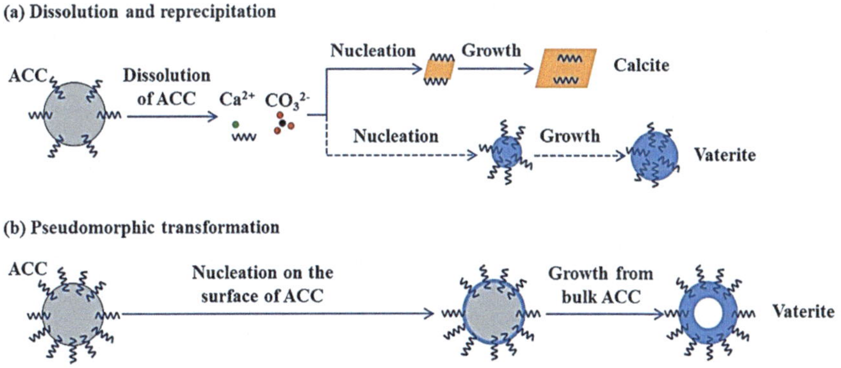

Fig. 7 Schematic illustration of ACC crystallization in solution. **a** In presence of a low concentration of poly(aspartic acid), calcite and vaterite form via a dissolution and reprecipitation process. **b** At a high concentration of poly(aspartic acid), stabilization of ACC for a longer time allows the vaterite formation via a pseudomorphic transformation process. Reprinted with permission from Zou et al. [70]. Copyright (2017), John Wiley and Sons

Table 2 Influence of additives on calcium carbonate phase and morphology

Additives	Phase	Morphology
Sr(II) and poly(acrylic acid) [56]	Calcite	Roundness in the rhombohedral calcite
Egg white and Cu^{2+} [57]	Vaterite	Spherical
Aspartic acid [58]	Calcite	Rhombohedral
Poly-lysine combined with electric field [60]	Calcite	Calcite with notches in the edges of the rhombohedra (at low concentration of additive) Calcite crystals elongated in the direction perpendicular to the substrate (at high concentration of additive) Cone-shaped (at very high concentration of additive)
Poly(styrene sulfonate) sodium and dextran sodium sulphate [21]	Calcite	Toroidal-like
N-stearoyl L-glutamic acid [61]	Calcite	Rhombohedral (crystal surfaces were rough which consisted of numerous nanoparticles)
Adipic acid [62]	Vaterite	Ellipsoidal and spherical
Whey protein [63]	Vaterite	Spherical
Curcumin [64] Quercetin [64]	Vaterite Calcite	Spherical Dumbbell-shaped
Tulsi [65] Green tea [65]	Vaterite, monohydrocalcite Calcite	Spherical Floral-shaped
Amino acids and dye molecules [66]	Calcite	Variation from the usual rhombohedral morphology
Poly(methacrylic acid) (PMAA) and Poly(benzyl methacrylate) (PBzMA) [67]	Calcite	Variation from the usual rhombohedral morphology
Co-polymer nanoparticles based on Poly(2-hydroxypropyl methacrylate) [68]	Calcite	Variation from the usual rhombohedral morphology
Alkanolamide 6502 [69]	Vaterite	Spherical
Poly(aspartic acid) [70]	Calcite (at low concentration of additive) Vaterite (at high concentration of additive)	Rhombohedral Spherical

the additive role during the crystallization. It has been observed that alcohols with a greater molecular weight than methanol exhibit stronger interaction with $CaCO_3$ than water. In this respect, as the hydrocarbon chain length increases, the interaction increases and accordingly, the product of crystallization will vary if not directly in terms of the phase; at least the properties will be different. It has been observed that as the hydrocarbon chain length increases, the hydrophobicity increases and prevents the crystal from further dissolution–recrystallization and renders hydrophobic properties to the crystal [69, 71]. Amino acids as additives in $CaCO_3$ formation have been studied extensively, and it has been observed that different amino acids result in the formation of crystals with different phases and morphologies, which is probably due to the fact that they vary in their role of action. Hou et al. studied the role of glycine and they reported the formation of vaterite, which takes place due to the lowering of activation energy of vaterite under the additive action [72]. They also reported that not just the phase but the morphology is also influenced by amino acids. The use of two different amino acids, glycine and aspartic acid resulted in a change in calcite morphology [73]. Aspartic acid being an acidic amino acid has two –COO– groups in a molecule which gives it the ability to influence two or more nuclei; on the other hand, glycine does not have this ability. Briegel et al. performed a study where a number of amino acids were subjected as additives during $CaCO_3$ formation [74]. The use of alanine resulted in calcite with the usual rhombohedral structure, cysteine gave vaterite with some calcite, leucine resulted in calcite with unusual structure and lysine resulted in calcite in accumulated form. They reported that it was possibly due to the difference in binding kinetics of the different amino acids with the nuclei that resulted in the formation of different polymorphs.

2.2.3 Role of Templates

The introduction of a substrate to influence the nucleation step during $CaCO_3$ crystallization is fundamentally what is referred to as heterogeneous nucleation. The substrate involved here is called the template and the effect induced by it is known as the templating effect. The process of $CaCO_3$ nucleation on templates is generally explained by three mechanisms: epitaxy, charge/polarization matching, and mutual templating [12].

In epitaxy, we generally talk about the complementarity in terms of the crystallographic structure of the nucleus and the template, where the crystallographic structure of the nucleus must fit within that of the substrate. This mechanism assumes the template to have a rigid structure and that the template and the forming crystal have a structural surface-repetitive, complimentary basis for pattern recognition [75]. Owing to these constraints, it becomes difficult to use this model to explain the nucleation process in all systems; hence the researchers have come up with other mechanisms. However, a close structural match between the template surface and the nucleating crystal enables a significant level of control over crystallization, where the growth of the crystal and eventually the overall morphology is controlled.

In the second mechanism, charge/polarization matching, the distribution of charges on the nucleus or the template is complemented by the opposite charge on the other. In addition to Coulomb charges, other interactions such as van der Waals, polarization, etc., are considered. As can be understood, the pre-requisite factors here determining the nucleation process are much more considerate of the interface structure as compared to the epitaxy mechanism. From the reported studies, we have examples where the charge/polarization matching gave the expected type of nucleus; however, there are instances where nucleation occurs, but the charge/polarization mechanism cannot be used to explain the process. As a result, another mechanism, the mutual templating mechanism, to comprehensively understand the templating process was developed.

In a mutual templating mechanism, the structure of the nucleus and the template mutually influence each other to achieve an optimum interface, both structurally and energetically. This phenomenon has been justified by molecular dynamics studies [76]. The experimental studies have shown that the growth rate is influenced by the flexibility of the substrate and low interface energy between the template and the nucleus promotes the growth rate [77].

2.2.4 Oriented Nucleation

So far, we have understood how the formation of a crystal ($CaCO_3$) takes place. In brief, the constituting ions assemble to form the pre-nucleation clusters, which lead to the formation of the nucleus upon further addition of ions, and the nucleus above the critical size further develops to generate a crystal. Therefore, it can be said that there are several steps leading to the generation of a matured $CaCO_3$ crystal. In the course of this whole process, the organic molecules and surfaces if introduced play a significant role where they influence the stabilization of a particular polymorph or a specific crystal face. This stabilization of a specific crystal face on a template surface is referred to as oriented nucleation and all the subsequently grown crystals of the mineral phase have this specific crystal plane oriented parallel to the template surface [12]. Therefore, describing the role of templates in $CaCO_3$ formation in a nutshell, we can say that the organic surface lent by the template during the crystallization process;

i. Encourages heterogeneous nucleation by lowering the surface free energy of the developing nuclei.
ii. Aids in the formation of a particular polymorph by stabilizing a specific crystal face.

3 Polymorph Control (Formation of Metastable Phases)

The factors responsible for the formation of metastable phases have already been discussed in detail in the previous sections regarding the mechanism of nucleation and crystal growth. The introduction of organic molecules (additives) or surfaces (templates) enables us to successfully isolate the metastable phase like spherical vaterite by terminating the growth of the crystals to the rhombohedral calcite. This is achieved by either of the two ways; the free energy of nucleation, ΔG_n, is lower for calcite than that of vaterite and under normal circumstances, continuous supersaturation will allow the nucleation of calcite, but the introduction of a suitable surface, as already discussed, raises the ΔG_n of calcite as compared to that of vaterite and ultimately facilitating the formation of the metastable vaterite [72, 78]. In a second way, the organic additives simply bind to specific crystal faces of the growing crystal post the nucleation process and prevent the crystal from further maturation. Similarly, these additives post binding to specific crystal faces directly control the morphology of the resulting crystal.

Here, we look at some of the recent studies on templating effects on $CaCO_3$ precipitation. Kanoje et al. studied the influence of template based on surfactant-polymer on the morphology of $CaCO_3$ [79]. The co-operative effect of the surfactant and the polymer contributed to the formation of different polymorphs and morphology. Three different polymers, poly(vinyl glycol), poly(vinyl alcohol), and poly(vinyl acetate) were taken and they were used individually in combination with the surfactant cetyltrimethylammonium bromide. As a result of the templating effect, spherical, bone-like, and cluster-like morphologies of $CaCO_3$ were obtained.

Abeywardena et al. studied the precipitation of $CaCO_3$ in which they extracted calcium from dolomite using sucrose solution [80]. In the next part of the study, they performed the crystallization of $CaCO_3$ using sucrate (Sucrose + OH^- = Sucrate$^-$) and sodium dodecyl sulphate individually and also in combination as the templates. The polymorphism exhibited as a result of varying pH and under the influence of the templates was reported. Previously, using the same source of calcium, dolomite, Somarathna et al. reported the role of polyethylene glycol templates on the morphology of $CaCO_3$; additionally, the role of temperature and pH in tandem with the templating-effect on the polymorphism was reported [81].

Rianasari et al. synthesized lamellar stacks of $CaCO_3$ single-crystal sheets under the influence of the template poly(acrylic) acid [82]. They had taken calcium acetate as the $CaCO_3$ precursor. The combination of the precursor and the template was chosen as such, based on the assumption that in the mixture of the two, Ca^{2+} and COO^- chelated complexes would be attached in a "net-like" structure, and the –OH groups would be bonded by H-bonding. The variation in the concentration of the precursor and the template (Ca^{2+}/COO^-) allowed the control in the morphology of $CaCO_3$ and produced molecular sheets of unit cell dimensions. Figure 8 schematically represents the formation of $CaCO_3$ molecular sheets using a two-step process.

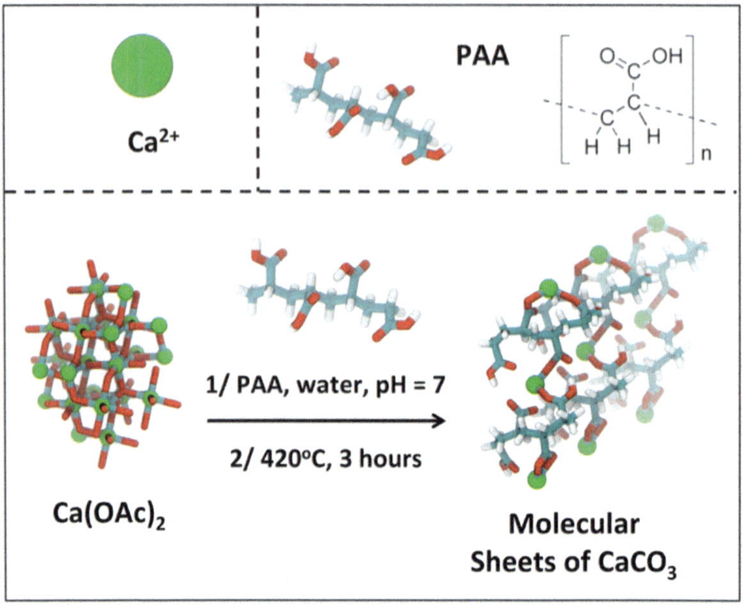

Fig. 8 Schematic representation of the molecular sheets of $CaCO_3$ formation using a two-step process. A nacre-like lamellar structure of molecular sheets of $CaCO_3$ interleaved with organic material, (amorphous carbon) is produced using a formulation of poly(acrylic) acid (PAA) and calcium acetate (Ca-Ac) using a chemical template technique and followed by calcination. Reprinted from Rianasari et al. [82] under the terms and conditions of the Creative Commons CC BY license. https://creativecommons.org/about/cclicenses/

Lopez-Berganza et al. studied the combined effects of additives and surface templates on $CaCO_3$ formation [83]. Two different poly-electrolyte films, poly(ethylene imine) and a two-layer film composed of poly(allylamine hydrochloride) and poly(acrylic acid), were subjected as templates and the organic additives used were amino acids and monosaccharides. The experimental observations followed by the kinetic studies on the precipitation process provided insights into understanding the mechanisms of mineralization, helping us to develop biomimetic models to understand biomineralization.

Different from the templates discussed thus far, which were either polymeric organic compounds or surfactants, by drawing a recent work performed by Wei et al. as an example, we can understand the role of biological cells as templates in the precipitation of $CaCO_3$ [84]. In this work, they subjected three types of microbial cells as templates; *Escherichia coli*, *Staphylococcus aureus*, and Beer Yeast. In one set of experiments, they used the cells directly as templates and in another set of experiments, they encapsulated the cells within polyelectrolyte and then used the encapsulated cells as the template. The precipitated $CaCO_3$ particles in the presence of the bare cells were a mixture of calcite and vaterite, with the ratio of the latter being higher. With the encapsulated cells, *Escherichia coli* and *Staphylococcus aureus* gave similar results, but with yeast, the ratio of calcite was more than that of vaterite. They reported with the justification that the difference in polymorphism observed was due to the differences in cell wall components of the cells used as templates.

Although both additives and templates are dictating factors that determine the formation of a particular polymorph during in vitro generation of $CaCO_3$, however, the difference lies in their mode of action during the various stages of crystal maturation. The additives, which are comprised of both organic and inorganic species, are believed to act during the crystal growth stage, whereby they function by either inhibiting the growth or by facilitating the growth along a particular crystal face. On the other hand, the template molecules behave as nucleators, directing the nucleation step during $CaCO_3$ formation (heterogeneous nucleation) and it is the surface chemistry of the template that induces oriented nucleation of the complementary crystal face.

4 Influence of Reaction Conditions During In Vitro $CaCO_3$ Formation

4.1 Influence of Temperature

The effect of reaction temperature on the formation of $CaCO_3$ in the absence of any additive has been studied and it was observed that the low temperatures, 20–30 °C, generally favor the formation of calcite. The high temperatures above 60 °C are suitable for the formation of aragonite and this can be explained by kinetic control due to higher growth rates of aragonite as compared to the other polymorphs. The metastable vaterite is usually stable at low temperatures, less than 10 °C [85, 86]. However, the introduction of additives during the bio-inspired mineralization of $CaCO_3$ at different temperatures influences the formation of polymorphs different from the additive-free mineralization. Paul et al. in their study, reported the specific formation of vaterite at 60 °C in the presence of the additive whey protein; they suggested that at a high temperature of 60 °C, β-lactoglobulin component of whey protein is broken down into smaller fragments [63]. Hence, a greater number of particles are available to influence the Ca^{2+} precipitation by adsorption on the $CaCO_3$ pre-clusters, influencing the formation of the least stable polymorph vaterite. The formation of calcite with morphologies different from the usual rhombohedral morphology was reported using the flavonoid quercetin, where at 10 °C, the particles had an ellipsoidal shape and at 80 °C, the particles exhibited a dumbbell shape, clearly suggesting the role of the additive during the mineralization of $CaCO_3$ [64].

4.2 Influence of Concentration

Prah et al. in their study, observed that it was the stable calcite phase that formed via the transformation of the amorphous precursor and the metastable vaterite and aragonite at low concentrations of the reactants [87]. However, at high concentrations, vaterite was stabilized, preventing its transformation to calcite. The concentrations of the reactants also play a crucial role in determining the size of the crystals. Calcium and carbonate ion concentrations are key parameters controlling particle size. Belkacemi et al. reported that when the concentration of Ca^{2+} is increased in comparison to CO_3^{2-}, the average size of the particles is decreased and the metastable phases like ACC and vaterite are preferably formed, whereas, when the concentration of CO_3^{2-} is more, the size of the particles increases and the formation of calcite are favored [88]. The supersaturation is directly dependent on the concentration ratio of the constituent ions involved in the reaction at any given precipitation time, $[Ca^{2+}]_t/[CO_3^{2-}]_t$. during the precipitation, a high concentration of Ca^{2+} involves a high value of R and, therefore, an increase in the supersaturation. Hence, the nucleation phase, corresponding to the formation of nuclei (small crystals), dominates the crystalline growth leading to the decrease of $CaCO_3$ particle size. Exactly the opposite mechanism dictates the formation of large crystalline particles when the concentration of CO_3^{2-} is more. Here, the value of R is lowered and consequently, the saturation level is lowered as a result, heterogeneous nucleation occurs where small crystallites form the core and aggregation of other crystals take place on them, leading to the formation of large-sized $CaCO_3$ particle.

In the synthesis involving additives, for a fixed concentration of the constituent ions Ca^{2+} and CO_3^{2-}, the concentration of additive used plays a crucial role in determining the phase, morphology, and mechanical properties of the resulting composites. For whey protein as an additive, it was observed that on lowering the concentration of the additive to 1.5 mg/mL, the percentage of formation of vaterite had gone down to ~70% as compared to ~90% when 2.5 mg/mL was used, the precipitation being performed under ambient conditions [63].

4.3 Influence of Ionic Strength

The effect of ionic strength on the yield of the polymorphs formed during $CaCO_3$ precipitation was studied by Tai et al. [89]. They used sodium chloride in order to vary the ionic strength of the solution and reported that with an increase in the ionic strength, the aragonite phase dominates the phase in conditions where the calcite phase is usually favored. However, they also made a statement based on their experimental findings that as compared to the other factors influencing the formation of $CaCO_3$, the ionic strength is not a very effective influencing factor. Paul et al. in their work on whey protein, studied the role of ionic strength and reported that in the presence of the additive whey protein and when the ionic strength of the solution was increased by the addition of 100 mM NaCl, the percentage of vaterite formed had slightly gone up to 93% as compared to the synthesis in the absence of NaCl where 90% vaterite was obtained [63]. In another work, they reported that when 100 mM NaCl was introduced to the reaction involving the additive curcumin, the percentage of vaterite had gone up to 98% as compared to 90% without NaCl [64].

5 Mechanical Properties

Amongst the several functions performed by the biominerals, skeletal support and protection stand out as the trademark feature of these biogenic materials [90]. In the invertebrates, it is the biomineral of our subject, calcium carbonate, which is responsible for these functions. Not just calcium carbonate but other biominerals such as calcium phosphates have shown such excellent mechanical properties that they can compete in terms of these properties with the synthetically engineered ceramic materials fabricated at drastic conditions of temperature and pressure. The complex and hierarchical structures of these composite biominerals are the essential features responsible for superior mechanical properties such as hardness and crack resistance [91–94]. As a result, material chemists have been fascinated and have always been in the quest of synthesizing materials that would match the properties of the biominerals such that they could be utilized for various applications [95]. To understand the mechanical properties of the biogenic composites, it is essential to take cognizance of the microstructure, properties, and morphologies of the individual constituents. Therefore, understanding the composition of the biominerals was of first-hand importance, whereby it was observed that in the organic–inorganic hybrid composites, the inorganic mineral phase is closely associated with organic molecules. These organic molecules were either present between the crystal units or embedded within single crystals, suggesting that the occlusion of generally very low levels of organic species contributed to the superior mechanical properties of the biominerals as compared to their synthetic inorganic counterparts [96].

The biogenic calcite single-crystal is reported to possess better mechanical strength as compared to the geologic calcite single-crystal available in the form of Iceland spar. Kunitake et al. performed a comparative study on the mechanical properties of biogenic calcite from the outer shell of the mollusk *Atrina rigida* and the geologic calcite from Iceland spar [97]. The shell of *Atrina rigida* is made up of two layers and each layer contains mineral crystals with neighboring organic matrices. The mineral in the inner layer is constituted by aragonite "tablets," and in the outer layer, it is calcite present as elongated single-crystal "prisms. Based on their study, they reported that these calcite prisms in *Atrina rigida* are harder than the geologic calcite in Iceland spar. They summarized that the difference in the hardness could be due to factors such as the occluded organic molecules, Mg^{2+}, residual defects arising out of the formation process, etc. Based on similar studies, which revealed the properties of the biogenic minerals and the possible reasons responsible for their superiority, the incorporation of organic molecules, especially with calcite single crystals has been a subject of considerable interest.

Kim et al. incorporated functionalized polystyrene within calcite single crystals and investigated the mechanical properties of the composite [98]. The extent of occlusion was >20 vol% and the modulus and stiffness of the resulting calcite composites were measured to be 4–5 units less and the hardness was 1.5 less than that of the Iceland spar. The real highlight of the incorporation was in terms of the improvement acquired in the fracture behavior attributed to the strong association of the polymer particles with the crystal. In another study, Kim et al. reported the role of incorporated amino acids on the hardness of the resulting calcite composite [99]. The two amino acids used were glycine and aspartic acid. The concentration of incorporation was varied and in addition to changes in morphology, a significant increase in the hardness of the crystals was observed. Calcite crystals with hardness almost twice as that of Iceland spar were fabricated.

6 Summary

In this chapter, we have discussed all the aspects related to $CaCO_3$ biomineralization and how this biogenic process is mimicked by material chemists to attain the biological standard. To understand the crystallization process, we have presented an insight study into the different stages of the crystal growth mechanism. To get a glimpse of the present-day scenario of bio-inspired synthesis research on $CaCO_3$, we have cited numerous recent examples of additives and templates used by researchers around the world. All possible types of additives, ranging from metal ions to low molecular weight organic compounds to high molecular weight species such as proteins and polymers, have been discussed in this chapter with an insight into their mode of action during the in vitro biomineralization of $CaCO_3$. This should provide the reader an ample amount of ideas to strategize their synthesis so as to attain the desired phase or morphology and also to generate $CaCO_3$ with enhanced mechanical properties.

References

1. L. Addadi, S. Weiner, A pavement of pearl. Nature **389**, 912 (1997)
2. M.D. Brasier, O.R. Green, A.P. Jephcoat, A.K. Kleppe, M.J. Van Kranendonk, J.F. Lindsay, A. Steele, N.V. Grassineau, Questioning the evidence for earth's oldest fossils. Nature **416**, 76 (2002)
3. A.-W. Xu, Y. Ma, H. Cölfen, Biomimetic mineralization. J. Mater. Chem. **17**, 415 (2007)
4. H.A. Lowenstam, S. Weiner, *On Biomineralization* (Oxford University Press, 1989)
5. S.I. Stupp, G.C. Mejicano, J.A. Hanson, Organoapatites: materials for artificial bone. II. Hardening reactions and properties. J. Biomed. Mater. Res. **27**, 289 (1993)
6. M. Viani, T.E. Scha, J.B. Thompson, N.A. Frederick, J. Kindt, P.K. Hansma, [Smith et Al 1999] Molecular mechanistic origin of the toughness of natural adhesive fibers and composites. **399**, 761 (1999)

7. S. Mann, Synthesis of barium sulfate from surfactant–inorganic nanoparticles reviews the chemistry of form. Angew. Chem. Int. Ed. **39**, 3392 (2000)
8. H. Du, U. Steiner, E. Amstad, Nacre-inspired hard and tough materials. Chimia (Aarau) **73**, 29 (2019)
9. P.U.P.A. Gilbert, M. Abrecht, B.H. Frazer, The organic-mineral interface in biominerals. Mol. Geomicrobiol. **59**, 157 (2018)
10. R.J.P. Williams, An introduction to biominerals and the role of organic molecules in their formation source. Philos. Trans. R. Soc. Lond. B **304**, 411 (1984)
11. L.B. Gower, Biomimetic model systems for investigating the amorphous precursor pathway and its role in biomineralization. Chem. Rev. **108**, 4551 (2008)
12. N.A.J.M. Sommerdijk, G. de With, Biomimetic $CaCO_3$ mineralization using designer molecules and interfaces. Chem. Rev. **108**, 4499 (2008)
13. K.K. Sand, C.S. Pedersen, S. Sjöberg, J.W. Nielsen, E. Makovicky, S.L.S. Stipp, Biomineralization: long-term effectiveness of polysaccharides on the growth and dissolution of calcite. Cryst. Growth Des. **14**, 5486 (2014)
14. J.W. Shen, C. Li, N.F.A. Van Der Vegt, C. Peter, Understanding the control of mineralization by polyelectrolyte additives: simulation of preferential binding to calcite surfaces. J. Phys. Chem. C **117**, 6904 (2013)
15. B.T. Kato, Polymer/calcium carbonate layered thin-film composites. Adv. Mater. **12**, 1543 (2010)
16. H. Du, E. Amstad, Water: how does it influence the $CaCO_3$ formation? Angew. Chem. Int. Ed. **59**, 1798 (2020)
17. L.B. Mao, H.L. Gao, H. Bin Yao, L. Liu, H. Cölfen, G. Liu, S. M. Chen, S.K. Li, Y.X. Yan, Y.Y. Liu, S.H. Yu, Synthetic nacre by predesigned matrix-directed mineralization. Science **354**, 107 (2016)
18. X.R. Xu, A.H. Cai, R. Liu, H.H. Pan, R.K. Tang, K. Cho, The roles of water and polyelectrolytes in the phase transformation of amorphous calcium carbonate. J. Cryst. Growth **310**, 3779 (2008)
19. D.B. Trushina, T.V. Bukreeva, M.V. Kovalchuk, M.N. Antipina, $CaCO_3$ vaterite microparticles for biomedical and personal care applications. Mater. Sci. Eng. C **45**, 644 (2014)
20. B. Cantaert, D. Kuo, S. Matsumura, T. Nishimura, T. Sakamoto, T. Kato, Use of amorphous calcium carbonate for the design of new materials. ChemPlusChem **82**, 107 (2017)
21. H. Bahrom, A.A. Goncharenko, L.I. Fatkhutdinova, O.O. Peltek, A.R. Muslimov, O.Y. Koval, I.E. Eliseev, A. Manchev, D. Gorin, I.I. Shishkin, R.E. Noskov, A.S. Timin, P. Ginzburg, M.V. Zyuzin, Controllable synthesis of calcium carbonate with different geometry: comprehensive analysis of particle formation, cellular uptake, and biocompatibility. ACS Sustain. Chem. Eng. **7**, 19142 (2019)
22. Y. Svenskaya, B. Parakhonskiy, A. Haase, V. Atkin, E. Lukyanets, D. Gorin, R. Antolini, Anticancer drug delivery system based on calcium carbonate particles loaded with a photosensitizer. Biophys. Chem. **182**, 11 (2013)
23. H. Yang, Y. Wang, T. Liang, Y. Deng, X. Qi, H. Jiang, Y. Wu, H. Gao, Hierarchical porous calcium carbonate microspheres as drug delivery vector. Prog. Nat. Sci. Mater. Int. **27**, 674 (2017)
24. S. Mofakhami, E. Salahinejad, Biphasic calcium phosphate microspheres in biomedical applications. J. Control. Release **338**, 527 (2021)
25. D. Paul, D. Sachan, G. Das, Silver nanoparticles embedded on in-vitro biomineralized vaterite: a highly efficient catalyst with enhanced catalytic activity towards 4-nitrophenol reduction. Mol. Catal. **504**, 111433 (2021)
26. D. Paul, G. Das, Efficient solid-state synthesis of biomineralized vaterite-derived pure $CaMnO_3$ perovskite for effective photocatalysis. CrystEngComm **23**, 4050 (2021)
27. L.A. Estroff, L. Addadi, S. Weiner, A.D. Hamilton, G.M. Whiteside, M. Science, W.Y. Investigator, Introduction: Biomineralization. **108**, 10 (2008)
28. J.W. Beck, R.L. Edwards, E. Ito, F.W. Taylor, J. Recy, F. Rougerie, P. Joannot, C. Henin, Sea-surface temperature from coral skeletal strontium/calcium ratios. Science **257**, 644 (1992)

29. Z. Zou, W.J.E.M. Habraken, G. Matveeva, A.C.S. Jensen, L. Bertinetti, M.A. Hood, C.-Y. Sun, P.U.P.A. Gilbert, I. Polishchuk, B. Pokroy, J. Mahamid, Y. Politi, S. Weiner, P. Werner, S. Bette, R. Dinnebier, U. Kolb, E. Zolotoyabko, P. Fratzl, A hydrated crystalline calcium carbonate phase: calcium carbonate hemihydrate. Science **363**, 396 (2019)
30. N.H. De Leeuw, S.C. Parker, Surface structure and morphology of calcium carbonate polymorphs calcite, aragonite, and vaterite: an atomistic approach. J. Phys. Chem. B **102**, 2914 (1998)
31. G. Wulff, XXV. Zur Frage Der Geschwindigkeit Des Wachsthums Und Der Auflösung Der Krystallflächen. Zeitschrift Für Krist. Cryst. Mater. **34**, 449 (1901)
32. B. Dickens, W.E. Brown, The crystal structure of calcium carbonate hexahydrate at ~ –120°. Inorg. Chem. **9**, 480 (1970)
33. H. Effenberger, Kristallstruktur und Infrarot-Absorptionsspektrum yon synthetischem Monohydrocalcit, $CaCO_3.H_2O$. Monatshefte fiir Chemic. **112**, 899 (1981)
34. D. Gebauer, A. Völkel, H. Cölfen, Stable prenucleation calcium carbonate clusters. Science **322**, 1819 (2008)
35. M.H. Nielsen, S. Aloni, J.J. De Yoreo, In situ TEM imaging of $CaCO_3$ nucleation reveals coexistence of direct and indirect pathways. Science **345**, 1158 (2014)
36. W.A. Deer, R.A. Howie, J. Zussman, *Introduction to the Rock Forming Minerals* (Longman, Harlow, UK, 1992)
37. E.S. Dana, *A Textbook of Mineralogy* (Wiley, New York, 1958)
38. F. Liendo, M. Arduino, F.A. Deorsola, S. Bensaid, Nucleation and growth kinetics of $CaCO_3$ crystals in the presence of foreign monovalent ions. J. Cryst. Growth **578**, 126406 (2022)
39. S. Mann, Molecular recognition in biomineralization. Nature **332**, 119 (1988)
40. H. Cölfen, Bio-inspired mineralization using hydrophilic polymers, in *Biomineralization II* (Springer Berlin Heidelberg, n.d.), pp. 1–77
41. N. Spanos, P.G. Koutsoukos, Kinetics of precipitation of calcium carbonate in alkaline PH at constant supersaturation. spontaneous and seeded growth. J. Phys. Chem. B **102**, 6679 (1998)
42. A. E. Nielsen, *Kinetics of Precipitation* (Oxford Pergamon Press, 1964)
43. X.Y. Liu, S.W. Lim, Templating and supersaturation-driven anti-templating: principles of biomineral architecture. J. Am. Chem. Soc. **125**, 888 (2003)
44. A.M. Travaille, E.G.A. Steijven, H. Meekes, H. Van Kempen, Thermodynamics of epitaxial calcite nucleation on self-assembled monolayers. J. Phys. Chem. B **109**, 5618 (2005)
45. P.C. Rieke, Application of Van Oss-Chaudhury-good theory of wettability to interpretation of interfacial free energies of heterogeneous nucleation. J. Cryst. Growth **182**, 472 (1997)
46. D. Gebauer, How Can Additives Control the Early Stages of Mineralisation? Minerals **8** (2018)
47. P. Bots, L.G. Benning, J.D. Rodriguez-Blanco, T. Roncal-Herrero, S. Shaw, Mechanistic insights into the crystallization of amorphous calcium carbonate (ACC). Cryst. Growth Des. **12**, 3806 (2012)
48. T. Ogino, T. Suzuki, K. Sawada, The rate and mechanism of polymorphic transformation of calcium carbonate in water. J. Cryst. Growth **100**, 159 (1990)
49. S.R. Dickinson, G.E. Henderson, K.M. McGrath, Controlling the kinetic versus thermodynamic crystallisation of calcium carbonate. J. Cryst. Growth **244**, 369 (2002)
50. S. Weiner, W. Traub, Macromolecules in mollusc shells and their functions in biomineralization. Philos. Trans. R. Soc. Lond. Ser. B, **304**, 425 (1984)
51. E. Weber, B. Pokroy, Intracrystalline inclusions within single crystalline hosts: from biomineralization to bio-inspired crystal growth. CrystEngComm **17**, 5873 (2015)
52. N. Watabe, Studies on shell formation. J. Ultrastruct. Res. **12**, 351 (1965)
53. F.C. Meldrum, S.T. Hyde, Morphological influence of magnesium and organic additives on the precipitation of calcite. J. Cryst. Growth **231**, 544 (2001)
54. H.H. Teng, P.M. Dove, C.A. Orme, J.J. De Yoreo, Thermodynamics of calcite growth: baseline for understanding biomineral formation. Science **282**, 724 (1998)
55. Y.-Y. Kim, C.L. Freeman, X. Gong, M.A. Levenstein, Y. Wang, A. Kulak, C. Anduix-Canto, P.A. Lee, S. Li, L. Chen, H.K. Christenson, F.C. Meldrum, The effect of additives on the early stages of growth of calcite single crystals. Angew. Chem. **129**, 12047 (2017)

56. H.Y. Kim, T. Yang, W. Huh, Y. Kwark, Y. Lee, I.W. Kim, The Combined effects of Sr(II) and poly (acrylic acid) on the morphology of calcite. Materials **12**, 3339 (2019)
57. X. Liu, K. Li, C. Wu, Z. Li, B. Wu, X. Duan, Y. Zhou, C. Pei, Influence of copper (II) on biomineralization of $CaCO_3$ and preparation of micron pearl-like biomimetic $CaCO_3$. Ceram. Int. **45**, 14354 (2019)
58. H. Du, C. Courrégelongue, J. Xto, A. Böhlen, M. Steinacher, C.N. Borca, T. Huthwelker, E. Amstad, Additives: their influence on the humidity- and pressure-induced crystallization of amorphous $CaCO_3$. Chem. Mater. **32**, 4282 (2020)
59. R. Stepić, L. Jurković, K. Klementyeva, M. Ukrainczyk, M. Gredičak, D.M. Smith, D. Kralj, A.S. Smith, Adsorption of aspartate derivatives to calcite surfaces in aqueous environment. Cryst. Growth Des. **20**, 2853 (2020)
60. N. Wada, N. Horiuchi, M. Nakamura, K. Nozaki, A. Nagai, K. Yamashita, Calcite crystallization on polarized single calcite crystal substrates in the presence of poly-lysine. Cryst. Growth Des. **18**, 872 (2018)
61. G. Zhu, S. Yao, H. Zhai, Z. Liu, Y. Li, H. Pan, R. Tang, Evolution from classical to non-classical aggregation-based crystal growth of calcite by organic additive control. Langmuir **32**, 8999 (2016)
62. R. Sun, T. Willhammar, E. Svensson Grape, M. Strømme, O. Cheung, Mesoscale transformation of amorphous calcium carbonate to porous vaterite microparticles with morphology control. Cryst. Growth Des. **19**, 5075 (2019)
63. D. Paul, S. Halder, G. Das, Whey protein directed in vitro vaterite biomineralization: influence of external parameters on phase transformation. Colloids Interface Sci. Commun. **36**, 100255 (2020)
64. D. Paul, G. Das, Curcumin and quercetin as templates in the in vitro biomineralization of $CaCO_3$: a comparative study on phase modulation. Ceram. Int. **47**, 12334 (2021)
65. D. Paul, D. Sachan, S. De, G. Das, Modulation of the $CaCO_3$ phase and morphology by tuning the sequence of addition: an insight into the formation of monohydrocalcite. New J. Chem. **45**, 18337 (2021)
66. B. Marzec, D.C. Green, M.A. Holden, A.S. Coté, J. Ihli, S. Khalid, A. Kulak, D. Walker, C. Tang, D.M. Duffy, Y.Y. Kim, F.C. Meldrum, Amino acid assisted incorporation of dye molecules within calcite crystals. Angew. Chem. Int. Ed. **57**, 8623 (2018)
67. Y.-Y. Kim, M. Semsarilar, J.D. Carloni, K.R. Cho, A.N. Kulak, I. Polishchuk, C.T. Hendley, P.J.M. Smeets, L.A. Fielding, B. Pokroy, C.C. Tang, L.A. Estroff, S.P. Baker, S.P. Armes, F.C. Meldrum, Structure and properties of nanocomposites formed by the occlusion of block copolymer worms and vesicles within calcite crystals. Adv. Funct. Mater. **26**, 1382 (2016)
68. Y.Y. Kim, L.A. Fielding, A.N. Kulak, O. Nahi, W. Mercer, E.R. Jones, S.P. Armes, F.C. Meldrum, Influence of the structure of block copolymer nanoparticles on the growth of calcium carbonate. Chem. Mater. **30**, 7091 (2018)
69. J. Wang, J. Song, Z. Ji, J. Liu, X. Guo, Y. Zhao, J. Yuan, The preparation of calcium carbonate with different morphologies under the effect of alkanolamide 6502. Colloids Surf. A Physicochem. Eng. Asp. **588** (2020)
70. Z. Zou, L. Bertinetti, Y. Politi, P. Fratzl, W.J.E.M. Habraken, Control of polymorph selection in amorphous calcium carbonate crystallization by poly(aspartic acid): two different mechanisms. Small **13**, 1 (2017)
71. D.V. Okhrimenko, J. Nissenbaum, M.P. Andersson, M.H.M. Olsson, S.L.S. Stipp, Energies of the adsorption of functional groups to calcium carbonate polymorphs: the importance of -OH and -COOH groups. Langmuir **29**, 11062 (2013)
72. W. Hou, Q. Feng, Morphology and formation mechanism of vaterite particles grown in glycine-containing aqueous solutions. Mater. Sci. Eng. C **26**, 644 (2006)
73. W. Hou, Q. Feng, Morphologies and growth model of biomimetic fabricated calcite crystals using amino acids and insoluble matrix membranes of *Mytilus edulis*. Cryst. Growth Des. **6**, 5 (2006)
74. C. Briegel, J. Seto, *Single Amino Acids as Additives Modulating $CaCO_3$ Mineralization*, Advanced Topics in Biomineralization (IntechOpen, London, 2012)

75. J.H. Van der Merwe, Misfit dislocations in epitaxy. Metall. Mater. Trans. A Phys. Metall. Mater. Sci. **33**, 2475 (2002)
76. D.M. Duffy, A.M. Travaille, H. Van Kempen, J.H. Harding, Effect of bicarbonate ions on the crystallization of calcite on self-assembled monolayers. J. Phys. Chem. B **109**, 5713 (2005)
77. S.J. Cooper, R.B. Sessions, S.D. Lubetkin, Role of interfacial tension in nucleation beneath monolayer films. Langmuir **13**, 7165 (1997)
78. H. Tong, W. Ma, L. Wang, P. Wan, J. Hu, L. Cao, Control over the crystal phase, shape, size and aggregation of calcium carbonate via a L-aspartic acid inducing process. Biomaterials **25**, 3923 (2004)
79. B. Kanoje, D. Patel, K. Kuperkar, Morphology modification in freshly precipitated calcium carbonate particles using surfactant-polymer template. Mater. Lett. **187**, 44 (2017)
80. M.R. Abeywardena, R.K.W.H.M.K. Elkaduwe, D.G.G.P. Karunarathne, H.M.T.G.A. Pitawala, R.M.G. Rajapakse, A. Manipura, M.M.M.G.P.G. Mantilaka, Surfactant assisted synthesis of precipitated calcium carbonate nanoparticles using dolomite: effect of PH on morphology and particle size. Adv. Powder Technol. **31**, 269 (2020)
81. Y.R. Somarathna, M.M.M.G.P.G. Mantilaka, D.G.G.P. Karunaratne, R.M.G. Rajapakse, H.M.T.G.A. Pitawala, K.G.U. Wijayantha, Synthesis of high purity calcium carbonate micro- and nano-structures on polyethylene glycol templates using dolomite. Cryst. Res. Technol. **51**, 207 (2016)
82. I. Rianasari, F. Benyettou, S.K. Sharma, T. Blanton, S. Kirmizialtin, R. Jagannathan, A chemical template for synthesis of molecular sheets of calcium carbonate. Sci. Rep. **6**, 1 (2016)
83. J.A. Lopez-Berganza, R.M. Espinosa-Marzal, Mechanistic approach to predict the combined effects of additives and surface templates on calcium carbonate mineralization. Cryst. Growth Des. **16**, 6186 (2016)
84. Y. Wei, H. Xu, S. Xu, H. Su, R. Sun, D. Huang, L. Zhao, Y. Hu, K. Wang, X. Lian, Synthesis and characterization of calcium carbonate on three kinds of microbial cells templates. J. Cryst. Growth **547**, 125755 (2020)
85. K.Y. Chong, C.H. Chia, S. Zakaria, Polymorphs calcium carbonate on temperature reaction, in *AIP Conference Proceedings*, vol. 1614 (2014), p. 52
86. E.M. Flaten, M. Seiersten, J.P. Andreassen, Polymorphism and morphology of calcium carbonate precipitated in mixed solvents of ethylene glycol and water. J. Cryst. Growth **311**, 3533 (2009)
87. J. Prah, J. Macek, G. Drazic, Precipitation of calcium carbonate from a calcium acetate and ammonium carbamate batch system. J. Cryst. Growth **324**, 229 (2011)
88. R. Babou-Kammoe, S. Hamoudi, F. Larachi, K. Belkacemi, Synthesis of $CaCO_3$ nanoparticles by controlled precipitation of saturated carbonate and calcium nitrate aqueous solutions. Can. J. Chem. Eng. **90**, 26 (2012)
89. C.Y. Tai, F.B. Chen, Polymorphism of $CaCO_3$ precipitated in a constant-composition environment. AIChE J. **44**, 1790 (1998)
90. F.C. Meldrum, H. Cölfen, Controlling mineral morphologies and structures in biological and synthetic systems. Chem. Rev. **108**, 4332 (2008)
91. S. Weiner, L. Addadi, H.D. Wagner, Materials design in biology. Mater. Sci. Eng. C **11**, 1 (2000)
92. M.A. Meyers, P.Y. Chen, A.Y.M. Lin, Y. Seki, Biological materials: structure and mechanical properties. Prog. Mater. Sci. **53**, 1 (2008)
93. S. Kamat, X. Su, R. Ballarini, A.H. Heuer, Structural basis for the fracture toughness of the shell of the conch *Strombus gigas*. Nature **405**, 1036 (2000)
94. J. Aizenberg, J.C. Weaver, M.S. Thanawala, V.C. Sundar, D.E. Morse, P. Fratzl, Materials science: skeleton of Euplectella Sp.: structural hierarchy from the nanoscale to the macroscale. Science **309**, 275 (2005)
95. S. Weiner, L. Addadi, Design strategies in mineralized biological materials. J. Mater. Chem. **7**, 689 (1997)
96. A. Berman, J. Hanson, L. Leiserowitz, T.F. Koetzle, S. Weiner, L. Addadi, Biological control of crystal texture: a widespread strategy for adapting crystal properties to function. Science **259**, 776 (1993)

97. M.E. Kunitake, L.M. Mangano, J.M. Peloquin, S.P. Baker, L.A. Estroff, Evaluation of strengthening mechanisms in calcite single crystals from mollusk shells. Acta Biomater. **9**, 5353 (2013)
98. Y.Y. Kim, L. Ribeiro, F. Maillot, O. Ward, S.J. Eichhorn, F.C. Meldrum, Bio-inspired synthesis and mechanical properties of calcite-polymer particle composites. Adv. Mater. **22**, 2082 (2010)
99. Y.Y. Kim, J.D. Carloni, B. Demarchi, D. Sparks, D.G. Reid, M.E. Kunitake, C.C. Tang, M.J. Duer, C.L. Freeman, B. Pokroy, K. Penkman, J.H. Harding, L.A. Estroff, S.P. Baker, F.C. Meldrum, Tuning hardness in calcite by incorporation of amino acids. Nat. Mater. **15**, 903 (2016)

Mineralization of Low-Carbon Bio-carbonate, Bio-phosphate, and Bio-composite Cements

Xiaoniu Yu and Qiyong Zhang

Abstract Ordinary Portland cement (OPC) emits a lot of carbon dioxide during the production process. The production of 1 ton of OPC emits about 0.8 tons of carbon dioxide. Microbial cement emits no or a small amount of carbon dioxide and belongs to low-carbon bio-cement. Hence, three types of bio-cements are prepared to partially replace OPC. Growth curves of carbonate-mineralization microbe (CMM) and phosphate-mineralization microbe (PMM) show that the breeding process of microbes becomes relatively stable and slow after 20 h. Size and morphology of bio-carbonate and bio-phosphate cements can be regulated by pure water, bacterial body, bacterial secretion, and bacterial solution. Ammonia/ammonium (NH_3/NH_4^+) can well be converted into environment-friendly carbonate and struvite precipitation when HPO_4^{2-} and Mg^{2+} are contained in the mixture solution of CMM and urea in the production process of the bio-composite cement. The shape of bio-$CaCO_3$, bio-$BaHPO_4$, and bio-composite cement crystals are mainly spherical, dumbbell, and irregular block structures with different sizes in bacterial solution, respectively. However, the shape of their crystals between sand particles is mainly ball, dumbbell, and irregular block cluster structures with different sizes, respectively. Therefore, the morphology of these crystals in bacterial solution is different from that in sand environment. Bacterial solution and sand particle play an important role in the nucleation, growth, and accumulation of crystals for bio-$CaCO_3$, bio-$BaHPO_4$, and bio-composite cements. Pores of loose sand columns can well be filled by bio-$CaCO_3$, bio-$BaHPO_4$, and bio-composite cements.

Keywords Carbonate-mineralization microbe · Phosphate-mineralization microbe · Bio-$CaCO_3$ · Bio-$BaHPO_4$ · Bio-composite cement

X. Yu
Jiangsu Key Laboratory of Construction Materials, Southeast University, Nanjing 211189, China
e-mail: xnyu09@163.com

Q. Zhang (✉)
College of Engineering and Technology, Southwest University, Chongqing 400715, China
e-mail: qyzhang22@swu.edu.cn

S. J. Ikhmayies (ed.), *Advances in Minerals Research*, Advances in Material Research and Technology, https://doi.org/10.1007/978-3-031-49175-7_8

1 Introduction

Minerals are important components for living systems, and they are found in all kinds of bacterium, plants, and animals, in which these minerals serve a multitude of crucial biochemical and structural functions. Structure, morphology, and size of mineral materials can be regulated by organic matrix such as polysaccharides, proteins, proteolipids, and proteoglycans. These organic matrices can control the nucleation, growth, and orientation of minerals in biological environments [1]. The biogenic mineral, which has the special multistage structure, is synthesized by organic and inorganic interactions in the interphase [2]. The orientation of minerals is adjusted by the above mineralization process. In addition, three-dimensional media such as metabolites and bacterial body have been applied as the template of the ingenious morphology for mineral particles in the bacterial solution. When bacterial body and metabolites are used as a template, the biogenic minerals are finally formed through the processes of molecular preassembly, interphase molecular recognition, growth modulation, and cell processing. The bacterial solution and biomacromolecules have been used as novel additives to adjust the morphology of mineral materials, such as $CaCO_3$ [3–6], $BaCO_3$ [7], $BaHPO_4$ [8, 9], $LiFePO_4/C$ [10], CdS [11], and Ag_2S [12].

In the process of biomineralization to produce cement, there is no CO_2 emission or a small amount of CO_2 emission, and it is a kind of low-carbon bio-cement. The CO_2 concentration will continue to increase in the global atmosphere because of the continued use of fossil fuels and cement production, and today its concentration is above 400 ppm [13]. The CO_2 is the main greenhouse gas compared with water vapor, nitrous oxide, methane, and other gases, which contributes 60% to the greenhouse effect according to literature reported [13]. Reducing CO_2 emissions can effectively reduce the greenhouse effect, the frequency, and intensity of climate disasters, and it can keep the safety of people's life and improve its properties [13]. It is necessary to achieve global carbon emission reduction through low-carbon emission technologies.

When sand particles are used as a template, the biomineralization process occurs in sand surface due to products of bio-phosphate, bio-carbonate, and bio-composite cements. Finally, loose sand particles are cemented by bio-carbonate, bio-phosphate, and bio-composite cements into a whole with the growth of bio-cement crystals. Bio-mineralization is broadly divided into bio-augmentation and bio-stimulation. The micro-environment conditions of bio-mineralization are often influenced by the bacterial metabolism and organic matrix. Bio-mineralization can occur in the solution and sand environment. The carbonate-mineralization microbe (CMM) and phosphate-mineralization microbe (PMM) are often selected as mineralizing bacteria due to the following four reasons: (1) the strain should be environmentally friendly; (2) the heritability of the strain should be good, and the strain can be stored for a long time; (3) enzymes that hydrolyze substrates can be produced by the strain; (4) the strains can survive in adverse environments. In this chapter, different types of low-carbon bio-cement can be prepared with gelling properties. Its morphology and size can be regulated by bacterial liquid, bacterial body, secretion, solvent, sandy soil,

Table 1 Medium composition of CMM

Nutrients	Mass (g)	Main effects
Yeast	15.0 or 20.0	Supply C and N
NH_4Cl/ NH_4SO_4 $NiCl_2$	10.0 0.02	Induction of urease Composition of urease
Deionized water	1000.0	Solvent

and so on. These low-carbon bio-cements are well bind to loose sandy-soil particles with a certain mechanical strength.

2 Methodology

2.1 Preparation Method of CMM

The composition of the growth medium for CMM is shown in Table 1. Figure 1 shows the schematic diagram of the cultivation process of CMM. The culture medium of *Sporosarcina pasteurii* (1 L, pH = 9) contains 15 or 20 g of yeast extract powder, 10 g of ammonium chloride, and 0.02 g of nickel chloride. The growth medium is subjected to the autoclave process for sterilization at 121 °C for 25 min. After cooling, 0.5–1% v/v of strains are inoculated to 1 L of the medium in a sterile fume hood and then grown at a temperature of 25–3 0 °C in an orbital shaker with a speed of 170–260 rad/min. After 24–72 h of the cultivation, the cultivated microorganisms are harvested at the stationary phase and stored in a refrigerator for 0–3 months.

2.2 Preparation Method of PMM

The medium of PMM (1 L, pH = 7) contains 5 g of peptone and 3 g of beef extract, then placed in an autoclave and sterilized at 121 °C for 25 min, as shown in Table 2. The medium solution becomes turbid after the cultivation, the harvested PMM also needs to be stored in the refrigerator and used within 3 months.

2.3 Preparation of Low-Carbon Bio-carbonate, Bio-phosphate, and Bio-composite Cements

Usually, CMM can hydrolyze urea to form carbonate and ammonium ions. PMM can hydrolyze organophosphate monoester (ROP) to form phosphate ions. Different morphologies and size of low-carbon bio-carbonate, bio-phosphate, and

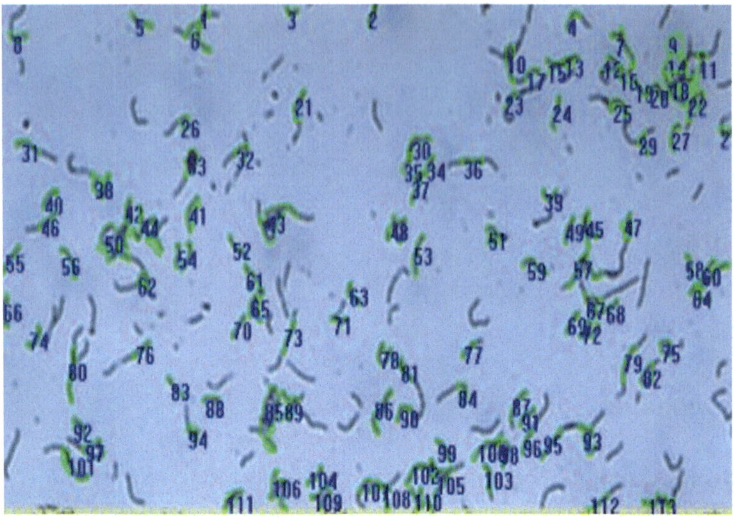

Fig. 1 Morphology of CMM [14]

Table 2 Medium composition of PMM

Nutrients	Mass (g)	Main effects
Peptone	5.0	Supply C and N
Beef extract	3.0	Supply C and N
Deionized water	1000.0	Solvent

bio-composite cements can be prepared by carbonate, phosphate, and ammonium ions reacting with calcium, barium, and magnesium/hydrogen phosphate ions respectively in bacterial liquid, bacterial body, secretion, water, sandy-soil and so on. These biocements can be poured into or mixed with loose sandy soil particles, and then they are cemented to a whole with a certain mechanical strength according to literatures reported.

3 The Mineralization of Bio-carbonate Cement

3.1 Bio-mineralization Mechanism

Figure 1 shows that CMM is a rod-shaped structure [14]. The urease produced by CMM only needs 1 microsecond to hydrolyze urea, while urea is hydrolyzed for about 3.6 years in the natural environment. Hydrolysis rate of urea in the urease is the 10^{14} times of natural condition [15]. Figure 2 shows the hydrolysis and mineralization mechanisms of urea catalyzed by urease [16]. First, water molecules coordinated

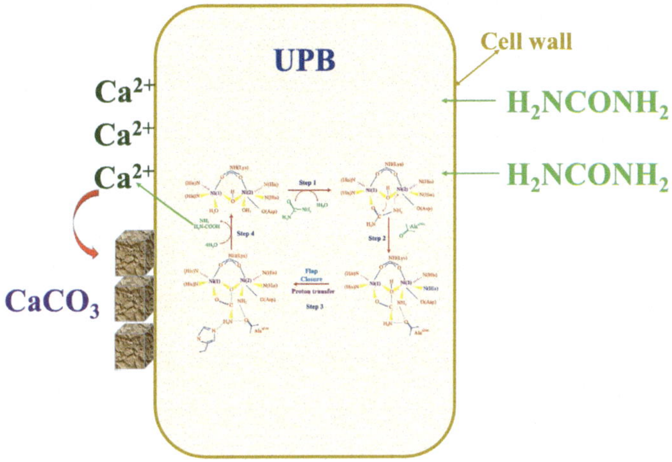

Fig. 2 Schematic diagram of preparation of bio-minerals

with Ni(1) are attacked by O atom of urea, and three water molecules are fixed. Then, the urea bridges Ni(1) and Ni(2) atoms, and a conjugated structure (OCO) is formed via hydroxyl oxygen attacking C = O. Next, the proton transfer and structural transformation occur in the urease. The urease-urea complex is unstable, and it is converted to urease, NH_3 and NH_2COOH. Finally, the unstable NH_2COOH can quickly be hydrolyzed into NH_3/NH_4^+ and CO_3^{2-} in the solution. Metal ions may react with CO_3^{2-}, which causes the precipitation of MCO_3, $M_2(CO_3)_3$, etc.

3.2 *Bio-mineralization in Solution*

X-ray diffraction (XRD) pattern confirms that the main components of the precipitates are calcite (PDF Card No. 86-2334) and vaterite (PDF Card No. 72-0506), as shown in Fig. 3 [17]. The densities of calcite and vaterite are 2.71 and 2.60 g/cm^3 according to the XRD index results, respectively. Bio-$CaCO_3$ crystals show different shapes and sizes when they are obtained under different reaction conditions. For the conditions of pure water and bacterial body, bio-$CaCO_3$ crystals are block structure and their size is in the range of 10–30 and 5–10 μm, respectively. The particles with bacteria imprint can be observed in the bacterial body. The spherical structure in bacterial secretion is observed, and its size is about 1–10 μm. However, the shape of bio-$CaCO_3$ crystals is block and spherical structure (10–100 μm) when it is precipitated in bacterial solution (Fig. 4) [17]. XRD patterns of seawater-based bio-carbonate cements show that the mixture of calcite magnesium, monohydrocalcite, and calcium carbonate is found in precipitates when the pH of the seawater-based urease-produced bacteria (UPB) is 8.01, as shown in Fig. 5 [18]. The particles with the bacteria imprint are also observed when the mixture of calcite magnesium and

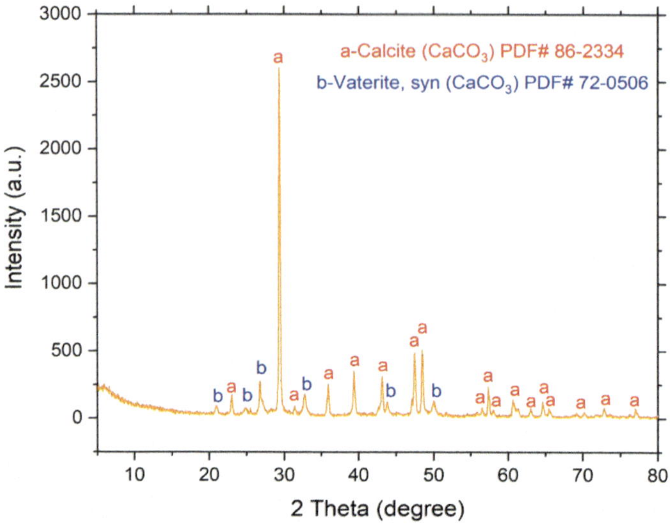

Fig. 3 XRD pattern of bio-$CaCO_3$ crystals [17]

Fig. 4 SEM images of bio-$CaCO_3$ crystals [17]

calcite precipitate at pH = 6.04 (Fig. 5) [18]. Scanning electron microscope (SEM) images show the morphology of seawater-based bio-carbonate cement which precipitates at seawater-based UPB with 6.04 pH (a) and 8.01 pH (b) ((a) pH = 6.04; (b) pH = 8.01), as shown in Fig. 6 [18]. The morphology of particles for the seawater-based bio-carbonate cement is mainly irregular spherical, and particles size is between 10 and 100 μm. Therefore, the pH of the seawater-based UPB cannot affect the shape and size of bio-carbonate particles.

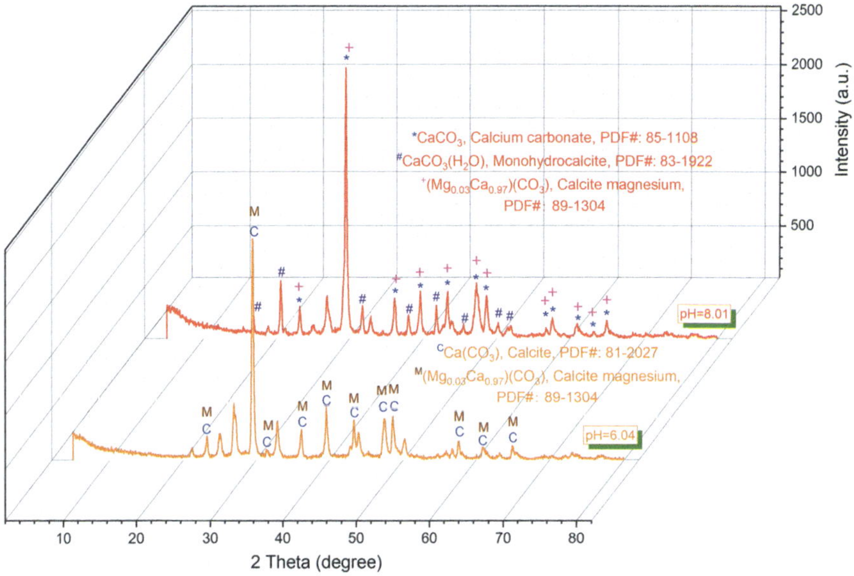

Fig. 5 XRD patterns of seawater-based biocarbonate cements [18]

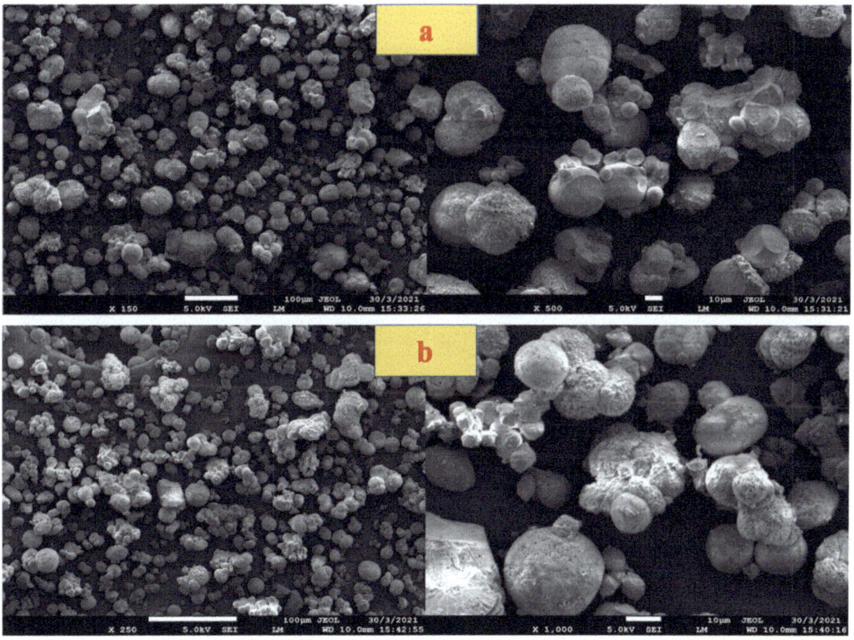

Fig. 6 Seawater-based UPB induced precipitation of carbonate (**a** pH = 6.04; **b** pH = 8.01) [18]

3.3 Bio-mineralization Between Loose Sand Particles

The mineralization process of bio-$CaCO_3$ cement between loose sand particles can be observed at different times [14]. Many bacterial bodies can be adsorbed on the surface of sand particles when sand particles are immersed in bacterial solution. The pores in sand can be filled by bio-$CaCO_3$ when the bacterial solution and the mixture solution of urea and $CaCl_2$ are grouted into sand column. The precipitation of $CaCO_3$ increases with the increase in the time, and loose sand particles are cemented by these precipitates before 5.0 h [19]. After that, the change in the effect of cementation is not significant, which may be caused by the lack of bacterial solution or cementation solution [19]. Therefore, the size and number of bio-$CaCO_3$ crystals do not change after 5.0 h. SEM images show that loose sand columns can be cemented by the bio-$CaCO_3$ with the strength, and the morphology of crystals is the sphere clusters with a size of 1–10 μm, as shown in Fig. 7.

SEM images indicate that pores between sand particles can well be filled by seawater-based bio-carbonate cement, as shown in Fig. 8 [18]. Due to the precipitation of the bio-$CaCO_3$, the size of pores decreases with the increase of cementation times. The morphology of particles for the bio-carbonate cement is mainly irregular spherical, and the size of particles is between 10 and 100 μm. When cementation times are 12, pores are almost completely filled. Therefore, the content of the

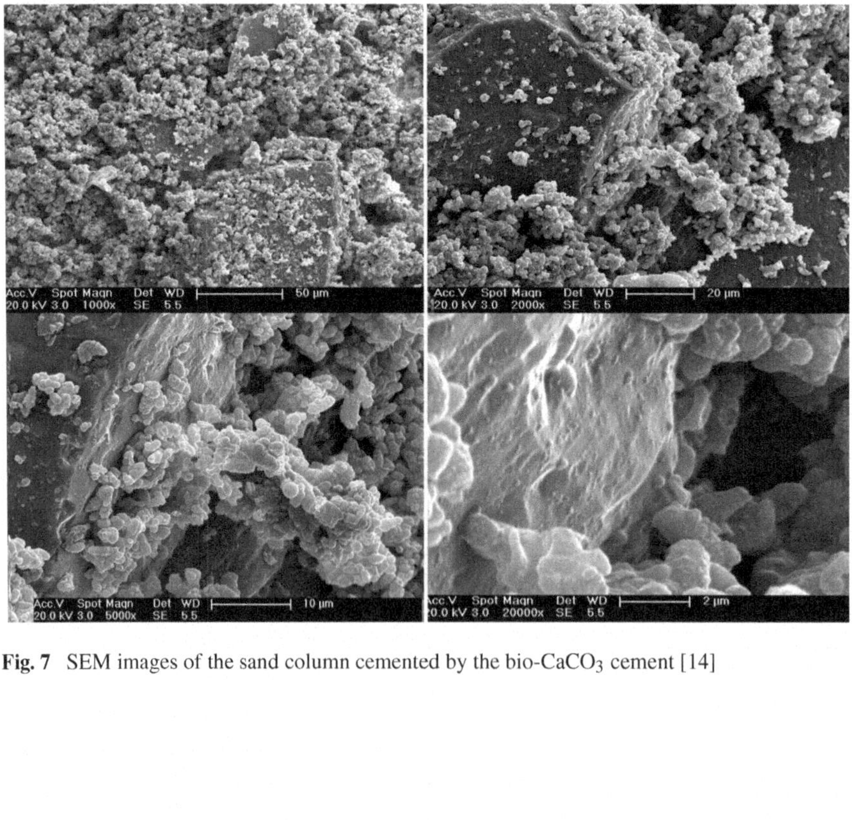

Fig. 7 SEM images of the sand column cemented by the bio-$CaCO_3$ cement [14]

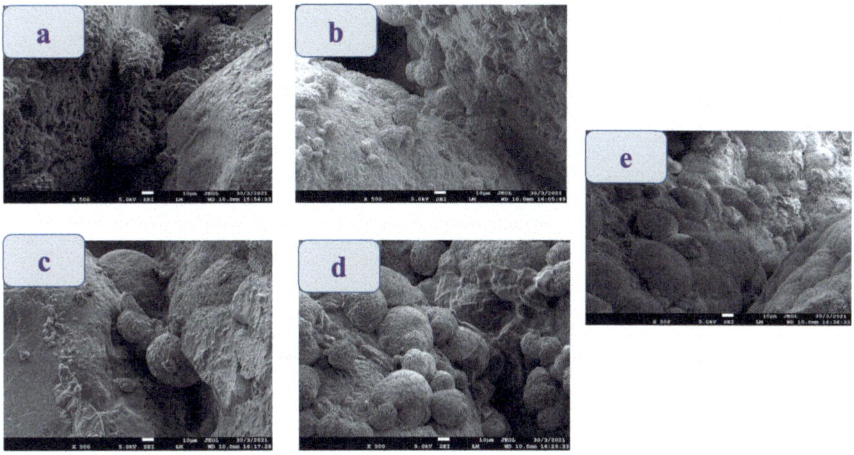

Fig. 8 SEM images of bio-blocks with different number of injections: **a** 1, **b** 3, **c** 6, **d** 9, **e** 12 [18]

bio-carbonate cement is largest, and the unconfined compressive strength is also highest.

4 The Mineralization of Bio-phosphate Cements

4.1 Bio-mineralization Mechanism of PMM

Figure 9 displays that the PMM is a rod-shaped structure [20]. The hydrolysis mechanism for the alkaline phosphatase (ALP) is indicated in Fig. 10 [20]. In the free ALP, the phosphate-binding site can be filled by three water (H_2O) molecules [21]. The hydroxyl group (OH) of the Ser102 can participate in a hydrogen bond with an Mg-coordinated hydroxide ion [21]. The ALP-substrate complex (ALP-ROP) is formed when two oxygen atoms of ROP coordinate with Zn(1) and Zn(2) [21]. The O of Ser102 attacks the P of ROP, and then the RO^- is lost in the ALP-ROP. The covalent of ALP-phosphate intermediate (ALP-P) can be formed due to the inversion of the phosphorus center. A nucleophilic hydroxide ion (OH^-) is coordinated with Zn1, and it attacks the covalent of ALP-P to form the non-covalent of the ALP-phosphate complex (ALP-P_i), which causes a second inversion of configuration at the phosphorus center [21]. The Mg is coordinated with H_2O, which can donate a proton to the O^r in the Ser102, and then the PO_4^{3-} is obtained. Finally, metal ions may react with PO_4^{3-}, which causes the precipitation of $MHPO_4$, $M_3(PO_4)_2$, MPO_4, etc.

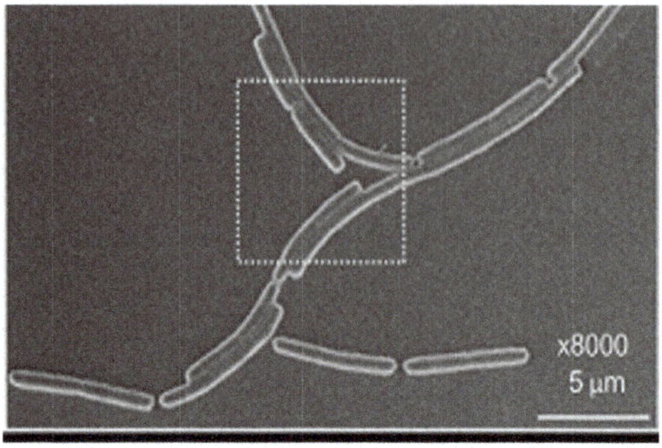

Fig. 9 Shape of PMM [20]

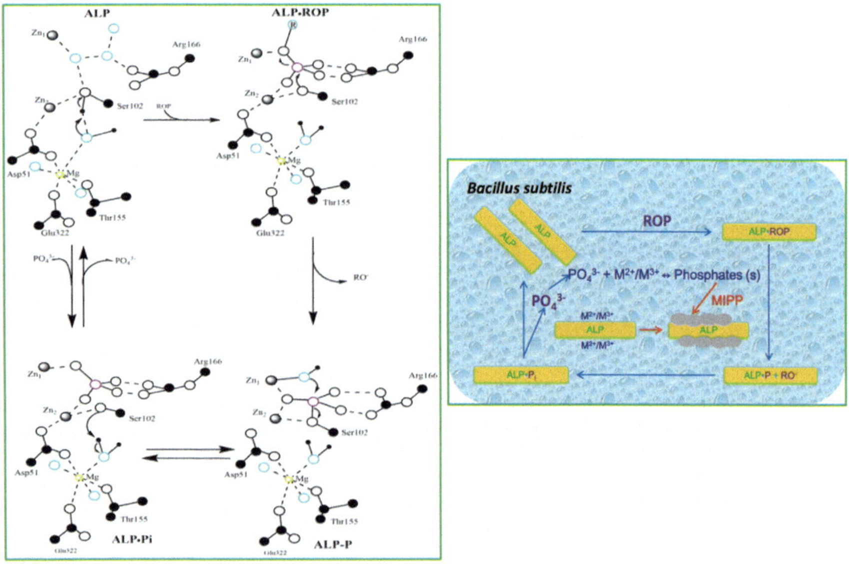

Fig. 10 The hydrolysis mechanism of ALP [20]

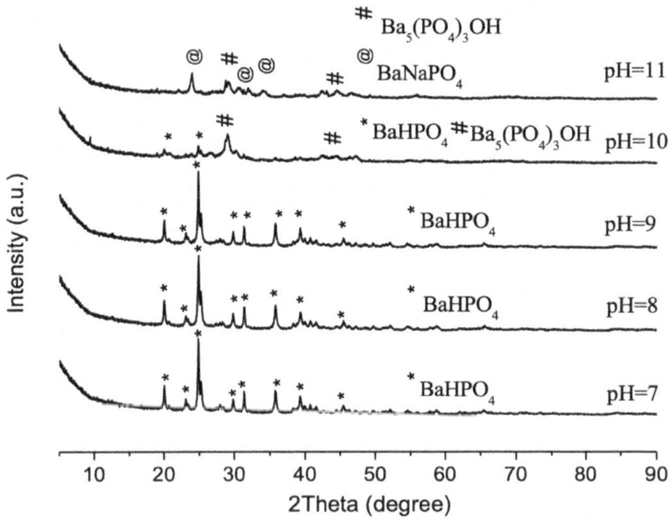

Fig. 11 XRD patterns of barium phosphates at different pH conditions [20]

4.2 Bio-mineralization of Biological Barium Phosphate Cement

4.2.1 Structure of Biological Barium Phosphates Generated at Different pH

XRD analysis is performed on the barium phosphate synthesized at different pH, as shown in Fig. 11 [20]. At pH < 10, diffraction peaks of samples show that crystalline compounds are mainly $BaHPO_4$. When the pH is 7, 8, and 9, corresponding standard cards of $BaHPO_4$ synthesized by the biomineralization are JCPDS Nos. 17-0929, 17-0929, and 72-1370, respectively. However, at pH = 10, the surface of $BaHPO_4$ crystals such as (034) disappears, and the barium hydroxyphosphate ($Ba_5(PO_4)_3OH$) including (112) and (211) appears. The above results display that the chemical component of precipitates is a mixture of $BaHPO_4$ and $Ba_5(PO_4)_3OH$. XRD patterns show that the above $BaHPO_4$ and $Ba_5(PO_4)_3OH$ correspond to standard JCPDS Nos. 72-1370 and 78-1141, respectively. When the pH is 11, the main component of the precipitate is a mixture of $Ba_5(PO_4)_3OH$ and $NaBaPO_4$, and corresponding standard cards are JCPDS Nos. 78-1141 and 33-1210, respectively.

4.2.2 Distribution Coefficient of Phosphate Ions at Different pH

The ion distribution coefficient (δ) can explain the formation of barium phosphate at different pH conditions [20, 22]. The δ of phosphate ions is related to the concentration of hydrogen ion (H^+). The δ of phosphate ions at different pH conditions can be calculated by the distribution coefficient equation. The distribution coefficient of $[HPO_4^{2-}]$ (δ_2) is larger than that of the distribution coefficient of $[PO_4^{3-}]$ (δ_3) when the pH is 7, 8, and 9, so precipitates are mainly $BaHPO_4$. When the pH is 10 and 11, the δ_3 is smaller than that of δ_2, which can be ignored according to the distribution coefficient equation. However, when the pH is 10 and 11, XRD results show that phosphate compounds are formed, so the δ_3 cannot be ignored. The differences in product types are mainly attributed to the pH of the solution after the reaction of the PMM with the substrate and the weakly acidic $BaCl_2 \cdot 2H_2O$. For example, when the pH is less than 10, all PO_4^{3-} ions can be converted to HPO_4^{2-} ions. At pH = 10, some PO_4^{3-} ions can be converted into HPO_4^{2-} ions. When the pH exceeds 10, all PO_4^{3-} ions cannot be converted to HPO_4^{2-} ions. Therefore, the PMM produces ALP during the growth process, which can hydrolyze substrates to produce PO_4^{3-} ions.

In a word, when the pH of the mixed solution after the reaction between the PMM and the substrate is adjusted to 7, 8 and 9, the chemical component of the mineralized product is $BaHPO_4$. When the pH is 10, OH^-, HPO_4^{2-} and PO_4^{3-} in the mixed solution react with Ba^{2+} to form a mixture of $BaHPO_4$ and $Ba_5(PO_4)_3OH$. However, when the pH is 11, OH^- and PO_4^{3-} ions in the mixed solution react with Ba^{2+} and Na^+ to form a mixture of $Ba_5(PO_4)_3OH$ and $NaBaPO_4$.

4.2.3 Morphology and Size of Barium Phosphate Particles at Different pH Conditions

At pH = 7, the $BaHPO_4$ crystal is an irregular structure, and the width and length of the particles are less than 20 μm [22]. When the pH is 8, the $BaHPO_4$ crystal is spherical and irregular, and the particle size changes from 1 to 20 μm [22]. At pH = 9, the $BaHPO_4$ crystal is a dumbbell-shaped structure, and the width and length of the particles are between 1 and 10 μm [22]. When the pH is 10, mixed crystal of $BaHPO_4$ and $Ba_5(PO_4)_3OH$ is an irregular spherical structure, and the particle size changes from 1 to 15 μm [22]. At pH = 11, the mixed crystal of $NaBaPO_4$ and $Ba_5(PO_4)_3OH$ is an irregular block structure, and the particle size changes from 1 to 30 μm [22].

4.2.4 Morphology of Barium Phosphate Particles at Different Solution

SEM images indicate that the shape and size of $BaHPO_4$ particles obtained at different reaction conditions [23]. The shape of $BaHPO_4$ is irregular at different reaction conditions. The dumbbell-shaped structure with a smooth surface can be observed in BHP-a particles [23]. The length of dumbbell is approximately 1.5 times the width. In the SEM images of BHP-b particles, aggregate clusters are found in specimens [23]. BHP-c particles shape is mainly the oval structure [23]. The length of particles is about equal to twice the width. However, when Na_2HPO_4 is dissolved in the bacterial solution, the spherical structures with some irregular shapes occur. The size distribution of spherical particles is non-uniform, and the diameter changes from 5 to 20 μm. In the bacterial secretion, the size range of irregular spherical particles is between 5 and 20 μm. The shape and size are similar with BHP-d. The irregular quadrilateral box is observed in the bacterial body, and its length is approximately the twice width.

4.2.5 Morphology of the Bio-$BaHPO_4$ with Different Standing Time

Yu et al. [24] showed SEM images of the bio-$BaHPO_4$ stayed in the mixture of PMM and substrate for 0, 12, 24, and 36 h. At 0 h, the shape of the bio-$BaHPO_4$ particles is an irregular structure, and the particle size is mainly distributed in 2–20 μm. At 12 h, the morphology of particles is mainly dumbbell-shaped and irregular spherical structures, and the particle size changes from 10 to 100 μm. When the bio-$BaHPO_4$ is stayed for 24 h, the shape of particles is mainly irregular spherical and dumbbell-shaped structures, and the particle size is mainly distributed in 5–50 μm. At 36 h, the main morphology of the bio-$BaHPO_4$ is also irregular spherical and dumbbell-shaped structures, and the particle size changes from 5 to 30 μm. In conclusion, the products at 0, 12, 24, and 36 h are irregular particles, irregular spherical aggregates, irregular spherical and dumbbell-shaped aggregates, and irregular spherical shapes, respectively. The particles size distribution is wide.

4.2.6 Bio-mineralization of the Bio-$BaHPO_4$ Between Loose Sand Particles

Figure 12 shows SEM images of the bio-$BaHPO_4$ mixed with sand at different standing times and dosages [25]. Particle sizes of the bio-$BaHPO_4$ in Fig. 12a, b are significantly smaller than those in Fig. 12c. For a given dosage, the number of the bio-$BaHPO_4$ decreases with the increase in standing time, which causes a lower porosity at 36 h for the sand column (36 h) compared with 12 and 24 h. The particles sizes of the bio-$BaHPO_4$ are basically the same in Fig. 12a, b, but the degree of aggregation between the particles in Fig. 12a is significantly lower than that in Fig. 12b. Therefore, the porosity of sand column at 12 h is lower than that of sand column at 24 h. The compressive strength of the sand column cemented by the bio-$BaHPO_4$ at 24 h is greater than that of sand column at 0 and 36 h.

SEM images of sand columns cemented by 40% of the bio-$BaHPO_4$ after 12, 24, and 36 h are shown in Fig. 12d–f. The effect of standing time on the size, number, porosity, and compressive strength of the bio-$BaHPO_4$ is similar to the above-mentioned dosage of 20%.

Figure 13 shows the microstructure of sand columns formed by the bio-$BaHPO_4$ at different dosages [25]. The size of pores between sand particles decreases with the increase in the bio-$BaHPO_4$ content for a given magnification. The above results are consistent with the change of porosity with the increase of the slurry content.

Fig. 12 SEM images of the bio-sandstones under different content and standing time: **a–c** 20% content and standing time 12, 24, and 36 h, **d–f** 40% content and standing time 12, 24, and 36 h [25]

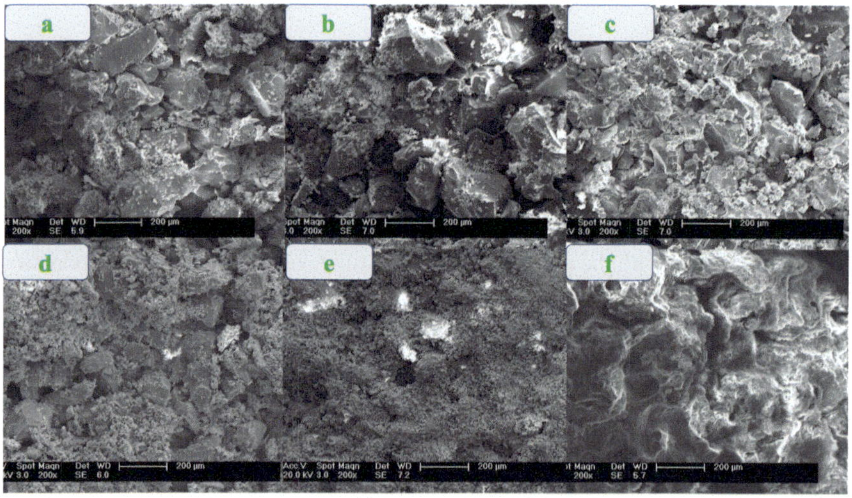

Fig. 13 SEM images of sand columns under different content: **a** 10%, **b** 20%, **c** 30%, **d** 40%, **e** 50%, **f** 60% [25]

4.3 Bio-mineralization of Biological Magnesium Phosphate Cement

4.3.1 Functional Groups of Biological Magnesium Phosphate

Fourier-transform infrared spectroscopy (FTIR) spectra show the functional groups of chemical/biological magnesium phosphate. Chemical/biological magnesium phosphate in the wavenumber range of 3000.00–3600.00 and 1631.94–1654.66 cm^{-1} is caused by water crystallization in magnesium phosphate [26]. The vibration absorption bands at 2366.89–2266.68, 1400.86–1406.07, and 871.10–874.32 cm^{-1} are carbonate ions, which may be caused by carbon dioxide in the air during synthesis or testing [26]. The strong vibrational peaks at 1027.96, and 1079.22 cm^{-1}, and the bending vibration peaks at 571.19, and 601.16 cm^{-1} are consistent with PO_4^{3-} reported in [27, 28]. The bending vibration absorption peaks of chemical/biological magnesium phosphate at 744.20–778.15 cm^{-1} are H–O(P) [29].

4.3.2 Chemical Components of Biological Magnesium Phosphate

Chemical components for different types of magnesium phosphate are analyzed by XRD patterns (Fig. 14). Index results indicate that chemical/biological magnesium phosphates are accurately retrieved to the reported structures. The standard card corresponds to chemical $MgHPO_4{\cdot}3H_2O$ is JCPDS No. 72-0023, as shown in Fig. 14a. The standard card corresponds to chemical $Mg_3(PO_4)_2{\cdot}5H_2O$ is JCPDS No. 35-0329 (Fig. 14c). The standard cards correspond to the mixture of $MgHPO_4{\cdot}1.2H_2O$ and $Mg_3(PO_4)_2{\cdot}5H_2O$ is JCPDS Nos. 49-0752 and 35-0329 respectively, as shown in Fig. 14b. The standard card corresponds to biological $Mg_3(PO_4)_2{\cdot}5H_2O$ is JCPDS No. 35-0329 (Fig. 14d).

4.3.3 Microstructure of Chemical/Biological Magnesium Phosphates

SEM images for chemical/biological magnesium phosphates are displayed in Ref. [20]. The chemical $MgHPO_4{\cdot}3H_2O$ is an irregular particle with cluster structures, and particles size is 1–3 μm. The mixture of $MgHPO_4{\cdot}1.2H_2O$ and $Mg_3(PO_4)_2{\cdot}5H_2O$ is also an irregular particle with cluster structures, and particles size is also between 1 and 3 μm, which is similar to the particle size of chemical $MgHPO_4{\cdot}3H_2O$. Similarly, the chemical $Mg_3(PO_4)_2{\cdot}5H_2O$ is also an irregular particle with cluster structures, but particle size changes from 1 to 5 μm. The morphology of biological $Mg_3(PO_4)_2{\cdot}5H_2O$ is the honeycomb structure with 1–2 μm of irregular particle clusters attached in their surface.

Figure 15 shows that electron diffraction (ED) and transmission electron microscope (TEM) images of chemical/biological magnesium phosphates [20]. The size of chemical $MgHPO_4{\cdot}3H_2O$ particles with irregular sheet-like structures is less than

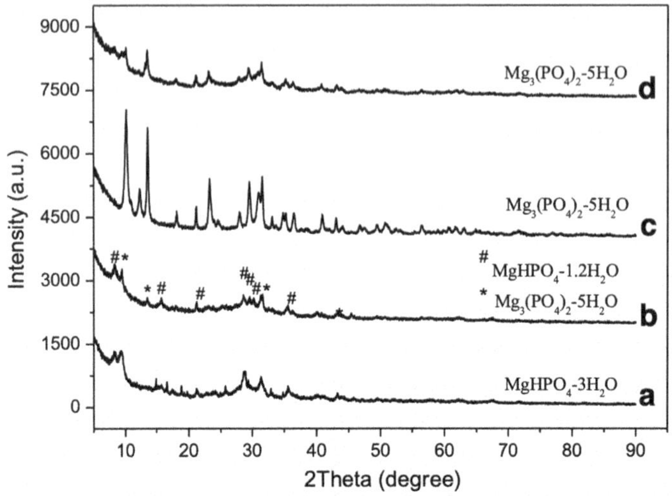

Fig. 14 XRD patterns of chemical (**a**, **c**) and biological (**b**, **d**) magnesium phosphates [20]

100 nm, and the crystallinity is good, as shown in Fig. 15a, b. Figure 15c is a TEM image of the mixture of $MgHPO_4 \cdot 1.2H_2O$ and $Mg_3(PO_4)_2 \cdot 5H_2O$. The size of particles with irregular flocculent structure is less than 100 nm. Diffraction peaks cannot be observed in the ED image, particles crystallinity is poor (Fig. 15d). Figure 15e shows that the size of chemical $Mg_3(PO_4)_2 \cdot 5H_2O$ with irregular spherical structures changes from 50 to 300 nm. Different sizes of light spots are shown in Fig. 15f, indicating good crystallinity of particles. The size of biological $Mg_3(PO_4)_2 \cdot 5H_2O$ with sheet-like structures changes from 50 to 200 nm (Fig. 15g). ED images show that different sizes of light spots are observed in Fig. 15h, which also indicates that the crystallinity of particles is good.

4.3.4 Thermal Properties of Chemical/Biological Magnesium Phosphate

The thermogravimetric-differential thermal (DSC-TGA) curves display thermal properties of chemical/biological magnesium phosphate [30, 31]. The TGA curve of the chemical $MgHPO_4 \cdot 3H_2O$ shows a sustained mass loss which is approximately 30.37% from room temperature to 800 °C due to the loss of crystal water and the decomposition of $MgHPO_4$. The DSC curve shows two endothermic peaks at 87.3 and 644.9 °C. The peak at 87.3 °C is the loss of crystal water, while the peak at 644.9 °C is the decomposition temperature of $MgHPO_4$.

The TGA curve of the biological mixture of $MgHPO_4 \cdot 1.2H_2O$ and $Mg_3(PO_4)_2 \cdot 5H_2O$ shows a sustained mass loss which is approximately 45.03% from room temperature to 800 °C, due to the loss of crystal water, the decomposition of organic matrix and $MgHPO_4$. The DSC curve shows two endothermic peaks 87.2

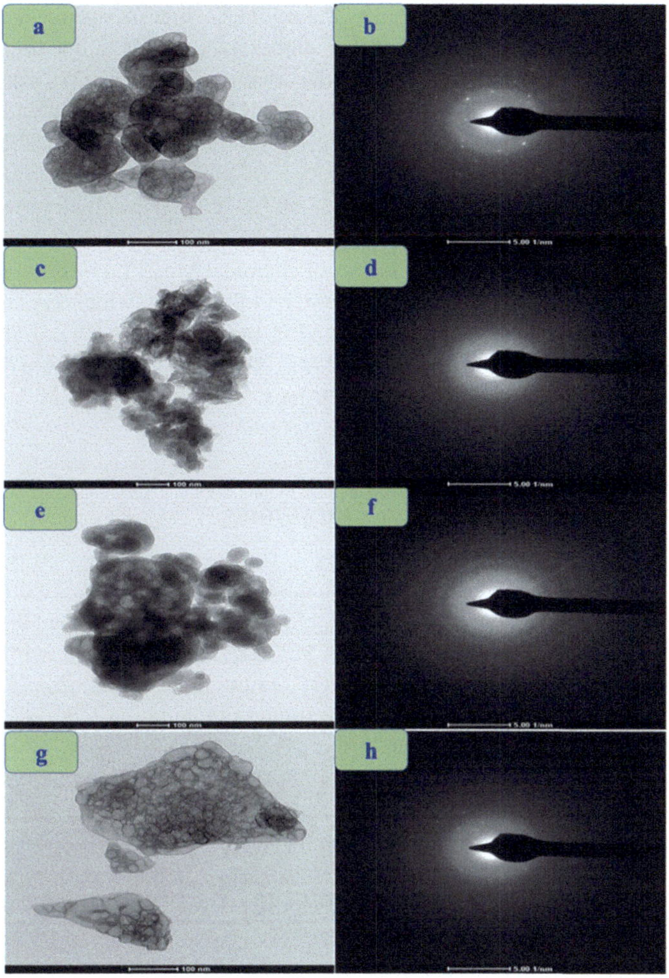

Fig. 15 TEM and ED images of chemical/biological magnesium phosphates: **a**, **b** chem-$MgHPO_4{\cdot}3H_2O$, **c**, **d** bio-mixture of $MgHPO_4{\cdot}1.2H_2O$ and $Mg_3(PO_4)_2{\cdot}5H_2O$, **e**, **f** chem-$Mg_3(PO_4)_2{\cdot}5H_2O$, **g**, **h** bio-$Mg_3(PO_4)_2{\cdot}5H_2O$ [20]

and 724.0 °C, respectively. The peak of 87.2 °C is the loss of crystal water, while the peak at 724.0 °C is the decomposition temperature of $MgHPO_4$ or the dehydration temperature of $Mg_3(PO_4)_2{\cdot}5H_2O$.

The TGA curve of the chemical $Mg_3(PO_4)_2{\cdot}5H_2O$ shows a sustained mass loss which is approximately 31.06% from room temperature to 800 °C, due to the loss of crystal water. The DSC curve indicates temperatures of three endothermic peaks are 85.0, 131.1, and 670.1 °C, respectively. The endothermic peaks at 85.0 and 131.1 °C represent the temperature of loss of crystal water at different stages, and 670.1 °C can be considered as the dehydration temperature of $Mg_3(PO_4)_2{\cdot}5H_2O$.

The TGA curve of the biological $Mg_3(PO_4)_2{\cdot}5H_2O$ shows a sustained mass loss which is approximately 32.55% from room temperature to 800 °C due to the loss of crystal water and thermal decomposition of the organic matrix. The DSC curve shows three endothermic peaks at 87.2, 226.2, and 698.7 °C, respectively. The endothermic peak at 87.2 °C represents the temperature of loss of crystal water. The peak of 226.2 °C may be caused by the decomposition of $Mg_3(PO_4)_2{\cdot}5H_2O$ with a small amount of organic matrix, and the peak of 698.7 °C is the dehydration temperature of $Mg_3(PO_4)_2{\cdot}5H_2O$. By comparing the DSC-TGA curves of chemical and biological $Mg_3(PO_4)_2{\cdot}5H_2O$, the mass loss of chemical $Mg_3(PO_4)_2{\cdot}5H_2O$ (31.06%) is lower than that of biological $Mg_3(PO_4)_2{\cdot}5H_2O$ (32.55%), which is caused by a small amount of organic matrix encapsulated in biological $Mg_3(PO_4)_2{\cdot}5H_2O$, and the crystallization temperature of biological $Mg_3(PO_4)_2$ after dehydration (698.7°C) is higher than that of chemical $Mg_3(PO_4)_2$ (670.1 °C).

4.4 *Comparison of Biological Phosphates*

PMM, which produces an alkaline phosphatase, can continuously hydrolyze the substrate, and thus PO_4^{3-} ions are obtained. Due to the presence of metal ions with different properties, the phosphates such as $M_3(PO_4)_2$, $MHPO_4$, and MPO_4 can be produced when acidic cation reacts with PO_4^{3-} in the mixed solution of PMM and substrate. The morphologies of barium phosphate precipitation induced by PMM can be imparted at different pH conditions. The average crystal sizes of barium phosphate deposited in the mixed solutions at pH = 7, 8, 9, 10, and 11 are 33.40, 29.37, 24.13, 47.76, and 96.53 μm, respectively. The results of XRD and calculated $[HPO_4^{2-}]$ (δ_2) distribution coefficient show that when the pH is less than 10, the main component is $BaHPO_4$. However, when the pH is 10, the main component is a mixture of $BaHPO_4$ and $Ba_5(PO_4)_3OH$. When the pH is 11, the main component is a mixture of $Ba_5(PO_4)_3OH$ and $BaNaPO_4$. The comparison of the morphology, size, and thermal properties of chemical $BaHPO_4$ and biological $BaHPO_4$ shows that the morphology of chemical $BaHPO_4$ is a trapezoidal structure, while the morphology of biological $BaHPO_4$ is irregular spherical cluster and dumbbell-shaped structures. Acidic $MgCl_2{\cdot}6H_2O$ is added to the mixed solution, which can generate a nanoscale mixture of $MgHPO_4{\cdot}1.2H_2O$ and $Mg_3(PO_4)_2{\cdot}5H_2O$. When the neutral anhydrous $MgCl_2$ is added, nanoscale $Mg_3(PO_4)_2{\cdot}5H_2O$ is obtained, and its morphology and thermal properties are different from that of chemical $Mg_3(PO_4)_2{\cdot}5H_2O$. The morphology of biological $Mg_3(PO_4)_2{\cdot}5H_2O$ is a honeycomb structure, while the chemical $Mg_3(PO_4)_2{\cdot}5H_2O$ is an irregular cluster structure and the crystallization temperature of $Mg_3(PO_4)_2$ after dehydration is lower than that of biological $Mg_3(PO_4)_2$. In conclusion, the biological $BaHPO_4$, the mixture of $MgCl_2{\cdot}6H_2O$ and $Mg_3(PO_4)_2{\cdot}5H_2O$ can be synthesized when acidic $BaCl_2{\cdot}2H_2O$ and $MgCl_2{\cdot}6H_2O$ are added to the mixed solution. The neutral anhydrous $MgCl_2$ is mixed with PO_4^{3-}, which generates the biological $Mg_3(PO_4)_2{\cdot}5H_2O$. Except for micron-sized biological $BaHPO_4$, the size of other biological phosphates is in nanometers. Therefore, metal ions with different properties react with PO_4^{3-} in the mixed solution to mainly generate nanoscale phosphates.

5 The Bio-mineralization of Bio-composite Cement

Urea can be quickly hydrolyzed by the urease, which is produced by CMM, and thus ammonium (NH_4^+) and carbonate (CO_3^{2-}) ions are obtained [32]. Theoretically, two moles of NH_4^+ can be produced by the hydrolysis of one mole of urea by CMM [32]. Ammonium can be converted into environment-friendly carbonate and struvite precipitation when HPO_4^{2-} and Mg^{2+} are contained in the CMM solution [33–37].

5.1 Selection of Magnesium Source

Figure 16a shows when 100 mL of the mixed solution of urea (1 mol/L) and $MgCl_2{\cdot}6H_2O$ (2 mol/L) is added to 100 mL of the mixed solution of $K_2HPO_4{\cdot}3H_2O$ (1 mol/L) and CMM, many white precipitates are observed. The dried product is mainly composed of $MgHPO_4(H_2O)_3$ and $MgNH_4PO_4(H_2O)_6$, the corresponding standard cards are JCPDS No. 75-1714 and 71-2089. After the mixed solution of urea (1 mol/L) and $MgCl_2{\cdot}6H_2O$ (2 mol/L) reacts with 100 mL of the mixed solution of $K_2HPO_4{\cdot}3H_2O$ (1 mol/L) and CMM, the product is mainly composed of $NH_4MgPO_4{\cdot}6H_2O$ (JCPDS No. 15-0762) and $MgCO_3(H_2O)_3$ (JCPDS No. 70-1433) (Fig. 16b). The results of the XRD show that the addition of $K_2HPO_4{\cdot}3H_2O$ can convert NH_3/NH_4^+ into struvite during the reaction process. The weakly acidic $MgCl_2{\cdot}6H_2O$ can react with CO_3^{2-} in the solution, which results in the decomposition of CO_3^{2-} so that the magnesium carbonate cannot be formed. Therefore, weakly acidic magnesium salts cannot be used as ideal precipitants. However, neutral $MgCl_2$ does not cause the decomposition of the CO_3^{2-} which reacts with magnesium ions to form magnesium carbonate. Therefore, the anhydrous $MgCl_2$ as the precipitant is used in the preparation of bio-composite cements.

5.2 Composition of NH_3/NH_4^+ Conversion Products

The XRD results of the products are shown in Fig. 17 [34]. When the mixture of $K_2HPO_4{\cdot}3H_2O$ (1 mol/L) and CMM, the mixture of 1 mol/L of urea and 2 mol/L of $MgCl_2$ are mixed, the main components of the product are the mixture of $MgCO_3(H_2O)_3$ and $NH_4MgPO_4{\cdot}6H_2O$. The corresponding standard cards are JCPDS No. 70-1433 and 15-0762 (Fig. 17a). When the mixture of $K_2HPO_4{\cdot}3H_2O$ (1.5 mol/L) and CMM, the mixture of 1 mol/L of urea and 2.5 mol/L of $MgCl_2$ are mixed, the main components of the product are the mixture of $MgNH_4PO_4(H_2O)_6$

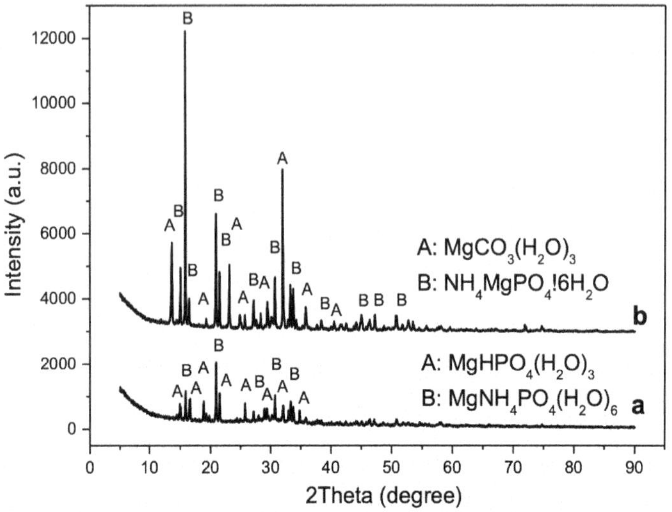

Fig. 16 Influence of the precipitant on the products: **a** the mixture solution of urea and $MgCl_2{\cdot}6H_2O$ as the precipitant, **b** the mixture solution of urea and $MgCl_2$ as the precipitant

(JCPDS No. 71-2089) and $MgCO_3{\cdot}2H_2O$ (JCPDS No. 18-0768), as shown in Fig. 17b. When the mixture of $K_2HPO_4{\cdot}3H_2O$ (2 mol/L) and CMM, the mixture of 1 mol/L of urea and 3 mol/L of $MgCl_2$ are mixed, the chemical component of the product is the mixture of $MgNH_4PO_4(H_2O)_6$, $MgCO_3{\cdot}2H_2O$, and $MgHPO_4(H_2O)_3$. Standard cards of the mixture are JCPDS Nos. 77-2303, 18-0768, and 70-2345 (Fig. 17c). The presence of $MgHPO_4(H_2O)_3$ indicates that $K_2HPO_4{\cdot}3H_2O$ is in excess in the XRD pattern. The above XRD results show that the presence of $K_2HPO_4{\cdot}3H_2O$ can effectively convert the NH_3/NH_4^+ generated by CMM hydrolyzing urea into struvite in the presence of Mg^{2+}, which can effectively reduce the emission of NH_3.

5.3 *Bio-mineralization in the Solution*

Figure 18 shows that SEM images for different types of the mixture of struvite and magnesium carbonate were obtained at different reaction conditions [34]. Figure 18a, b are the images of the mixture of $NH_4MgPO_4{\cdot}6H_2O$ and $MgCO_3(H_2O)_3$. The size of irregular particles is mainly distributed between 10 and 100 μm, and the surface of particle is relatively rough. SEM images of the mixture of $MgNH_4PO_4(H_2O)_6$ and $MgCO_3{\cdot}2H_2O$ are shown in Fig. 18c, d. The shape of particles is also irregular, and the size of particles is mainly distributed between 10 and 100 μm. Figure 18e, f show the microstructures of the mixture of $MgNH_4PO_4(H_2O)_6$, $MgCO_3{\cdot}2H_2O$, and $MgHPO_4(H_2O)_3$, which exists in the form of irregular particle clusters with relatively rough surfaces and sizes ranging from 100 to 300 μm. From

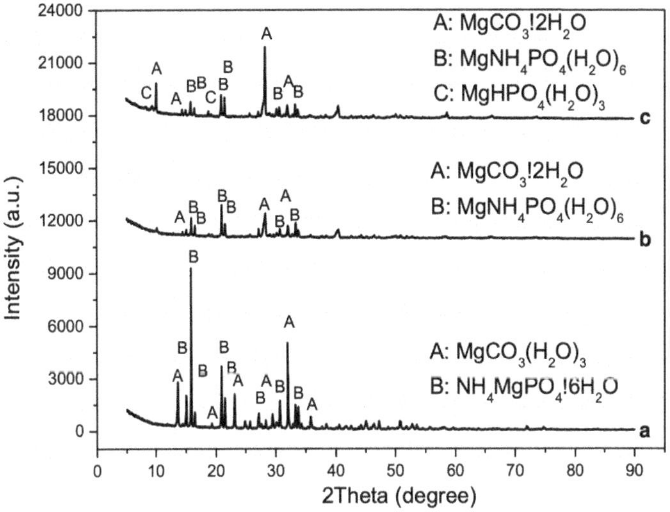

Fig. 17 XRD patterns of conversion products of NH_3/NH_4^+: **a** the mixture solution of urea (1 mol/L) and $MgCl_2$ (2 mol/L) as the precipitant, **b** the mixture solution of urea (1 mol/L) and $MgCl_2$ (2.5 mol/L) as the precipitant, **c** the mixture solution of urea (1 mol/L) and $MgCl_2$ (3 mol/L) as the precipitant [34]

the above results, the morphology of the mixture of struvite and magnesium carbonate induced by microorganisms is mainly irregular, and the average particle size of the mixture of $MgNH_4PO_4(H_2O)_6$, $MgCO_3{\cdot}2H_2O$, and $MgHPO_4(H_2O)_3$ is larger than that of the mixture of $MgCO_3{\cdot}2H_2O$ and $MgNH_4PO_4(H_2O)_6$, $MgCO_3(H_2O)_3$ and $NH_4MgPO_4{\cdot}6H_2O$.

5.4 Bio-mineralization Between Loose Sand Particles

The X-ray computed tomography (XCT) is used to observe the evolution of the three-dimensional (3D) pore structure for sand columns at cementation times 2, 4 and 6, as shown in Fig. 19 [33]. The VG Studio Max 2.0 software of the XCT equipment can apply gray threshold algorithm and defect analysis module to get the hole information according to the literature reported [19]. The size of the hole can be represented by the color scale, in which red to blue represents from large holes to small holes. With the increase of cementation times, the defect size in the sand column decreases. The maximum defect size of the sand column after 2 cementation times is 1401 mm^3, and it decreases to 738 mm^3 after 6 cementation times. The product of microbial cement increases with the increase of cementation times, which causes the filling of defects inside sand column.

With the increase of cementation times, the defect size gradually decreases. The defects are filled by the microbial cement with the increase of cementation times

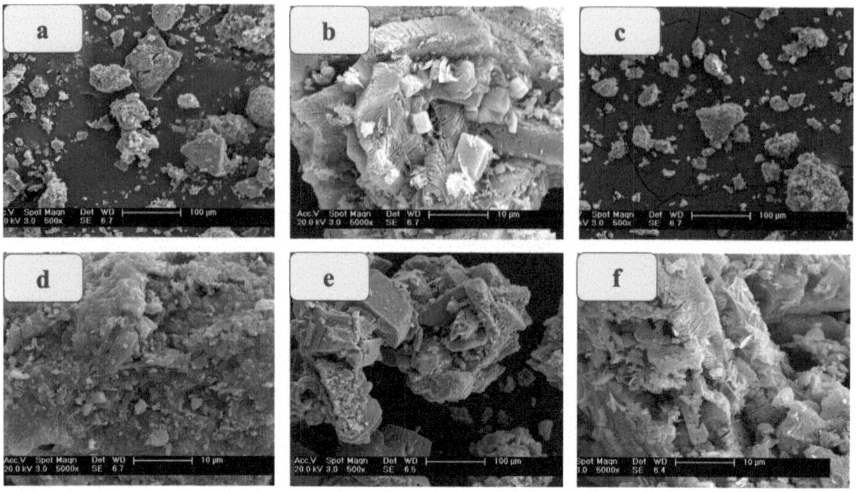

Fig. 18 SEM images of different types of mixture for struvite and magnesium carbonate: **a**, **b** the mixture of $NH_4MgPO_4{\cdot}6H_2O$ and $MgCO_3(H_2O)_3$, **c**, **d** the mixture of $MgNH_4PO_4(H_2O)_6$ and $MgCO_3{\cdot}2H_2O$, **e**, **f** the mixture of $MgNH_4PO_4(H_2O)_6$, $MgCO_3{\cdot}2H_2O$, and $MgHPO_4(H_2O)_3$ [34]

which decreases the number and size of defects. The internal defects of loose sand columns in the initial state are unevenly distributed, so their overall defect distribution is non-uniform [19], as shown in Fig. 19. XCT images show that the average porosity of sand columns is 36.20, 32.21, and 20.21% when the cementation times are 2, 4, and 6, respectively [33]. The above results are similar to average porosity of sand columns (34.62, 29.55, and 25.15%) measured by paraffin drainage method in cementation times 2, 4, and 6.

SEM images of sand columns with different cementation times are shown in Fig. 20 [33]. Particles shape is mainly irregular sheet-like structure at 2, 4, and 6 cementation times. Figure 20a, b show that the pore between sand particles cannot be fully filled due to insufficient bio-composite cement when cementation times are 2. When cementation times are 4, the amount of bio-composite cement is larger than that of 2, and the pores between sand particles can be partially filled, as shown in Fig. 20c, d. Figure 20e, f show that when cementation times are 6, the amount of bio-composite cement exceeds 2 and 4, and the pores between sand particles can be fully filled. Loose sand particles are cemented into a whole with good mechanical properties.

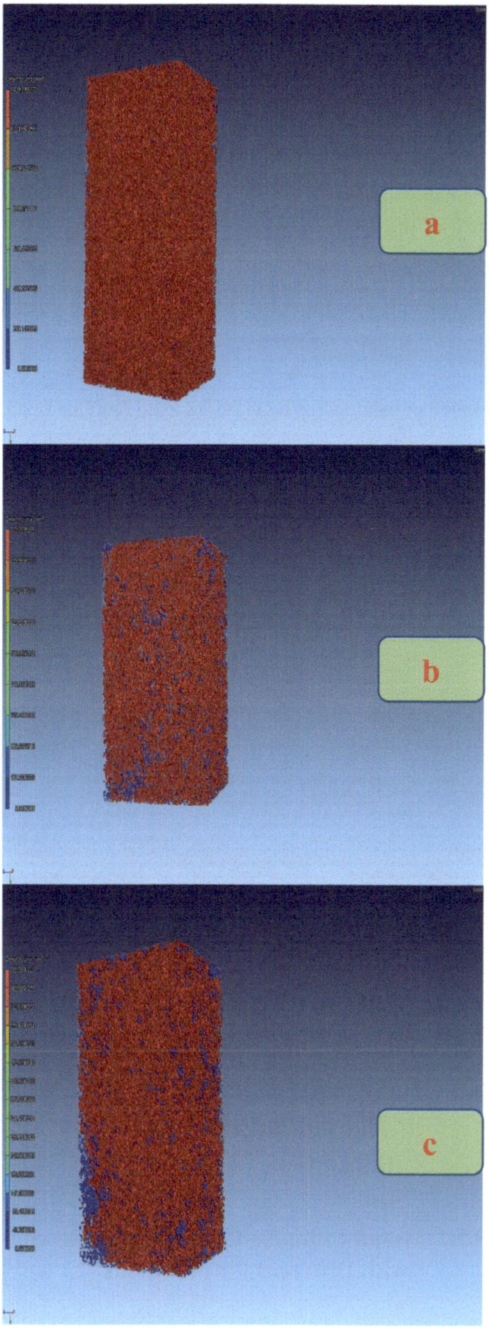

Fig. 19 3D pore structure evolution of sand columns under different cementation times: **a** 2, **b** 4, **c** 6 [33]

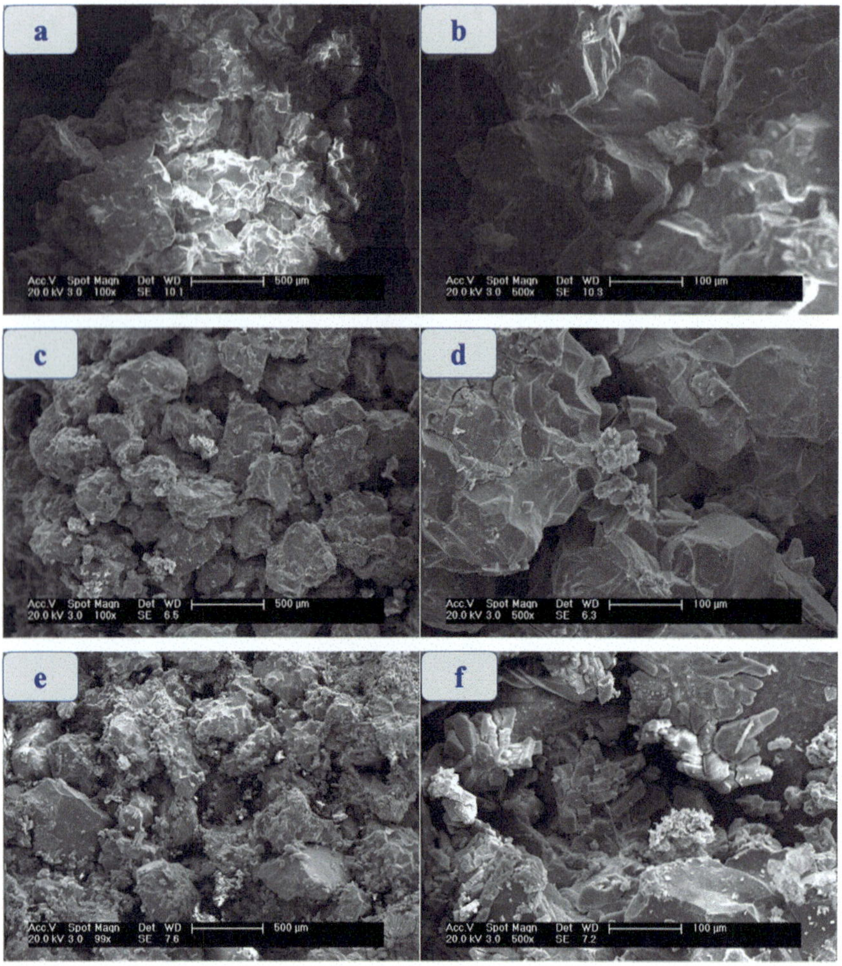

Fig. 20 SEM images of sand columns under different cementation times: **a**, **b** 2, **c**, **d** 4, **e**, **f** 6 [33]

5.5 Bio-mineralization of Different Formulations of Bio-composite Cement and Cementation of Sand Columns

5.5.1 Sand Columns Cemented by Bio-composite Cements with Different Formulations

The XRD peaks of sand column cemented by the CJ1 show that the main components are $MgNH_4PO_4(H_2O)_6$ (JCPDS No. 71-2089), quartz (JCPDS No. 46-1045), and $Mg_5(CO_3)_4(OH)_2{\cdot}4H_2O$ (JCPDS No. 25-0513) (Fig. 21a). The

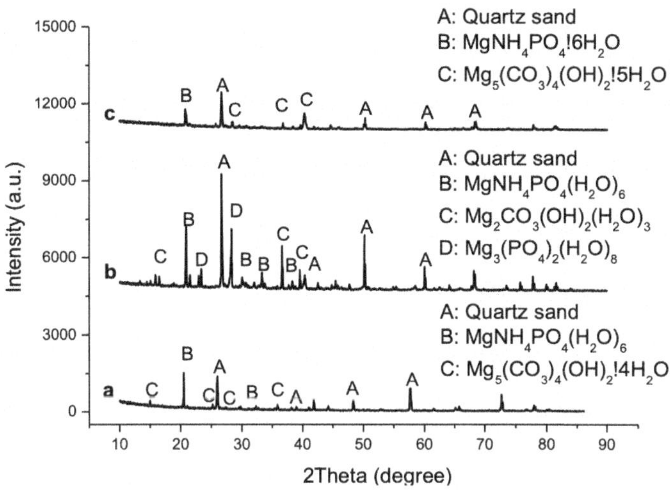

Fig. 21 XRD patterns of sand columns cemented by different formulas of bio-composite cements: **a** CJ1, **b** CJ1.5, **c** CJ2 [34]

XRD peaks of sand column cemented by the CJ1.5 show that the main components are $MgNH_4PO_4(H_2O)_6$ (JCPDS No. 71-2089), quartz (JCPDS No. 86-2237), $MgCO_3(OH)_2(H_2O)_3$ (JCPDS No. 70-0591), and $Mg_3(PO_4)_2(H_2O)_8$ (JCPDS No. 84-1148), as shown in Fig. 21b. The XRD peaks of sand column cemented by the CJ2 show that the main components are $MgNH_4PO_4 \cdot 6H_2O$ (JCPDS No. 15-0762), quartz (JCPDS No. 89-1961), and $Mg_5(CO_3)_4(OH)_2 \cdot 5H_2O$ (JCPDS No. 23-1218) (Fig. 21c). After CJ1, CJ1.5, and CJ2 cementing sand columns, chemical components of cementitious materials are a mixture of hydromagnesite and struvite. Therefore, the NH_3/NH_4^+ emitted during above cementation process can be converted into struvite, which is an environmentally friendly substance, in the presence of $K_2HPO_4 \cdot 3H_2O$ and Mg^{2+}.

5.5.2 Internal State of Sand Columns Cemented by Different Formulations of Bio-composite Cement

SEM images of sand columns are shown in Fig. 22 [34]. Figure 22a, b show the internal state of sand column cemented by the CJ1. The shape of the bio-composite cement is mainly an irregular sheet-like structure, and the pore between sand grains can be fully filled by the product of these sheet-like structures. The internal state of sand columns cemented by the CJ1.5 shows that the bio-composite cement has an irregular particle cluster structure, and the pore between sand grains can be fully filled by the product of these cluster structures (Fig. 22c, d). The internal state of sand column cemented by the CJ2 indicates that the shape of the bio-composite cement

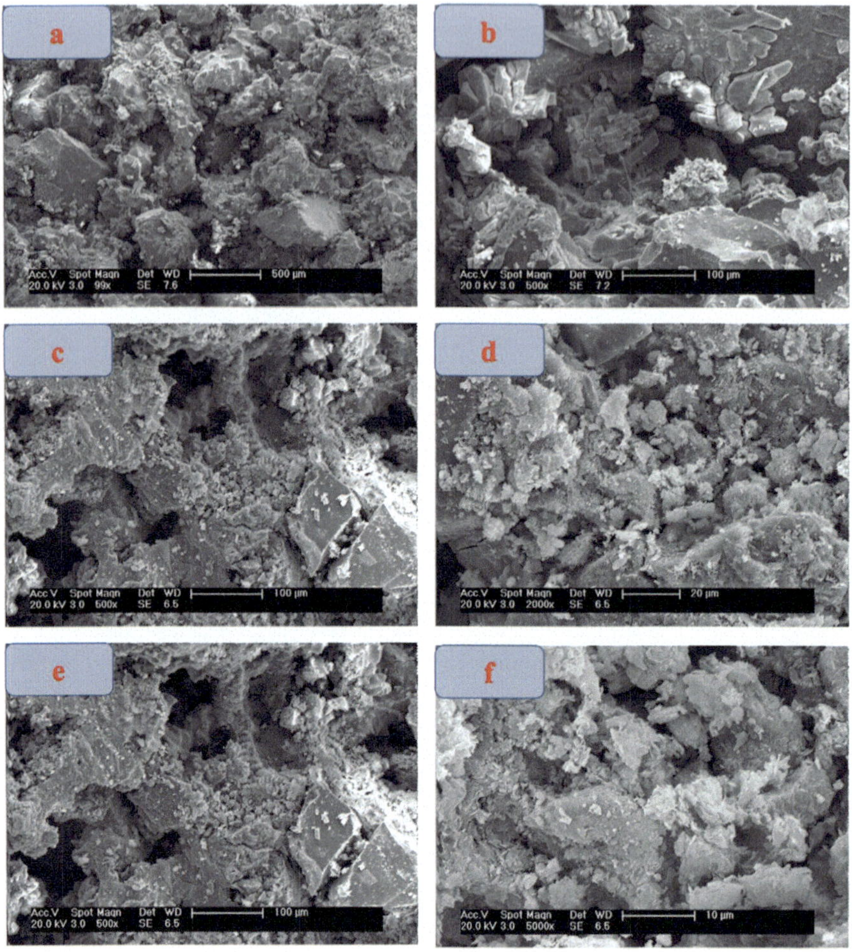

Fig. 22 SEM images of sand columns cemented by different formulas of bio-composite cement: **a**, **b** CJ1, **c**, **d** CJ1.5, **e**, **f** CJ2 [34]

is also an irregular particle cluster structure, and it plays a similar role in the pore between sand particles, as shown in Fig. 22e, f.

6 Conclusions

In this chapter, urease and alkaline phosphatase secreted by CMM and PMM are used to hydrolyze the substrate, and finally CO_3^{2-} and PO_4^{3-} ions are generated. In the presence of metal ions with different properties, bio-phosphate, bio-carbonate, and bio-composite cement are generated, and their composition, microstructure, cementation properties, etc., are discussed. The diverse shape and size of bio-phosphate, bio-carbonate, and bio-composite cement can be obtained in the presence of pure water, bacterial body, bacterial secretion, and bacterial solution. The shapes of bio-carbonate crystals are block, spherical, and block structures, corresponding to mineralization solution of pure water, bacterial body, bacterial secretion, and bacteria, respectively. The shape of bio-$BaHPO_4$ crystals is hexagonal, pentagonal, spherical, and dumbbell, corresponding to mineralization solution of pure water, bacterial body, bacterial secretion, and bacteria, respectively. The morphology of bio-carbonate, bio-phosphate, and bio-composite cements between sand particles is mainly ball, dumbbell, and irregular block cluster structures with different sizes, respectively. Therefore, pure water, bacterial body, bacterial secretion, bacterial solution, and sand particles play an important role in the mineralization process of bio-phosphate, bio-carbonate, and bio-composite cements.

Acknowledgements The research described in this chapter was financially supported by the Opening Funds of Jiangsu Key Laboratory of Construction Materials (Grant No. CM2018-02).

References

1. A.A. Veis, A window on biomineralization. Science**307**, 1419–1420 (2005)
2. S. Mann, Molecular recognition in biomineralization. Nature **332**, 119–124 (1988)
3. G. Fu, S. Valiyaveettil, B. Wopenka, D.E. Morse, CaCO3 Biomineralization: acidic 8-kDa proteins isolated from aragonitic abalone shell nacre can specifically modify calcite crystal morphology. Biomacromolecules **6**, 1289–1298 (2005)
4. Y.F. Ma, L. Qiao, Q.L. Feng, In-vitro study on calcium carbonate crystal growth mediated by organic matrix extracted from fresh water pearls. Mater. Sci. Eng. C **32**, 1963–1970 (2012)
5. S. Mann, B.R. Heywood, S. Rajam, J. Derek Birchall, Controlled crystallization of CaCO3 under stearic acid monolayers.Nature **334**, 692–695 (1988)
6. L. Cheng, C.X. Qian, R.X. Wang, J.Y. Wang, Study on the mechanism of calcium carbonate formation induced by carbonate-mineralization microbe. Acta Chim. Sin. **65**, 2133–2138 (2007)
7. W.K. Zhu, X.G. Luo, C. Zhang, J. Zhou, Microbiological precipitation of barium carbonate. Chin. J. Inorg. Chem. **27**, 2053–2060 (2011)
8. F. Wang, G.Y. Xu, Z.Q. Zhang, The effect of pH on morphological control of barium hydrogen phosphate crystal by a double-hydrophilic copolymer. Mater. Lett. **59**, 808–812 (2005)
9. X.N. Yu, C.X. Qian, X. Wang, Microbiological precipitation, morphology and thermal behavior of barium hydrogen phosphate. J. Chil. Chem. Soc. **60**, 2885–2887 (2015)
10. X.D. Zhang, W. He, Y.Z. Yue, R.M. Wang, J.X. Shen, S.J. Liu, J.Y. Ma, M. Li, F.X. Xu, Bio-synthesis participated mechanism of mesoporous LiFePO4/C nanocomposite microspheres for lithium ion battery. J. Mater. Chem. **22**, 19948–19956 (2012)
11. J. Lin, E. Cates, P.A. Bianconi, A synthetic analogue of the biomineralization process: controlled crystallization of an inorganic phase by a polymer matrix. J. Am. Chem. Soc. **116**, 4738–4745 (1994)
12. J. Chen, Y.F. Kong, J.J. Ji, J. Ruan, K. Wang, F. Gao, D. Cui, Protein-induced structural evolution of silver sulfide at the nanoscale: from hollow particles to solid spheres. Nanoscale **4**, 4455–4458 (2012)

13. C.X. Qian, X.N. Yu, T.W. Zheng, Y.Q. Chen, Review on bacteria fixing CO_2 and bio mineralization to enhance the performance of construction materials. J. CO_2 Util. **55**, 101849 (2022)
14. X.N. Yu, J.G. Jiang, Mineralization and cementing properties of bio-canbonate cement, bio phosphate cement, and bio-canbonate/phosphate cement: a review. Environ. Sci. Pollut. Res. **25**, 21483–21497 (2018)
15. A.M. Barrios, S.J. Lippard, Decomposition of alkyl-substituted urea molecules at a hydroxide bridged dinickel center. Inorg. Chem. **40**, 1250 (2001)
16. S. Benini, W.R. Rypniewski, K.S. Wilson, S. Ciurli, S. Mangani, Structure-based rationalization of urease inhibition by phosphate: novel insights into the enzyme mechanism. J. Biol. Inorg. Chem. **6**, 778–790 (2001)
17. X.N. Yu, Z.H. He, X.Y. Li, Bio-cement modified construction materials and their performances. Environ. Sci. Pollut. Res. **29**, 11219–11231 (2022)
18. X.N. Yu, H. Rong, Seawater based MICP cements two-phase/one-phase cemented sand blocks. Appl. Ocean Res. **118**, 102972 (2022)
19. H. Rong, Preparation and binding mechanism of microbe cement. Dissertation, Southeast University, 2014
20. X.N. Yu, J.G. Jiang, J.W. Liu, W.Li, Review on potential uses, cementing process, mechanism and syntheses of nano-phosphate cementitious materials by the microbial mineralization method. Constr. Build. Mater. **273**, 121113 (2021)
21. B. Stec, K.M. Holtz, E.R. Kantrowitz, A revised mechanism for the alkaline phosphatase reaction involving three metal ions. J. Mol. Biol. **299**, 1303–1311 (2000)
22. X.N. Yu, C.X. Qian, X. Wang, Microbially induced deposition of barium phosphates and its ingredient, morphology, and size under different pH values. J. Southeast Univ. (Engl. Ed.) **31**, 506–510 (2015)
23. X.N. Yu, C.X. Qian, X., Morphology of barium hydrogen phosphate formation induced by phosphate-mineralization microbe. J. Wuhan Univ. Technol. Mater. Sci. Ed. **31**, 227–230 (2016)
24. X.N. Yu, C.X. Qian, Z.H. He, L.Z. Sun, Removal of heavy metal Ba (II) by phosphate mineralization microbe under different standing time. Fresenius Environ. Bull. **26**, 859–863 (2017)
25. X.N. Yu, C.X. Qian, B. Xue, X. Wang, The influence of standing time and content of the slurry on bio-sandstone cemented by biological phosphates. Constr. Build. Mater. **82**, 167–172 (2015)
26. M. Rajkumar, N.M. Sundaram, V. Rajendran, Preparation of size controlled, stoichiometric and bioresorbable hydroxyapatite nanorod by varying initial pH, Ca/P ratio and sintering temperature. Dig. J. Nanomater. Biostruct. **6**, 169–179 (2011)
27. R.L. Frost, M.L. Weier, Raman spectroscopic study of vivianites of different origins. Neues Jahrbuch fuer Mineralogie-Monatshefte **2004**, 445–463 (2004)
28. E. Walpersdorf, C.B. Koch, L. Heiberg, D.W. O'Connell, C. Kjaergaard, H.C.B. Hansen, Does vivianite control phosphate solubility in anoxic meadow soils?. Geoderma **193–194**, 189 (2013)
29. A. Bensalem, M. Ahluwalia, Synthesis of amorphous MgHPO4 center dot x(R) [R = ethanol; ethylene glycol] in anhydrous media. Mater. Res. Bull. **32**, 1473 (1997)
30. X.N. Yu, C.X. Qian, X. Wang, Biosynthesis of magnesium phosphates and its thermal property. Sci. Adv. Mater. **7**(9), 1730–1733 (2015)
31. X.N. Yu, C.X. Qian, L.Z. Sun, Chemosynthesis of nano-magnesium phosphates and its characterization. Dig. J. Nanomater. Biostruct. **11**(4), 1099–1103 (2016)
32. M.M. Rahman, R.N. Hora, I. Ahenkorah, S. Beecham, M.R. Karim, A. Iqbal, State-of-the-art review of microbial-induced calcite precipitation and its sustainability in engineering applications. Sustainability **12**, 6281 (2020)
33. X.N. Yu, C.X. Qian, L.Z. Sun, The influence of the number of injections of bio-composite cement on the properties of bio-sandstone cemented by bio-composite cement. Constr. Build. Mater. **164**, 682 (2018)
34. X.N. Yu, C.X. Qian, B. Xue, Loose sand particles cemented by different bio-phosphate and carbonate composite cement. Constr. Build. Mater. **113**, 571 (2016)

35. X.N. Yu, C.X. Qian, J.G. Jiang, Desert sand cemented by bio-magnesium ammonium phosphate cement and its microscopic properties. Constr. Build. Mater. **200**, 116–123 (2019)
36. X.N. Yu, Q.W. Zhan, C.X. Qian, J.J. Ma, Y. Liang, The optimal formulation of bio-carbonate and bio-magnesium phosphate cement to reduce ammonia emission. J. Clean. Prod. **240**, 118156 (2019)
37. X.N. Yu, J. Chu, Y. Yang, C.X. Qian, Reduction of ammonia production in the biocementation process for sand using a new biocement. J. Clean. Prod. **286**, 124928 (2021)

Advances in Mineral-Based Form Stable Phase Change Materials

Xiaobin Gu and Kaijun Dong

Abstract The utilization efficiency of solar energy is low because of its intermittency and discontinuity. The issue can be solved by phase change material (PCM) based energy storage technology, therefore improving the utilization efficiency of solar energy. However, PCM's leakage problem has largely affected its practical application. To address the leakage problem, the preparation of form-stable PCMs (FSPCMs) using minerals, has been widely studied. For FSPCMs, mineral characteristics largely affect the preparation, properties, and application of mineral-based FSPCMs. This chapter, therefore, summarized the influences of minerals on the preparation, properties, and application of mineral-based FSPCMs. Furthermore, research prospects are also given in this chapter. This summary provides a clearer insight into the preparation, properties, and application of mineral-based FSPCMs and will help promote the development of mineral-based FSPCMs.

Keywords Form-stable phase change material · Minerals · Supporting material · Thermal energy storage · Property effect

1 Introduction

Nowadays, human society is constantly developing and the population is growing, so the consumption of fossil fuels is also increasing. Due to the limited resource reserves of fossil fuels and their catastrophic damage to the environment, the research on solar energy is getting more and more attention [1]. The intermittency and instability of solar energy result in low efficiency and greatly hinder its practical application. To

X. Gu (✉) · K. Dong
Guangzhou Institute of Energy Conversion, Chinese Academy of Sciences, Guangzhou 510640, China
e-mail: guxb@ms.giec.ac.cn

CAS Key Laboratory of Renewable Energy, Guangzhou, China

Guangdong Provincial Key Laboratory of New and Renewable Energy Research and Development, Guangzhou, China

S. J. Ikhmayies (ed.), *Advances in Minerals Research*, Advances in Material Research and Technology, https://doi.org/10.1007/978-3-031-49175-7_9

enhance its utilization efficiency and increase its use, energy storage technology is crucial [2]. In the middle of all kinds of energy storage technologies, PCM storage has attracted widespread attention and research from researchers, as it is a material with almost constant phase transition temperature and large latent heat, it is used to alleviate the drawbacks of solar energy [3]. PCM can be divided into three types according to their chemical composition, inorganic PCM, organic PCM, and eutectic PCM [4].

Despite all their advantages, phase change materials still suffer from leakage problems, which limit their application in many fields like building protection structures [5], temperature management of electronic devices [6], and temperature regulation of photovoltaic (PV) panels [7]. In order to solve the leakage problem of phase change materials and increase its applications field and scope, using minerals to make form-stable phase change materials (FSPCM) has been proven to be a major solution to solve the PCM leakage problem, owing to minerals having the characteristics of low price, comparatively large surface area, chemical inertness, and large porosity. Figure 1 indicates a schematic diagram of mineral-based FSPCM and raw materials required for its manufacturing, mainly including graphite, attapulgite, perlite, kaolinite, diatomite, sepiolite, montmorillonite and palyclite [8, 9]. Typically, natural minerals are first purified by an acid/alkali or calcination process. Compared to natural minerals, purified ore has more internal space for PCM packaging. As a result, the phase change materials used will be more easily melted and sorbed into the internal pores of the purified minerals. Because minerals themselves have many pores and a large specific surface area, they have good PCM adsorption capacity, in addition, the relevant consequences show, that mineral-based FSPCM has excellent thermal properties, so it has an excellent application prospect.

Minerals have important effects on the synthesis and properties of mineral-based FSPCM. Even though the main synthetic method of mineral-based FSPCMs is similar, the specific synthesis steps may be different due to the fact that the minerals all have some different properties. Each mixing method has its advantages and disadvantages, and the choices of particular mineral-based FSPCM preparation methods are

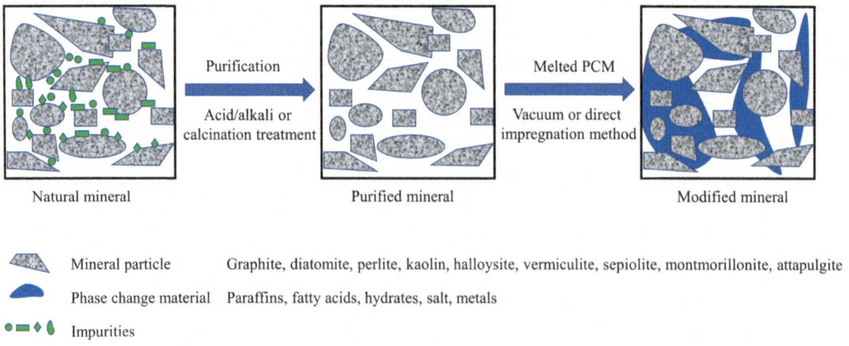

Fig. 1 Preparation of typical mineral-based FSPCM

significant for price and performance. In addition, due to the different intrinsic properties of minerals, the PCM load and the thermal performance of different mineral-based FSPCM may be significantly different. For instance, graphite-based FSPCM can absorb 90wt.% of phase change materials, but perlite can only absorb up to 60wt% of phase change materials. Moreover, by adding graphite, the thermal conductivity of FSPCM is increased, but perlite makes its thermal conductivity decrease. Therefore, by recognizing the influence of all kinds of minerals on mineral-based FSPCM performance, the mineral-based FSPCM was prepared for specific functions or fields.

According to the collected data, the research progress of mineral-based stable PCM is rarely summarized. Herein, the review of minerals is mainly to summarize the preparation, performance, and application of multifarious minerals. In addition, this article also reviews minerals and their research progress. This review focuses on the following topics. The influence of vacuum impregnation and direct impregnation on mineral-based FSPCM properties were summarized. The influence of all kinds of minerals on mineral-based FSPCM properties was investigated and summarized. The characteristic application of mineral-based FSPCM is introduced, and its development prospect is forecasted. This chapter intends to gain a more detailed comprehension of the synthesis, properties, the applications of multifarious mineral-based FSPCMs, and to boost the growth of mineral-based FSPCMs.

The following are the other parts of this chapter. Section 2 introduces the preparation process of mineral-based FSPCM, it investigates the influence of vacuum impregnation and direct impregnation on FSPCM properties. In Sects. 3, 4, and 5, the development of FSPCM around the three mineral types studied is introduced: graphite with the best heat transfer performance and phase change material adsorption, diatomaceous earth with medium thermal conductivity and relatively large phase change material adsorption, and pearlite with poor heat transfer performance and medium phase change material adsorption. We have conducted in-depth research on the influence of multifarious minerals on the performance of mineral-based FSPCMs. Section 6 summarizes the influence of the rest minerals on mineral-based FSPCM properties and highlights a number of emblematic consequences. Section 7 provides the application of mineral-based FSPCM. Finally, Sect. 9 introduces the conclusions and prospects of mineral-based FSPCM.

2 Typical Preparation Method of Mineral-Based FSPCMs

2.1 Preparation of Mineral-Based FSPCMs

The main preparation procedures of all mineral-based FSPCM are similar. In addition, Fig. 2 shows the schematic diagram of the program. As shown in Fig. 6. Because impurities occupy the pore space of the natural mineral, thereby reducing the adsorption capacity of PCM, pretreatment is usually required to purify the original mineral. Therefore, chemical treatment methods, such as acid/base treatment, are often used

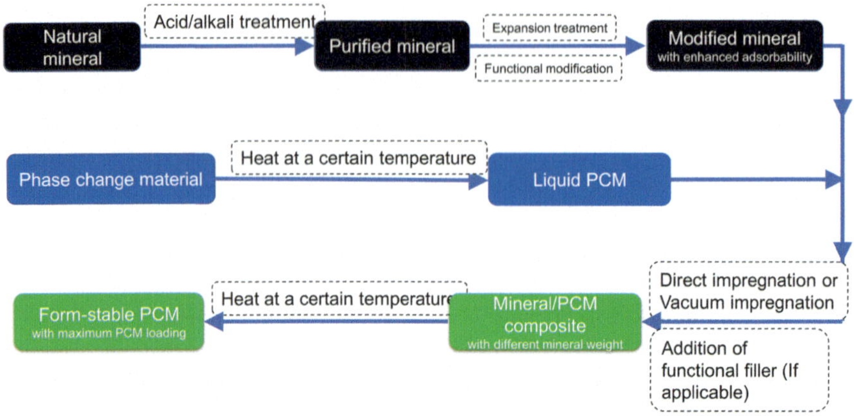

Fig. 2 Schematic of the preparation procedure of mineral-based FSPCMs

to clean up impurities from original minerals. Then, before they can be used as support materials for PCM, some minerals, like perlite, vermiculite, and graphite, must be expanded to improve their ability to adsorb PCM. In order to modify the above minerals, traditional high-temperature heating [10] and microwave radiation methods [11] are often used. Moreover, minerals are sometimes modified to adjust their surface wettability or to equip them with new functional groups.

Before proceeding with the impregnation step, the phase change materials must be heated until they become liquid. The modified minerals and liquid phase change materials use specific methods for combination. Common mixing approaches include vacuum impregnation and direct impregnation. In this mixing process, some functional fillers, such as nucleating agents, thickeners, and thermal conductivity enhancers can be added to the mixture to enhance the performance of the composite. After the mixing step, mineral-based composites with different mineral weights are obtained. The leakage of mineral-based composites is then studied by heating the phase change material in the composite material to melt. And then FSPCM is the material with a high phase change material load and no leakage.

2.2 *How Direct and Vacuum Impregnation Methods Affect the Properties of FSPCMs*

As we know, direct and vacuum impregnation are common combining methods for mineral-based FSPCM. The results indicate that direct impregnation and vacuum impregnation has distinct influences on PCM loading for different minerals. The effect is directly related to the pore size of minerals. When using minerals with relative pores, the impact of direct impregnation and vacuum impregnation on PCM load is almost the same, while when using minerals with smaller pores, the impact of direct impregnation and vacuum impregnation on PCM load is significantly different. The reason is that their adsorption principles are different. During the impregnation process, liquid phase change material enters the pores of minerals under the action of capillary force, and the atmospheric pressure present in the pores prevents its adsorption. So, when the atmospheric pressure in the pores is greater than the capillary force, the pores of minerals cannot absorb PCM. In general, smaller diameter pores have greater internal air pressure, so impregnation of PCM is challenging. In addition, if PCM does not completely fill the pores, the obtained PCM composite material has inherent leakage problems probably, as the air in the pore expands significantly at high temperatures [12].

2.2.1 Influence on PCM Loading in Graphite-Based FSPCMs

Specifically, graphite is an ordinary mineral that is commonly used for encapsulating PCM. Therefore, we chose graphite to contrast the influence of direct impregnation and vacuum impregnation on the phase change material load of graphite-based FSPCMs. In Table 1, the phase change material adsorption capacity of graphite by direct impregnation and vacuum impregnation are shown. There are almost identical phase change material load between direct impregnation and vacuum impregnation. Direct impregnation is more widely used than vacuum impregnation. In general, vacuum impregnation needs more complex experimental instruments (including vacuum pumps) and requires more energy than direct immersion input. Many researchers believe that direct impregnation is a more efficient way. However, it can be seen from Table 1 that the phase change material load of FSPCMs prepared using the vacuum impregnation method is not significantly different from that of FSPCM prepared directly impregnation method. In conclusion, because of the low price and uncomplicated equipment, it is recommended to prepare graphite-based FSPCM by direct impregnation method.

2.2.2 Influence on PCM Loading in Other Mineral-Based FSPCMs

Different from the research progress of graphite, the research of other minerals on the effect of FSPCM on PCM is not enough. According to the available studies, the PCM loading between direct impregnation and vacuum impregnation has considerable variation for diatomite and perlite. For vermiculite and perlite, it is known from previous studies that there is little change between direct and vacuum impregnation (Table 2), and vacuum impregnation is more suitable when the base materials have relatively small pores in diameter. The influence of direct impregnation and vacuum impregnation on the phase change material load of every mineral is different and needs to be further studied.

Table 1 Comparison of the PCM loadings of graphite-based FSPCM by the direct impregnation and vacuum impregnation

Preparation method	Mixing time (h)	Mixing temperature (°C)	Mixing speed (rpm)	Features	PCM	Loading	Refernces
Vacuum impregnation	6	60	–	**Advantages**: Removes the air in the pores of minerals and facilitates the PCM adsorption **Disadvantages**: Needs a vacuum pump, consumes electricity, and costs more	Stearic acid	90	[13]
	12	85	–		Polyethylene glycol	90	[14]
	1	80	30		Octadecane	76.9	[15]
	0.5	50	–		$Na_2SO_4 \cdot 10H_2O$	93	[16]
	0.5	80	–		$CaCl_2 \cdot 6H_2O$	60	[17]
	–	–	–		Sebacic acid	85	[18]
Direct impregnation	24	65	1 time/8 h	**Advantages**: Costs less and no complex equipment is required **Disadvantages**: Cannot remove the air in the pores of minerals, which may hamper the PCM adsorption	Myristic–palmitic–stearic acid	92.86	[19]
	1	100	–		Stearic acid	75	[20]
	–	65	–		Paraffin	85.6	[21]
	1	80	–		Stearic acid-benzamide	88	[22]
	–	160	–		$MgCl_2 \cdot 6H_2O$	83.33	[23]
	5 min	50	–		$CaCl_2 \cdot 6H_2O$	90	[24]

Table 2 Comparison of PCM loading by direct and vacuum impregnation in some mineral-based FSPCMs

Mineral	Mixing time (min)	Mixing temperature (°C)	Mixing speed (rpm)	PCM loading (wt. %)		Change in PCM loading (wt. %)	Ref
				DI	VI		
Diatomite	10–120	70–120	–	50.0	70.0	+ 20.0	[12]
Perlite	60	60	–	35.0	50.0	+ 15.0	[25]
Vermiculite	–	70	–	71.2	70.7	−0.5	[26, 27]
Sepiolite	–	–	–	25.4	25.2	−0.2	[28]

Note DI refers to direct impregnation and VI refers to vacuum impregnation

3 Graphite-Based Form Stable PCMs

3.1 Graphite Characteristics

Graphite has the highest PCM absorption and thermal conductivity and is therefore considered to be the best candidate for supporting materials in minerals used for the preparation of FSPCM. In more detail, the morphology of graphite is characterized by a carbon-based mineral with a hexagonal structure. The structure of natural graphite is layered, but after expansion, its structure is worm-like. The specific surface area of graphite is large and there are many pores, and graphite has good adsorption capacity after expansion. Expanded graphite (EG). In addition, due to the typical layered structure of graphite single crystals [29], minimal phonon scattering occurs in graphite flakes with perfect base surfaces; Therefore, the thermal conductivity of EG is prominent in the radial direction. Conversely, due to the low lattice vibration [30], the thermal conductivity perpendicular to the radial direction is reduced by a factor of 200, which results in the anisotropy of the thermal conductivity of the graphite. Its thermal conductivity was reported to be as high as 4180W/m·K radially, but the average thermal conductivity of commercial graphite is much lower (390W/m·K radially) [30]. It is a good performance, equivalent to high-conductivity copper (about 480W/m·K), silver (450W/m·K), beryllium oxide (260W/m·K), and aluminum nitride (200W/m·K); It is also much higher than alumina (25W/m·K) [31].

3.2 Influence of Graphite on the Properties of FSPCMs

3.2.1 Influence of Graphite on the Form Stability of FSPCMs

The formal stability of FSPCMs mainly depends on the capacity of materials to retain their shape, although the ambient temperature is higher than its phase transition temperature (PCM converted to liquid), the stability of FSPCM will not be affected. When PCM is encapsulated with porous materials, the formal stability of FSPCM is better when there is a higher PCM load. The phase change material load in FSPCMs represent the maximum load of the phase change material and no leakage occurs. Because of the large specific surface area and increased porosity after graphite expansion, the phase change material load of graphite-based FSPCM is generally between 80–90 wt.% (Table 3), and the prepared FSPCM has good morphological stability. Specifically, the PCM load in graphite-based FSPCM is affected by the specific surface area of the graphite. Moreover, the surface wettability between the base material and the PCM is also one of the influencing factors.

(a) Influence of the specific surface area of graphite

The higher specific surface area of EG generally indicates greater adsorption capacity of PCM. EG with a larger specific surface area can offer more space for PCM adsorption. For example, Kim et al. [15] produced EG/octadecane FSPCM where EG has a specific surface area of 37.6 m^2/g. The mass fraction of PCM in FSPCMs is about 76.9%. Li et al. [13] reported EG/stearate FSPCM; The EG they prepared has a larger specific surface area (76.0 m^2/g) and correspondingly higher PCM load. Therefore, the phase change material load capacity of graphite-based FSPCM increases with the increase of the specific surface area of graphite. Therefore, the phase change material load of graphite-based FSPCM increases as the graphite-specific surface area increases.

(b) Influence of surface wettability between the graphite and PCM

The shape stability of graphite-based FSPCMs can be improved by enhancing the surface wettability between expanded graphite and phase change material. Because the enhanced wettability can lead to a strong bond between EG and PCM, this promotes the progress of adsorption and restricts PCM to the graphite pore. Zou et al. [34] reported that the hydrophilic modification of EG could increase the inorganic PCM load from 51.1% to 82.6%. Xie et al. [35] applied A1203 for the first time to improve the hydrophilicity of EG, and their results showed that A1203EG could load up to 80.71wt.%PCM, which was 70wt.% higher than the original EG.

3.2.2 Influence of Graphite on the Phase Change Temperature of FSPCMs

Some research suggested that graphite can reduce the undercooling of inorganic phase change material and effectively enhance the phase transition behavior. Due to

Table 3 Effect of graphite on the PCM loadings and thermal properties of FSPCMs

	PCM types	PCM loading	Pure PCM		Graphite-based FSPCM		Specific surface area of carrier (m^2/g)	Hydrophobicity of carrier	Change in Temperature (ΔT, °C)	Percentage change in Latent heat ($\Delta H/H_p$, %)
			Temperature (T_p, °C)	Latent Heat (H_p, J/g)	Temperature (T_f, °C)	Latent Heat (H_f, J/g)				
Organic PCM	Myristic–palmitic–stearic acid [19]	**92.86**	42.38	159.8	42.99	151.4	–	–	+0.6	−5.3
	Succinic- adipic acid [32]	**90**	133.1	212.7	134.5	201.0	–	–	+1.4	−5.5
	Stearic acid [13]	**90**	67.54	225.67	68.29	192.77	76	–	+0.8	−14.6
	Stearic acid-benzamide [22]	**88**	63.16	200.04	64.54	174.02	–	–	+1.4	−13.0
	Sebacic acid [18]	**85**	129.0	214.2	128.0	188.0	–	–	−1.0	−12.2
	Palmitic acid [33]	**80**	61.53	192.59	60.81	149.66	–	–	−0.7	−22.3
	Octadecane [15]	**76.9**	26.5	219.2	25.0	160.7	37.6	–	−1.5	−26.7
Inorganic PCM	$Na_2SO_4 \cdot 10H_2O$ [16]	**93**	4.93	117.9	23.02	105.5	53.45	hydrophobic	+18.1	−10.5
	$MgCl_2 \cdot 6H_2O$ [23]	**83.33**	81.9	123.2	93.6	120.5	–	–	+11.7	−2.2

Note T_p-Solidification temperature of PCM, H_p-Solidification enthalpy of PCM, T_f-Solidification temperature of FSPCM, H_f-Solidification enthalpy of FSPCM

the large number of active sites in graphite, it promotes non-uniform nucleation of inorganic phase change material. A novel EG/$MgCl_2 \cdot 6H_2O$ FSPCM was prepared by Song et al. [23]. According to the report, adding 16.67 wt.% EG can reduce the undercooling of phase change material by 11.7 °C. Fu et al. [36] also reported on EG. It has the capacity to alleviate the undercooling of sodium acetate hydrate (more than 15 °C), while the synthetic graphite-based FSPCMs have lower undercooling (1.54 °C). For organic PCM that is not significantly supercooled, the phase transition temperature only changed due to graphite. Sari et al. [33] reported that after graphite adsorption of PCM, the phase transition temperature of PCM in FSPCMs increased a little compared to the original PCM. The reason for this phenomenon may be that the phase transition points measured by differential scanning calorimetry (DSC) have little hysteresis because of the large cooling rate of DSC, and this slight hysteresis becomes smaller because the thermal conductivity of graphite can increase the heat transfer rate of FSPCMs; Therefore, it has a slightly higher phase transition temperature.

3.2.3 Influence of Graphite on the Latent Heat of FSPCMs

The overall latent heat of FSPCM is reduced by the addition of graphite, and there are two possible reasons for this phenomenon. First, the storage and release of heat energy is a property of the PCM itself in the composite material. When graphite is added, due to its porous structure, the mass fraction of PCM in FSPCMs decreases, which causes the decrease of latent heat of FSPCM. Secondly, the free movement of liquid phase change material is limited by the pore structure in graphite, so the crystallinity of PCM is reduced, and the whole latent heat of FSPCM is also reduced. As shown in Table 3, when graphite is added, the latent heat decreases rapidly. For instance, Fu et al. [36] prepared a range of FSPCM with different mass fractions, among which the mass fractions of EG/$CH_3COONa \cdot 3H_2O$EG were 0, 3, 5, 8, 10, and 12 wt.%, and the latent heat characterized by DSC were 240.4, 238.6, 231.9, 223.1, 206.1, and 199.1 J/g, respectively. With the increase of EG mass fraction, related latent heat of FSPCMs will decrease accordingly. Li et al. [13] reported new FSPCMs with a mass fraction of 10 wt.% for EG. Unlike the original phase change material, the latent heat of the FSPCMs are reduced from 225.67 to 192.77 J/g. Compared with the original phase change material, the latent heat decline rate of graphite-based FSPCMs is 14.58%. In addition, in addition to 10wt%, the reduction of PCM in FSPCM, the restriction of Na free movement, and the latent heat of graphite-based FSPCMs are further reduced because of the inhibition of PCM crystallization through the EG pore structure. The graphite-based FSPCMs prepared by Gu et al. [37] and Li et al. [16] are the same as this result, and similar reductions in PCM crystallinity have been reported in their research.

3.2.4 Influence of Graphite on the Thermal Conductivity of FSPCMs

Due to the fact that the thermal conductivity determines the charging and discharging rate of the energy storage system, the thermal conductivity of PCM is very important in application. To greatly enhance the thermal conductivity of FSPCMs, graphite with large thermal conductivity is added to it. The addition of graphite provides a higher thermal conductivity passage in FSPCMs and improves their thermal conductivity. In addition, the thermal conductivity of FSPCM is also significantly related to the inherent properties of graphite, for example, anisotropy, interlayer distance, and FSPCM density.

(a) Influence of the role of graphite

In graphite-based FSPCM, graphite has two roles. On the one hand, when the amount of graphite added increases, the thermal conductivity of FSPCM significantly increases. On the other hand, as the mass fraction of graphite in FSPCM increases, the thermal conductivity of FSPCM will further improve. A thermal conductivity enhancement rate (TCER) was proposed, which is the ratio of the increase in thermal conductivity to the graphite load and was used to evaluate the effect of graphite on the thermal conductivity of FSPCM. As shown in Fig. 3, there is the result of using graphite-based FSPCM as both support materials and filler. Most TCERs do not exceed 20 when graphite is used as a filler [38–45]. In another case, the tier range of FSPCMs are mainly between 40 and 100. When graphite is used as the support material [13, 16, 18, 19, 22, 46–50], the TCER is larger than the sample using graphite as the filler. This may be because the three-dimensional net structure of expanded graphite can offer consecutive passages for the heat conduction in FSPCMs when it is used as a supporting material, while EG as a filler may agglomerate in FSPCM, resulting in uneven distribution, thus destroying the conduction passage and reducing thermal conductivity [51].

(b) Influence of the inherent properties of graphite

The thermal properties of graphite are closely related to its inherent properties like anisotropy and interlayer distance. Firstly, for the anisotropy of graphite, we know that the thermal conductivity of FSPCM containing EG also has obvious anisotropy characteristics, which is related to the anisotropy of EG. For instance, Wu et al. [20] The relationship between the thermal conductivity of expanded graphite and FSPCM was deeply studied. As the filling density of FSPCM reached 700 kg/m, the axial and radial thermal conductivity of FSPCM were 7.51 W/m·K^3 and 12.53 W/m·K^3, respectively, and the radial thermal conductivity was significantly higher than the axial thermal conductivity. Secondly, about the interlayer distance of graphite. According to relevant reports, the thermal conductivity of FSPCM increases with the increase of the distance between graphite layers. For instance, Lee et al. [52] According to the report, the thermal conductivity of FSPCMs reaches the highest value (3.56 W/m·K) when the graphite interlayer distance increases to the maximum (3.418 A), which is about 401% higher than that of FSPCM with EG interlayer distance of 3.340 A.

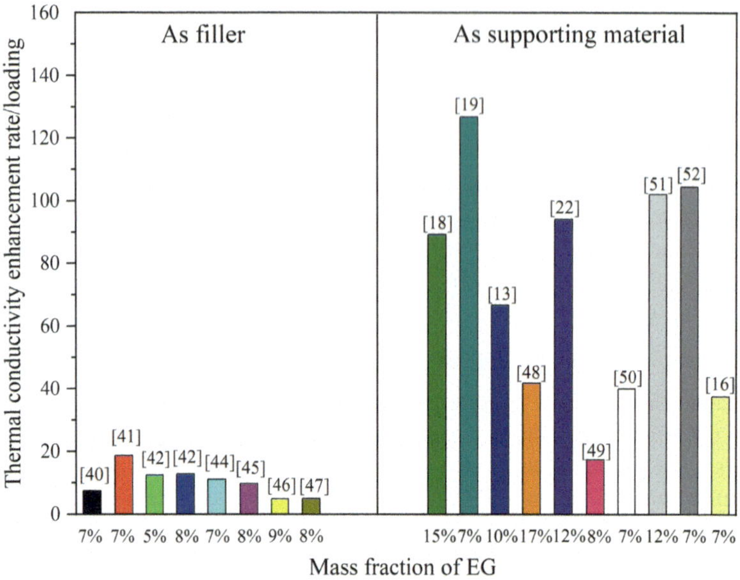

Fig. 3 Thermal conductivity of graphite as a support material and filler

(c) Influence of the microstructure of graphite affected by the density of FSPCM

When the density of FSPCM is low, the thermal conductivity of FSPCMs can be enhanced by increasing its apparent density to build a microscopic large thermal conductivity passage of internal graphite [53]. But there is also a density threshold for its apparent density. With the increase of the density of FSPCM, its thermal conductivity also increases, but when the optimal value is exceeded, the consecutive large thermal conductivity passage established by graphite is blocked by phase change material extruded from the graphite inner hole, and the thermal conductivity of FSPCM will gradually decrease with the continuous increase of its density. Wu et al. [20] deeply studied the relationship between the thermal conductivity of FSPCM and the microstructure of expanded graphite. At first, when the apparent density of FSPCMs increases, their thermal conductivity increases. When the packing density reaches the threshold, the thermal conductivity reaches the maximum. After that, the thermal conductivity decreases gradually as the apparent density continues to increase. This evolutionary process is able to be classified into four stages: Firstly, EG has a disordered worm-like microstructure, and incompact expanded graphite cannot come into close contact. At this stage, due to the discontinuity of the loose EG, the heat transfer is always interrupted, so the thermal conductivity is the smallest. Secondly, using EG to adsorb liquid phase change material to prepare EG-based FSPCMs. Then, press the inattentive FSPCM into fixed-shape blocks. At this stage, microscopic large thermal conductivity channels begin to form, and due to the low density of FSPCM, a large number of residual micropores inside the graphite lead to a high degree of discontinuity in the microscopic channels. Therefore, both axial and radial thermal conductivity can be enhanced compared with the first stage; These numbers are not big enough. Thirdly, as the degree of FSPCM compression increases, its porosity gradually decreases. As a result, the microscopic large thermal conductivity channels constituted by EG are continuous. The thermal conductivity gradually increases to its maximum value. Fourth, If the compression level continues to increase, the pores inside EG will become smaller than the volume of phase change material adsorbed in it. So, the phase change material in the pore of EG is squeezed out and the consecutive large thermal conductivity passage is interrupted, reducing the whole thermal conductivity of the FSPCMs.

3.2.5 Influence of Graphite on the Thermal Stability of FSPCMs

Thermal stability is defined as the capacity of FSPCMs to resist heat action. The influence of graphite on the thermal stability of FSPCM can be twofold. To begin with, the pore structure of EG can trap PCM in the cavity, and the interaction between graphite and PCM is able to delay the degradation of PCM and improve thermal stability. In addition, the high thermal conductivity of graphite also tends to cause degradation earlier. Therefore, different researchers have reported different effects of graphite on the thermal stability of FSPCMs. Liu et al. [32] used EG to adsorb eutectic mixtures of dicarboxylic acids. The initial weight loss temperature of FSPCM is higher than the original phase change material. Therefore, When adding EG, the thermal stability of FSPCM is improved. This is due to the interaction between graphite and PCM, which improves the thermal stability of PCM. In contrast, Wu et al. [20] studied the thermal stability of graphite/stearic acid FSPCM. Compared with pure stearic acid, the initial degradation temperature of graphite/stearic acid FSPCM was lower, indicating that its thermal stability was weakened. They attribute this change to graphite's enhanced thermal conductivity, which promotes the evaporation of stearic acid.

3.2.6 Influence of Graphite on the Thermal Reliability of FSPCMs

The reliability of FSPCMs represents the capacity of FSPCMs to remain stable over a long period of time, which is often characterized by accelerated thermal cycling tests. The results show that the FSPCM based on graphite has good thermal reliability, Despite testing, it was not found that graphite would significantly affect thermal reliability. It can be found that phase change material in FSPCMs keeps its thermal reliability after multiple thermal cycles, mainly because the adsorption of PCM in expanded graphite is a physical process and its chemical compatibility is good. Li et al. [16] produced EG/Na2SO4FSPCM and applied it to floor heating. In order to obtain the thermal reliability of FSPCM, 50 thermal cycles were carried out before and after. It is known from the test data that the phase transition points of melting and freezing only change by -1.55 and 0.63 °C. It is reasonable that enthalpy values declined by 9.16% and 0.69% respectively.

3.2.7 Influence of Graphite on Other Properties of FSPCMs

Graphite also has other excellent properties, such as good light absorption, strong electric conductivity, and good fire resistance, which also enhance the overall performance of graphite-based FSPCM. For example, to improve the photothermal conversion ability of PCM in composite materials, we can choose to add EG. EG can be used as a good light absorption part in FSPCMs to improve their light absorption performance. For instance, Zhang et al. [54] The solar energy memory of EG/RT-100 FSPCM was designed. The EG/RT-100 FSPCM is able to change the radiation into thermal energy and store it in phase-change material when it is exposed to solar radiation. Under the same lighting conditions, the temperature growth rate of graphite-based FSPCM is much faster than that of the original PCM. The thermal conductivity and electric conductivity of expanded graphite are excellent, which can not only be used to improve the photothermal conversion characteristics but also improve the electrothermal conversion performance of PCM. A kind of FSPCM prepared by Li et al. [55] uses expanded graphite as base material. When the mass fraction of expanded graphite in FSPCM is 15%, the FSPCM shows an excellent electric heating storage efficiency (65.7%) at 2.1 V voltage, thus improving the electric heating performance of PCM. In addition, due to the addition of expanded graphite, the fire resistance performance of FSPCMs is also improved. This is because the addition of EG forms a porous carbon layer under elevated temperature conditions and prevents the volatilization of paraffin, thus preventing the movement of fuel, oxygen, and heat between the block and the environment [56].

4 Diatomite-Based Form Stable PCMs

4.1 Diatomite Characteristics

Diatomaceous earth, a powdery mineral consisting of fossil wreckage, has three characteristics compared to other minerals used in the manufacture of FSPCM, namely correspondingly large PCM absorption and modest heat conductivity coefficient and can control humidity. First, diatomaceous earth has good adsorption capacity, which depends on its vesicular structure and biggish surface area [57]. Secondly, diatomite has high thermostability, and the heat conductivity coefficient is between 0.07 and 0.4 W/m·K [43, 58]. In addition, diatomaceous earth also has the ability to control humidity, which is determined by its inherently large surface area.

4.2 *Influence of Diatomite on the Properties of FSPCMs*

4.2.1 Influence of Diatomite on the Form Stability of FSPCMs

Diatomaceous earth has the characteristics of porous structure and low density, so the PCM load of diatomaceous earth-based FSPCM can be relatively high. Table 4 lists the PCM load and heat characteristics of diatomaceous earth-based FSPCM reported in the document. As shown in the following table, the PCM load range of most diatomaceous earth-based FSPCM is 60–70%, although a small number of researchers report that the PCM load is less than 50%, in general, the overall data show that the specific surface area and particle size of diatomaceous earth have a significant effect on the PCM load of diatomaceous earth-based FSPCM. That is, diatomite has good morphological stability for FSPCMs. The PCM load of diatomaceous earth-based FSPCM increases with the increase of the specific surface area of diatomaceous earth because, with the increase of the specific surface area, diatomaceous earth can provide more space to encapsulate PCM [58]. In addition, the adsorption capacity of diatomite for PCM is also related to its particle size, and the smaller the particle size, the higher the adsorption capacity of PCM. For example, Lee et al. [67] Three different particle sizes of diatomite were used to synthesize FSPCMs. The average diatomite diameters used are 4.276, 623.605, and 709.298 μm, respectively, and the corresponding PCM loads are 50, 38, and 35%, respectively. The experimental data show that the smallest particle size diatomite has the maximal PCM absorptivity. The writer analyzed that the disparate adsorption mechanism of diatomite may lead to the study results. When the agglomerant is lesser than the solid grain, a lot of bonding materials are distributed on the surface of particles during nucleation. When the agglomerant is greater than the solid grain, nucleation produces a nucleus with saturated pores because smaller particles are immersed in the larger binder material. The best diatomite adsorption principle is immersion, while the other two diatomite adsorption mechanism is coalescence, resulting in differences in PCM load in diatomite-based FSPCM.

4.2.2 Influence of Diatomite on the Phase Change Temperature of FSPCMs

The phase transition temperature of inorganic PCM and organic PCM is different under the influence of diatomite. As for the undercooling of inorganic PCM, diatomite as a nucleation catalyst can inhibit the undercooling of PCM because it can increase its crystallization rate by providing out-of-phase nuclear sites for inorganic PCM. Rao et al. [66] A kind of diatomaceous earth/$MgNO_3 \cdot 6H_2O$ FSPCM was studied. The experiment found that the undercooling capacity reduced from 7 °C to only 0.3 °C due to the addition of diatomaceous earth. Different from inorganic PCM, the phase transition temperature of organic PCM is almost steady after adding diatomite because of the good compatibility between organic PCM and diatomite. Xu et al.

Table 4 Relationship between PCM load and thermal properties of FSPCM and diatomite

	PCM types	PCM loading	Pure PCM		Diatomite-based FSPCM		Specific surface area of carrier (m^2/g)	Hydrophobicity of carrier	Change in temperature (ΔT, °C)	Percentage Change in enthalpy($\Delta H/H_p$, %)
			Temperature (T_p, °C)	Latent Heat (H_p, J/g)	Temperature (T_f, °C)	Latent Heat (H_f, J/g)				
Organic PCM	lauric acid-stearic acid [59]	72	28.9	166.9	30.0	114.5	68.9325	–	**+1.1**	**−31.4**
	Myristic acid [60]	72	53.0	173.6	53.3	119.4		–	**+0.3**	**−31.2**
	Capric-palmitic acid [61]	63	20.5	149.8	21.9	90.0		–	**+1.4**	**−39.9**
	Erythritol tristearate [62]	62	28.8	207.2	30.0	114.8		–	**+1.2**	**−44.6**
	Capric acid [63]	54	27.7	168.0	28.3	88.0	5.5108	–	**+0.6**	**−47.6**
	Lauric acid [64]	30	40.1	142.3	38.7	57.2		–	**−1.9**	**−59.8**
Inorganic PCM	Na_2SO_4 [65]	65	879.0	165.2	877.6	102.3	24.05	–	**−1.4**	**−38.1**
	$LiNO_3$ [65]	60	244.5	322.2	243.6	190.3	24.05	–	**−0.9**	**−40.9**
	$Mg(NO_3)\cdot 6H_2O$ [66]	45	81.4	163.1	88.4	74.2	4.441	–	**−7.0**	**−54.5**

Note T_p-Solidification temperature of PCM, H_p-Solidification enthalpy of PCM, T_f-Solidification temperature of FSPCM, H_f-Solidification enthalpy of FSPCM

[68] The diatomite/paraffin/carbon nanotube (CNT) FSPCM was synthesized, and the melting temperatures of the original PCM and diatomite-based FSPCM were found to be 27.47 °C and 27.12 °C, respectively, indicating that such changes could be ignored in practical applications.

4.2.3 Influence of Diatomite on the Latent Heat of FSPCMs

The potential heat of FSPCM is reciprocally ratio to the mass fraction of diatomite. The higher the mass fraction of diatomite is, the lower the potential heat of FSPCM is. The addition of diatomaceous earth reduces the mass fraction of PCM in FSPCM. Since PCM in FSPCM can store heat energy within a given temperature scope, the entire potential heat of diatomaceous earth-based FSPCM is also reduced. For example, Wen et al. [42] A diatomaceous earth/fatty acid FSPCM was prepared. According to the experimental data, the latent heat of pure PCM and diatomaceous earth-based FSPCM were 141.5 and 87.33 J/g, respectively.

4.2.4 Influence of Diatomite on the Thermal Conductivity of FSPCMs

According to the reported literature, the heat conductivity of diatomite ranges from 0.07 to 0.4 W/m·K [57, 58]. The heat conductivity of diatomaceous earth-based FSPCM is greatly affected by the different sources and densities of diatomaceous earth and any modifications to it. When the thermal conductivity of diatomite is greater than that of PCM, the holistic heat conductivity of diatomite-based FSPCM grows with the increase of diatomite content. When its heat conductivity is smaller than PCM, the addition of diatomite makes the thermal conductivity of diatomite FSPCM lower than the original PCM. In general, when the heat conductivity of diatomaceous earth-based FSPCMs are not enough to satisfy the admissions requirements; Thier heat conductivity can be improved by adding accessional heat conductivity enhancers such as silver [58], EG [67], carbon nanotubes [68], or graphene [69]. For instance, Sari et al. [70] produced a diatomaceous earth/carbon nanotube/PEG FSPCM. It is known that the thermal conductivity of the original PCM, diatomaceous earth, and diatomaceous earth-based FSPCM are 0.22, 0.08, and 0.15 W/m·K, respectively, indicating that the thermal conductivity of diatomaceous earth is significantly lower than that of the original PCM. The experimental results show that the thermal conductivity of FSPCM based on diatomaceous earth is also lower than that of the original PCM. The thermal conductivity of the FSPCM can then be improved by using carbon nanotubes with high thermal conductivity, which increased by 93% to 0.29 W/m·K when 2.5 wt.% carbon nanotubes were used. In another case, Qian et al. prepared a terpolymer FSPCM containing diatomaceous earth, polyethylene glycol, and silver [58] on the condition that the thermal conductivity of diatomaceous earth was higher than PCM, and the heat conductivity of FSPCM with diatomaceous earth was increased from less than 0.1 W/m·K to about 0.3 W/m·K, showing remarkable results. With the addition of silver, the heat conductivity increases sharply, further increasing to 0.82W/m·K, which is another major advance in the research.

4.2.5 Influence of Diatomite on the Thermal Stability of FSPCMs

Diatomaceous earth has good thermal stability, so with the addition of diatomaceous earth, the heat stability of diatomaceous earth-based FSPCM is better than the original PCM. The entirety heat stability of diatomaceous earth-based FSPCM is improved. Wen et al. prepared an FSPCM with significantly improved thermal stability [42]. In the experiment, capsaico-lauric acid (CA-LA) mixture was used as PCM and diatomite as sustentacular material. The initial degradation temperatures of CA-LA and diatomaceous earth/CA-LA are 172.5 °C and 178.3 °C respectively, which are increased by 5.8 °C severally. The consequence shows that the thermal stability of FSPCM is significantly improved with the addition of diatomite.

4.2.6 Influence of Diatomite on the Thermal Reliability of FSPCMs

The PCM in diatomaceous earth-based FSPCM can maintain its inherent good thermal reliability. This is because the impact of diatomaceous earth on the heat reliability of FSPCMs is negligible. Klarman et al. [43] proposed a diatomaceous earth/PEG FSPCM that could be used for thermal energy storage. The DSC curve is the same after 1000 thermal cycles. The experimental results show that after 1000 cycles, the thermal reliability of FSPCM is high, and the latent heat attenuation is only 1.1%. It is proved that diatomaceous earth has little effect on the heat reliability of FSPCM.

4.2.7 Influence of Diatomite on Other Properties of FSPCMs

Among the other properties of diatomaceous earth, its correlation with humidity is more prominent. Diatomaceous earth can also improve the humidness control capability of FSPCM due to its correspondingly large specific surface area and large poriness. For instance, Chen et al. [71] prepared an alkane mixture/SiO_2/diatomaceous earth FSPCM with good humidity control properties. The moisture absorption of diatomaceous earth is better than 8%, and PCM has a good ability to control humidity because of diatomaceous earth as a humidity control component.

5 Perlite-Based Form Stable PCMs

5.1 Perlite Characteristics

Although perlite has moderate phase change material adsorption capacity and low thermal conductivity, its low cost and good chemical inertness still make the preparation of FSPCM increasingly attractive. Specifically, perlite is a natural mineral that has been saturated with water for a long time, mainly composed of volcanic obsidian. It will transform into expanded perlite at 850–900 °C, which has a larger specific surface area and smaller density than perlite. The water in the perlite structure is evaporated during heating, causing it to expand; Compared to the unheated volume, it expands by 7–16 times; Therefore, this also makes the adsorption performance of expanded perlite good[72]. Due to its low thermal conductivity (approximately 0.09 W/m K [73]), perlite is usually a good choice for preparing insulation materials. Moreover, thanks to the good chemical inertness of perlite; Although it is added to other materials, chemical reactions will not occur. Pearlite also has a high melting temperature (950–1050 °C) [74], which allows it to maintain long-term stability in high-temperature environments.

5.2 Influence of Perlite on the Properties of FSPCMs

5.2.1 Influence of Perlite on the Form Stability of FSPCMs

In Sect. 3.2.1, it was mentioned that higher PCM loads represent better shape stability for mineral-based FSPCM. Compared with other mineral-based FSPCMs, perlite-based FSPCMs have moderate PCM adsorption capacity. Table 5 lists the load capacity and thermal performance of pearl rock-based FSPCM.As shown in the table, the PCM load of the pearl rock-based FSPCM is generally maintained at around 55%. The reason for this result is that the specific surface area of expanded perlite is relatively large and its density is low, and the expanded perlite itself is a layered structure with a large number of interconnected network channels inside. The capillaries in these channel pores have binding forces that can adsorb a certain amount of phase change material without leakage.

In addition, the surface wettability between phase change materials and perlite is one of the important factors affecting the PCM loading of perlite-based FSPCMs. The two components with good wettability have stronger binding forces, so better wettability usually represents a larger PCM loading amount. Ramakrishnan et al. [25] investigated the relationship between the shape stability and surface wettability of perlite-based FSPCMs. In the research process, the first step is to modify the surface of perlite to obtain hydrophobic perlite. The experimental results show that the contact angle between modified expanded perlite (EPO) and water is as high as 110.88°, which is 40.02° higher than the contact angle between raw expanded

Table 5 Effect of perlite on the PCM loadings and thermal properties of FSPCMs

PCM types	PCM loading	Pure PCM		Perlite-based FSPCM		Specific surface area of carrier(m^2/g)	Hydrophobicity of carrier	Change in Temperature (ΔT, °C)	Percentage change in Latent heat ($\Delta H/H_p$, %)
		Temperature (T_p, °C)	Latent Heat (H_p, J/g)	Temperature (T_f, °C)	Latent Heat (H_f, J/g)				
Tetradecanol [75]	77	35.8	204.6	35.9	158.1	–	–	+0.1	−22.7
Capric acid–palmitic acid [76]	64	23.7	137.9	25.7	88.4	–	Hydrophobic	+2.0	−35.9
Paraffin [77]	62.5	48.3	181.3	48.5	105.4	233	–	+0.2	−41.9
Paraffin [78]	60	29.0	152.6	28.9	88.7	–	–	−0.1	−41.9
Paraffin [79]	60	32.9	268.1	34.3	155.3	–	–	+1.4	−42.1
Lauric–palmitic–stearic acid [80]	55	30.4	150.6	30.3	81.3	1.294	–	−0.1	−46.0
Paraffin (hydrophobic EP) [25]	50	25.5	132.1	24.6	61.8	–	hydrophobic	−0.9	−53.2
Paraffin (hydrophilic EP) [25]	30	25.5	132.1	24.3	32.8	–	hydrophobic	−1.2	−75.2

Note T_p-Solidification temperature of PCM, H_p-Solidification enthalpy of PCM, T_f-Solidification temperature of FSPCM, H_f-Solidification enthalpy of FSPCM

perlite (EPW) and water. Therefore, the surface modification of perlite improves its hydrophobicity. Then, hydrophilic and hydrophobic perlite/paraffin composite materials were prepared, with PCM loading controlled at 35, 40, and 50 wt%. By using vacuum impregnation to adsorb PCM, it was found that the maximum loading capacities of hydrophilic and hydrophobic perlite were 35 and 50 wt%, respectively, without leakage. The author believes that paraffin, as a typical organic PCM, the strengthening bond between paraffin and hydrophobic perlite determines the difference in PCM loading. At present, research on inorganic phase change materials is still limited.

5.2.2 Influence of Perlite on the Phase Change Temperature of FSPCMs

The chemical compatibility between pearlite and PCM is good, and after adding pearlite, the change in phase transition temperature of the PCM is very small. As shown in Table 5, the temperature variation of the perlite/paraffin FSPCM manufactured by Zhang et al. [77] is only within 1.0°. After adding perlite, the melting temperature of paraffin decreased from 44.32 °C to 44.15 °C, and the freezing temperature increased from 48.30 °C to 48.51 °C. From the results, it is possible to conclude that the melting temperature and freezing temperature of paraffin changed by 0.17 and 0.21° C, respectively, which can be ignored in practical applications.

5.2.3 Influence of Perlite on the Latent Heat of FSPCMs

As shown in Table 5 and Fig. 4, as the mass fraction of perlite decreases, it will lead to an increase in the latent heat of FSPCMs, and there is a certain linear relationship between them [81, 82]. Because within a specific temperature range, only phase change materials in FSPCM can provide latent heat, and adding perlite reduces the mass fraction of phase change material, resulting in a decrease in total latent heat of FSPCMs.

5.2.4 Influence of Perlite on the Thermal Conductivity of FSPCMs

In addition, the thermal conductivity of perlite is lower than most phase change materials, so when perlite is added to it, the heat transfer performance of perlite-based FSPCMs will show a downward trend. However, the poor thermal conductivity of perlite-based FSPCMs is not conducive to practical applications, so the method of adding large thermal conductivity fillers is usually used to improve its thermal conductivity. Karaipekli et al. [79] investigated the thermal conductivity of perlite/paraffin FSPCMs. The thermal conductivities of pure paraffin, expanded perlite, and the perlite/paraffin FSPCMs were 0.22 and 0.15 W/m·K, respectively. Adding perlite reduced the thermal conductivity of paraffin by 0.07 W/m K. So as to enhance its thermal conductivity, carbon nanotubes (CNTs) were used as thermal conductivity

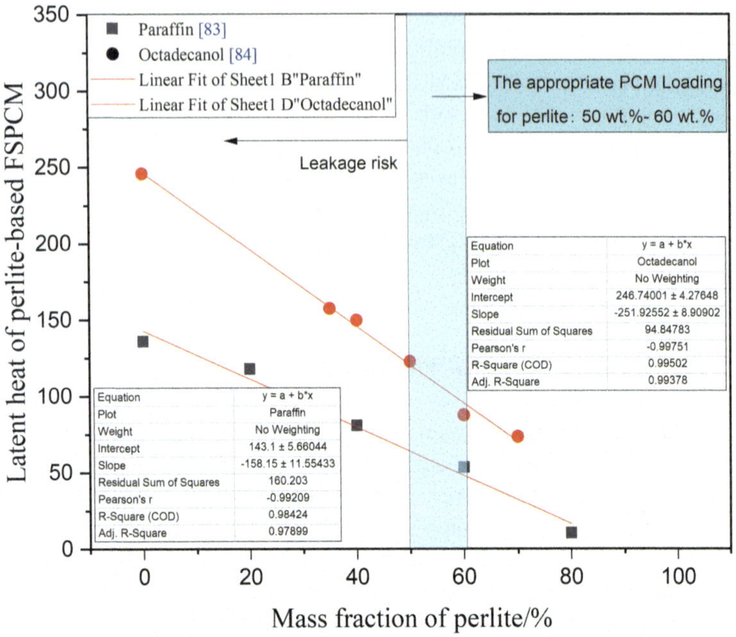

Fig. 4 Relationship between the mass fraction of perlite and latent heat of the FSPCM [81, 82]. *Note* the appropriate PCM loading is obtained from Table 5

enhancers, and the results showed that adding 1wt.% CNTs can enhance the thermal conductivity from 0.15 to 0.32 W/m K.

5.2.5 Influence of Perlite on the Thermal Stability of FSPCMs

Due to the excellent chemical inertness of perlite and the interaction between PCM and the inner surface of perlite, the thermal stability of perlite-based FSPCM is much higher than that of the original PCM. Increasing the initial degradation temperature of FSPCM using thermogravimetric analysis (TGA) can reflect the improvement effect of material thermal stability. Zuo et al. [83] prepared a perlite/paraffin FSPCM for thermal energy storage and systematically studied its thermal performance. The degradation of pure PCM is approximately 170 °C, while FSPCM based on perlite only begins to experience weight loss when the temperature reaches 200 °C, indicating a significant improvement in its thermal stability.

5.2.6 Influence of Perlite on the Thermal Reliability of FSPCMs

Pearlite itself has excellent chemical inertness and will not have a fundamental impact on the thermal reliability of FSPCM, and its thermal reliability remains within an acceptable level after multiple thermal cycles. For example, Zuo et al. [83] explored the thermal reliability of perlite-based FSPCMs and found that the latent heat of perlite-based FSPCMs before and after 200 thermal cycles were 112.7 and 110.1 J/g, respectively, a decrease of only 2.3%.

6 Other Mineral-Based Form Stable PCMs

Apart from the three minerals mentioned above, there is little attention and research on other mineral-based FSPCMs. However, currently known researches have shown that minerals, for example, vermiculite [84], kaolin [85], sepiolite [86], halloysite [87], attapulgite [88], montmorillonite [89], mullite [40], and palygorskite [90] can all be used as raw materials for synthesizing FSPCM.Because the properties of different minerals and their interaction with PCM vary, the properties of mineral-based FSPCM may be affected differently by each mineral.

6.1 *Influence of Other Minerals on the Form Stability of FSPCMs*

Due to the interaction between phase change materials and minerals, this may affect the PCM loading capacity of mineral-based FSPCM. Particularly, surface wettability is affected by interactions between phase change materials and minerals, and good surface wettability is one of the important factors in improving the phase change material loading capacity of halloysite-based FSPCM. For example, Liang et al. [91] used polydimethylsiloxane to modify the surface of halloysite and combined it with paraffin in a vacuum environment to obtain superhydrophobic halloysite-based FSPCM. The consequences indicated that compared to the original halloysite, the adsorption capacity of superhydrophobic halloysite for PCM increased by 11.4 wt.%, with a loading capacity of up to 62.9 wt.%. The strong affinity of superhydrophobic halloysite not only enhances the adsorption capacity of PCM but also reduces PCM leakage; Therefore, superhydrophobic halloysite has a greater PCM loading capacity than hydrophilic halloysite.

6.2 *Influence of Other Minerals on the Phase Change Temperature of FSPCMs*

If certain inorganic phase change materials have a high undercooling, minerals can serve as nucleating agents, promoting non-uniform nucleation and reducing the undercooling of inorganic phase change materials. Deng et al. [92] prepared vermiculite/$Na_2HPO_4\cdot 12H_2O$ FSPCM. The undercooling of this vermiculite-based FSPCM is 4 °C lower than the pure phase change material. On the contrary, organic phase change materials and minerals have very good chemical compatibility; Therefore, the interaction between minerals and organic PCM hardly affects the phase transition temperature[93], which makes the phase transition temperature of organic PCM tend to stabilize.

6.3 *Influence of Other Minerals on the Latent Heat of FSPCMs*

Usually, the latent heat of mineral-based FSPCM is affected by the mass fraction of PCM in it and is proportional, as only PCM in FSPCM has the function of storing and releasing heat, which contributes to the latent heat [94]. In addition, under the same mass fraction of PCM, the latent heat of mineral-based FSPCM is also affected by interactions between minerals and phase change materials. For example, Shi et al. [88] modified attapulgite with N—(2-aminoethyl)-3-aminopropyltrimethoxysilane (KH792). Then, three types of phase change materials (paraffin, polyethylene glycol, and lauric acid) were synthesized using modified attapulgite-based FSPCM. The results showed that when polyethylene glycol and paraffin were used as phase change materials with a mass fraction of 40%, the latent heat of FSPCM increased by 27.8% and 31.5% after modification with attapulgite, respectively. When lauric acid was used as PCM, the latent heat of the modified attapulgite-based FSPCMs was lower than the original attapulgite-based FSPCMs. The reasons for the above differences may be the presence of many hydroxyl groups on the surface of the original palygorskite, as well as the intense interaction between the original palygorskite and phase change materials such as polyethylene glycol and paraffin. After being modified by KH792, the number of hydroxyl groups on the surface of attapulgite significantly decreased, and then amino groups replaced some of the absorption sites of attapulgite. Phase change materials such as paraffin and polyethylene glycol weaken their interaction with modified attapulgite due to this displacement, and the latent heat and crystallinity of the two modified attapulgite-based FSPCMs increase accordingly [95, 96]. Because lauric acid has various functional groups, the interaction between attapulgite and lauric acid was not weakened by surface modification; Therefore, compared with the original attapulgite, the modified attapulgite has no increase in latent heat.

6.4 Influence of Other Minerals on the Thermal Conductivity of FSPCMs

For most mineral-based FSPCMs, if the heat transfer ability of the PCM is weaker than that of the mineral, its thermal conductivity is stronger than the original phase change material, and vice versa. However, the thermal conductivity of most mineral-based FSPCM does not reach the standards for applications; Adding high thermal conductivity enhancers is an effective method to enhance thermal conductivity. Because high thermal conductivity fillers form continuous channels within mineral-based FSPCMs, which play a positive role in increasing heat conduction, high thermal conductivity enhancers can improve thermal conductivity.

6.5 Influence of Other Minerals on the Thermal Stability of FSPCMs

Minerals mainly affect the thermal stability of mineral-based FSPCM from two aspects. To begin with, the pore structure of minerals provides storage space for PCM, and the interaction between phase change materials and minerals helps to increase the initial degradation temperature of PCM, thereby improving its thermal stability. In addition, some minerals with relatively high thermal conductivity often cause PCM to degrade earlier. Therefore, the positive[85] and negative[97] effects of minerals on thermal stability have been reported.

6.6 Influence of Other Minerals on the Thermal Reliability of FSPCMs

The inherent thermal reliability of the original phase change material determines the thermal reliability of mineral-based FSPCMs, and due to the good chemical compatibility between PCM and minerals, minerals have no significant impact on thermal reliability. For example, Shen et al. [98] synthesized a sepiolite/lauric acid FSPCM for thermal energy storage and studied its thermal reliability using thermal cycling tests. The consequences showed that the latent heat of FSPCM based on sepiolite decreased by only 3% after 300 thermal cycles, which is equivalent to pure lauric acid.

7 Applications of Mineral-Based Form Stable PCMs

7.1 *Effect of Mineral-Based FSPCMs on Building Performance*

Because mineral-based FSPCMs have the advantages of low cost and large heat storage capacity, it is widely used to reduce indoor temperature fluctuations in buildings. For example, mineral-based FSPCM can be integrated into buildings using direct mixing methods [99] and independent PCM panels [100]. In addition, multiple parts of the building, including (a) walls [24, 101, 102], (b) roofs [103, 104], and (c) floors [105, 106], all use the mineral-based FSPCMs and their impact on building performance was studied through experiments and numerical calculations [24].

7.1.1 Influence of Incorporating Mineral-Based FSPCMs into Buildings

Existing research has demonstrated that the combination of mineral-based FSPCMs is able to boost the thermal inertness of building materials and decrease indoor temperature fluctuations in buildings. Abden et al. [107] applied diatomite-based FSPCM to building ceilings. In their research, a diatom-based FSPCM was prepared to address the leakage issue of phase change materials, and its PCM loading capacity was 40%. Then, a gypsum board based on FSPCM was synthesized and a little laboratory with FSPCM-integrated gypsum board ceiling was constructed and modeled. The thermal/energy performance was estimated under actual environmental conditions and the economic benefits of traditional gypsum boards without FSPCM were analyzed. The consequences showed that the peak temperature of this laboratory without FSPCMs was significantly greater than the laboratory with FSPCMs, and the laboratory with FSPCM decreased by 8.9, 8.8, and 3.4 °C on days 1, 2, and 3, respectively. After incorporating FSPCM into the building, it can save 16.2% of the cooling load, and the investment payback period is just 1.7 years.

7.1.2 Influence of the Position and Thickness of Mineral-Based FSPCMs

Under the same conditions as FSPCMs, the thickness of phase change material panels and their position in building walls are important factors affecting the thermal performance of buildings. Ye et al. [24] applied graphite-based FSPCM panels to buildings to study the influence of FSPCM position on indoor temperature. Replace one of the four gypsum boards on each wall with the prepared FSPCM panel. The results indicate the indoor temperature changes in the external environment, data room, phase change material room, and testing room with four PCM panels installed in different locations. From the figure, it can be seen that the temperature amplitude of all test rooms using PCM panels is significantly lower than that of the reference room. In addition, when the PCM panel moves inward, the maximum temperature inside the phase change material continuously decreases from 30.7 °C to 29.6 °C, while the minimum temperature increases from 21.7 °C to 22.6 °C. The temperature change is minimal in the laboratory where the FSPCM panel is closest to the inner layer (farthest from the external environment). The reason for this result is that the heat consumption from the phase change material panel to the external environment decreases as the PCM panel moves from the outside to the inside; Therefore, when the FSPCM panel is at the innermost layer, the thermal performance of laboratory is the best. In addition, numerical methods were used to study the relationship between panel thickness, PCM type, and indoor temperature in the building, and the consequences showed that the thickness with the best effect was 8–10 mm.

7.2 *Influence of Mineral-Based FSPCMs on Battery Temperature*

In order to keep the high performance and energy ability of the battery, the operating temperature is supposed to be maintained within 60 °C. Because it has great thermal capacity and few operating costs, mineral-based FSPCM is superior to other temperature regulation methods. For example, Zhang et al. [6] mixed graphite with kaolinite-based FSPCMs to improve its thermal conductivity and loading capacity. The prepared FSPCM is applied to the battery management system. The consequences showed that the mineral-based FSPCM is able to effectively suppress the temperature increase of the battery. As the ambient temperature, in the beginning, was 26–28 °C, under the action of mineral-based FSPCM, the temperature of the battery was maintained below 45 °C, and the highest temperature was reduced by 27.74%. Moreover, the temperature of the battery with the addition of mineral-based FSPCMs was more uniform than that without mineral-based FSPCMs. As the discharge rate was 4 °C, the temperature change of the battery was just limited to 4.04 °C, which is 54.62% lower than that of batteries without FSPCM.

7.3 *Influence of Mineral-Based FSPCMs on the Performance of Photovoltaic Panels*

The impact of operating temperature on the performance of photovoltaic panels is very sensitive. Specifically, the photoelectric conversion efficiency of photovoltaic panels decreases by 0.4–0.65% when the temperature increases by 1 °C[108]. The phase change zone of PCM is narrow, and their heat storage capacity is large, which can effectively manage the surface temperature of photovoltaic panels and enhance their performance. For example, Luo et al. [7] proposed a PV-PCM system that combines photovoltaic panels with graphite-based FSPCM. The temperature of the PCM-PV system can be maintained below 50 °C for 200 min, which is 146 min longer than traditional photovoltaic panels. In addition, within 192 min, the output voltage of the PV-PCM panel was consistently higher than that of traditional photovoltaic panels, with the biggest increment of 1.11 V and an average increment of 0.75 V. Throughout the entire process, the total output power increased by 7.28%. Therefore, integrating PCM into photovoltaic panels can significantly improve its performance.

7.4 Influence of Mineral-Based FSPCMs on the Thermal Regulation of Asphalt Binders

When the ambient temperature is extremely high or low, the asphalt mixture is easy to cause rutting damage and thermal cracking [109]. Due to the narrow range of phase transition temperature and high latent heat, PCM is able to be applied to improve the performance of asphalt, as a suitable material. Zhang et al. [110] synthesized EG/ polyethylene glycol composite phase change material for regulating the temperature of asphalt binder and studied its feasibility in adjusting the temperature of asphalt mixture. To study the influence of FSPCM-modified asphalt binder on temperature regulation, the time taken for each sample to rise to 65 °C was compared. The results show that graphite-based FSPCM could reduce the thermal sensitivity of asphalt binder and improve its performance.

7.5 Influence of Mineral-Based FSPCMs on Other Applications

Researchers have reported various applications of mineral-based FSPCM in other fields, too. Such as solar distillers [111, 112], refrigerators [113, 114], waste heat recovery [115, 116], and interface solar steam generation [117]. The results show that mineral-based FSPCMs could enhance the performance of the above systems to a certain extent. Cheng et al. [111] applied graphite-based FSPCMs to enhance the performance of solar distillers. The consequences indicated that the daily water production using graphite-based FSPCM reached 3.41 L/m^2, which increased by 43.3% compared to the system without FSPCM. The graphite-based FSPCMs collected solar energy and stored it during the day, and were used to heat seawater at night, thereby extending the time for heating seawater and improving the daily productivity of water.

8 Conclusions and Future Works

This chapter systematically reviews and discusses the latest developments in mineral-based FSPCM. In particular, the influence of direct impregnation and vacuum impregnation on the performance of mineral-based FSPCM was compared and analyzed. This article reviews the influence of multiple minerals on the performance of mineral-based FSPCM. The typical applications of mineral-based FSPCM were introduced, and some future work was proposed.

(1) The main synthesis methods of mineral-based FSPCMs are similar, and direct and vacuum impregnation are the two most common synthesis methods. For graphite with larger pores, the use of direct impregnation and vacuum impregnation has almost no effect on the phase change material load in mineral-based FSPCM. On the contrary, for perlite with small pores, the phase change material load of FSPCMs prepared by direct impregnation and vacuum impregnation is significantly increased compared to the latter.

(2) Due to the different microstructures of each mineral, most minerals have good PCM adsorption properties. Diatomaceous earth and graphite can increase the phase transition temperature of PCM and alleviate the supercooling phenomenon of inorganic PCM, but the effect of residual minerals on this performance has not been reported. Usually, with the addition of minerals, the mass fraction of PCM in mineral-based FSPCMs decreases, resulting in a decrease in the latent heat of mineral-based FSPCM. Furthermore, due to the limitations of minerals like EG, the movement and crystallization of PCM are inhibited, further reducing its latent heat. Because of the good thermal conductivity of expanded graphite, the addition of expanded graphite can greatly enhance the thermal conductivity of FSPCM. The thermal conductivity of graphite is generally greater than that of other minerals; Therefore, it is necessary to add high thermal conductivity enhancers to enhance the whole thermal conductivity of else mineral-based FSPCMs.

    On most occasions, due to the excellent thermal stability of minerals themselves, the overall thermal stability of FSPCM can be improved. Minerals have no substantial impact on thermal reliability, making PCM keep its intrinsic great thermal reliability.

(3) By improving the performance of mineral-based FSPCMs and addressing their leakage issues, mineral-based FSPCM is able to be widely used to enhance the thermal inertia of buildings, improve the thermal management of batteries, improve thermal regulation of asphalt adhesives, and promote temperature control of photovoltaic panels. In addition, applications like solar distillers, refrigerators, and waste heat recovery have also been proven viable.

(4) Due to the shortage of guidance on preparation methods, direct impregnation and vacuum impregnation have been blindly used to prepare mineral-based FSPCM. It is recommended to conduct additional research to determine the impact of direct impregnation and vacuum impregnation on every mineral-based FSPCM. In addition, the physical performance of minerals and their interactions

with phase change material play a key part in the performance of mineral-based FSPCM, but there is relatively little research on it. Therefore, a more comprehensive preparation study is needed to explore the relationship between the physical performance of minerals and the performance of mineral-based FSPCM. In addition, even if graphite-based FSPCM has good thermal conductivity, a more efficient method is needed to improve the thermal conductivity of other mineral-based FSPCM to reach the requirements of applications. In addition, to explore its value-added applications, there is a need to develop new mineral modification methods, such as electrothermal conversion and photothermal conversion. Previous research has mostly focused on the preparation and characterization of low-temperature mineral-based FSPCM, and it is recommended to further focus on the development and practical application of medium to high-temperature FSPCM.

Acknowledgements This work was jointly supported by the Guangzhou Science and Technology Project (No. 2019A1515111177, 202206010026, and 202201010108), and the National Natural Science Foundation of China (42272042, 41872039).

References

1. K. Yuan, J. Shi, W. Aftab, M. Qin, A. Usman, F. Zhou, Y. Lv, S. Gao, R. Zou, Engineering the thermal conductivity of functional phase-change materials for heat energy conversion, storage, and utilization. Adv. Funct. Mater. **30**, 1904228 (2020). https://doi.org/10.1002/adfm.201904228
2. J. Liu, Z. Lin, Thermodynamic analysis of a novel dual-temperature air-source heat pump combined ejector with zeotropic mixture R1270/R600a. Energy Convers. Manag. **15**, 113078 (2020). https://doi.org/10.1016/j.enconman.2020.113078
3. M. Li, B. Mu, Effect of different dimensional carbon materials on the properties and application of phase change materials: a review. Appl. Energy. **15**, 695–715 (2019). https://doi.org/10.1016/j.apenergy.2019.03.085
4. L.F. Cabeza, A. Castell, C.D. Barreneche, A.D. Gracia, A.I. Fernández, Materials used as PCM in thermal energy storage in buildings: a review. Renew. Sust. Energy. Rev. **15**, 1675–1695 (2011). https://doi.org/10.1016/j.rser.2010.11.018
5. D. Kumar, M. Alam, J. Sanjayan, M. Harris, Comparative analysis of form-stable phase change material integrated concrete panels for building envelopes. Case Stud. Constr. Mater. **18**, e01737 (2023). https://doi.org/10.1016/j.cscm.2022.e01737
6. X. Zhang, C. Liu, Z. Rao, Experimental investigation on thermal management performance of electric vehicle power battery using composite phase change material. J. Clean. Prod. **201**, 916–924 (2018). https://doi.org/10.1016/j.jclepro.2018.08.076
7. Z. Luo, Z. Huang, N. Xie, X. Gao, T. Xu, Y. Fang, Z. Zhang, Numerical and experimental study on temperature control of solar panels with form-stable paraffin/expanded graphite composite PCM. Energy Convers. Manag. **149**, 416–423 (2017). https://doi.org/10.1016/j.enconman.2017.07.046
8. P. Lv, C. Liu, Z. Rao, Review on clay mineral-based form-stable phase change materials: preparation, characterization and applications. Renew. Sust. Energ. Rev. **68**, 707–726 (2017). https://doi.org/10.1016/j.rser.2016.10.014

9. S. Rostami, M. Afrand, A. Shahsavar, M. Sheikholeslami, R. Kalbasi, S. Aghakhani, M.S. Shadloo, H.F. Oztop, A review of melting and freezing processes of PCM/nano-PCM and their application in energy storage. Energy **211**, 118698 (2020). https://doi.org/10.1016/j.energy.2020.118698
10. A. Sarı, A. Karaipekli, Thermal conductivity and latent heat thermal energy storage characteristics of paraffin/expanded graphite composite as phase change material. Appl. Therm. Eng. **27**, 1271–1277 (2007). https://doi.org/10.1016/j.applthermaleng.2006.11.004
11. Z. Zhang, N. Zhang, J. Peng, X. Fang, X. Gao, Y. Fang, Preparation and thermal energy storage properties of paraffin/expanded graphite composite phase change material. Appl. Energy. **91**, 426–431 (2012). https://doi.org/10.1016/j.apenergy.2011.10.014
12. T. Qian, J. Li, Y. Deng, Pore structure modified diatomite-supported PEG composites for thermal energy storage. Sci. Rep. **6**, 32392 (2016). https://doi.org/10.1038/srep32392
13. D. Li, X. Cheng, Y. Li, H. Zou, G. Yu, G. Li, Y. Huang, Effect of MOF derived hierarchical Co_3O_4/expanded graphite on thermal performance of stearic acid phase change material. Sol. Energy **171**, 142–149 (2018). https://doi.org/10.1016/j.solener.2018.06.062
14. Y. Yang, Y. Pang, Y. Liu, H. Guo, Preparation and thermal properties of polyethylene glycol/expanded graphite as novel form-stable phase change material for indoor energy saving. Mater. Lett. **216**, 220–223 (2018). https://doi.org/10.1016/j.matlet.2018.01.025
15. D. Kim, J. Jung, Y. Kim, M. Lee, J. Seo, S.B. Khan, Structure and thermal properties of octadecane/expanded graphite composites as shape-stabilized phase change materials. Int. J. Heat Mass Transf. **95**, 735–741 (2016). https://doi.org/10.1016/j.ijheatmasstransfer.2015.12.049
16. C. Li, B. Zhang, B. Xie, X. Zhao, J. Chen, Tailored phase change behavior of $Na_2SO_4 \cdot 10H_2O$/expanded graphite composite for thermal energy storage. Energy Convers. Manag. **208**, 112586 (2020). https://doi.org/10.1016/j.enconman.2020.112586
17. Z.J. Duan, H.Z. Zhang, L.X. Sun, Z. Cao, F. Xu, Y.J. Zou, $CaCl_2 \cdot 6H_2O$/Expanded graphite composite as form-stable phase change materials for thermal energy storage. J. Therm. Anal. Calorim. **115**, 111–117 (2014). https://doi.org/10.1007/s10973-013-3311-0
18. S. Wang, P. Qin, X. Fang, Z. Zhang, S. Wang, X. Liu, A novel sebacic acid/expanded graphite composite phase change material for solar thermal medium-temperature applications. Sol. Energy **99**, 283–290 (2014). https://doi.org/10.1016/j.solener.2013.11.018
19. X. Yang, Y. Yuan, N. Zhang, X. Cao, C. Liu, Preparation and properties of myristic–palmitic–stearic acid/expanded graphite composites as phase change materials for energy storage. Sol. Energy **99**, 259–266 (2014). https://doi.org/10.1016/j.solener.2013.11.021
20. S. Wu, T.X. Li, T. Yan, Y.J. Dai, R.Z. Wang, High performance form-stable expanded graphite/stearic acid composite phase change material for modular thermal energy storage. Intl. Int. J. Heat Mass Transf. **102**, 733–744 (2016). https://doi.org/10.1016/j.ijheatmasstransfer.2016.06.066
21. Z. Zhang, X. Fang, Study on paraffin/expanded graphite composite phase change thermal energy storage material. Energy Convers Manag. **47**, 303–310 (2006). https://doi.org/10.1016/j.enconman.2005.03.004
22. G. Ma, J. Sun, Y. Zhang, Y. Jing, Y. Jia, Preparation and thermal properties of stearic acid-benzamide eutectic mixture/expanded graphite composites as phase change materials for thermal energy storage. Powder Technol. **342**, 131–140 (2019). https://doi.org/10.1016/j.powtec.2018.09.074
23. Z. Song, Y. Deng, J. Li, H. Nian, Expanded graphite for thermal conductivity and reliability enhancement and supercooling decrease of $MgCl_2 \cdot 6H_2O$ phase change material. Mater. Res. Bull. **102**, 203–208 (2018). https://doi.org/10.1016/j.materresbull.2018.02.024
24. R. Ye, W. Lin, K. Yuan, X. Fang, Z. Zhang, Experimental and numerical investigations on the thermal performance of building plane containing $CaCl_2 \cdot 6H_2O$/expanded graphite composite phase change material. Appl. Energy. **193**, 325–335 (2017). https://doi.org/10.1016/j.apenergy.2017.02.049
25. S. Ramakrishnan, J. Sanjayan, X. Wang, M. Alam, J. Wilson, A novel paraffin/expanded perlite composite phase change material for prevention of PCM leakage in cementitious composites. Appl. Energy. **157**, 85–94 (2015). https://doi.org/10.1016/j.apenergy.2015.08.019

26. J.A. Costa, A.E. Martinelli, R.M. Nascimento, A.M. Mendes, Microstructural design and thermal characterization of composite diatomite-vermiculite paraffin-based form-stable PCM for cementitious mortars. Constr. Build. Mater. **232**, 117167 (2020). https://doi.org/10.1016/j.conbuildmat.2019.117167
27. H. Wei, X. Li, Preparation and characterization of a lauric-myristic-stearic acid/Al_2O_3-loaded expanded vermiculite composite phase change material with enhanced thermal conductivity. Sol. Energy Mater. Sol. Cells **166**, 1–8 (2017). https://doi.org/10.1016/j.solmat.2017.03.003
28. Y. Konuklu, O. Ersoy, F. Erzin, Development of pentadecane/diatomite and pentadecane/sepiolite nanocomposites fabricated by different compounding methods for thermal energy storage. Int. J. Energy Res. **43**, 6510–6520 (2019). https://doi.org/10.1002/er.4534
29. Q. Wang, D. Zhou, Y. Chen, P. Eames, Z. Wu, Characterization and effects of thermal cycling on the properties of paraffin/expanded graphite composites. Renew. Energy **147**, 1131–1138 (2020). https://doi.org/10.1016/j.renene.2019.09.091
30. H.O. Pierson, Handbook of carbon, graphite, diamonds and fullerenes: processing, properties and applications. (Albuquerque, New Mexico, 1994), p. 55
31. S.K. Nayak, S. Mohanty, S.K. Nayak, Fundamental and innovative approaches for filler design of thermal interface materials based on epoxy resin for high power density electronics application: a retrospective. Multiscale Multidiscip. Model. Exp. Des. **3**, 103–129 (2020). https://doi.org/10.1007/s41939-019-00064-z
32. J. Liu, M. Xie, Z. Ling, X. Fang, Z. Zhang, Novel $MgCl_2$-KCl/expanded graphite/graphite paper composite phase change blocks with high thermal conductivity and large latent heat. Sol. Energy **159**, 226–233 (2018). https://doi.org/10.1016/j.solener.2017.10.083
33. A. Sarı, A. Karaipekli, Preparation, thermal properties and thermal reliability of palmitic acid/expanded graphite composite as form-stable PCM for thermal energy storage. Sol. Energy Mater. Sol. Cells **93**, 571–576 (2009). https://doi.org/10.1016/j.solmat.2008.11.057
34. T. Zou, W. Fu, X. Liang, S. Wang, X. Gao, Z. Zhang, Y. Fang, Hydrophilic modification of expanded graphite to develop form-stable composite phase change material based on modified $CaCl_2 \cdot 6H_2O$. Energy **190**, 116473 (2020). https://doi.org/10.1016/j.energy.2019.116473
35. N. Xie, Z. Li, X. Gao, Y. Fang, Z. Zhang, Preparation and performance of modified expanded graphite/eutectic salt composite phase change cold storage material. Int. J. Refrig. **110**, 178–186 (2020). https://doi.org/10.1016/j.ijrefrig.2019.10.008
36. W. Fu, T. Zou, X. Liang, S. Wang, X. Gao, Thermal properties and thermal conductivity enhancement of composite phase change material using sodium acetate trihydrate–urea/expanded graphite for radiant floor heating system. Appl. Therm. Eng. **138**, 618–626 (2018). https://doi.org/10.1016/j.applthermaleng.2018.04.102
37. X. Gu, S. Qin, X. Wu, Y. Li, Y. Liu, Preparation and thermal characterization of sodium acetate trihydrate/expanded graphite composite phase change material. J. Therm. Anal. Calorim. **125**, 831–838 (2016). https://doi.org/10.1007/s10973-016-5444-4
38. M. Li, Z. Wu, J. Tan, Properties of form-stable paraffin/silicon dioxide/expanded graphite phase change composites prepared by sol–gel method. Appl. energy. **92**, 456–461 (2012). https://doi.org/10.1016/j.apenergy.2011.11.018
39. X. Zhang, C. Zhu, G. Fang, Preparation and thermal properties of n-eicosane/nano-SiO_2/expanded graphite composite phase-change material for thermal energy storage. Mater. Chem. Phys. **240**, 122178 (2020). https://doi.org/10.1016/j.matchemphys.2019.122178
40. X. Gu, P. Liu, L. Bian, H. He, Enhanced thermal conductivity of palmitic acid/mullite phase change composite with graphite powder for thermal energy storage. Renew. Energy **138**, 833–841 (2019). https://doi.org/10.1016/j.renene.2019.02.031
41. N. Tan, T. Xie, Y. Feng, P. Hu, Preparation and characterization of erythritol/sepiolite/exfoliated graphite nanoplatelets form-stable phase change material with high thermal conductivity and suppressed supercooling. Sol. Energy Mater. Sol. Cells **217**, 110726 (2020). https://doi.org/10.1016/j.solmat.2020.110726
42. R. Wen, X. Zhang, Z. Huang, M. Fang, S. Huang, Preparation and thermal properties of fatty acid/diatomite form-stable composite phase change material for thermal energy storage. Sol. Energy Mater Sol. Cells. **178**, 273–279 (2018). https://doi.org/10.1016/j.solmat.2018.01.032

43. S. Karaman, A. Karaipekli, A. Sarı, A. Bicer, Polyethylene glycol (PEG)/diatomite composite as a novel form-stable phase change material for thermal energy storage. Sol. Energy Mater. Sol. Cells **95**, 1647–1653 (2011). https://doi.org/10.1016/j.solmat.2011.01.022
44. S. Liu, J. Han, Q. Gao, W. Kang, R. Ren, L. Wang, D. Chen, D. Wu, Lauric acid/bentonite/flake graphite composite as form-stable phase change materials for thermal energy storage. Mater. Express **10**(2), 214–224 (2020). https://doi.org/10.1166/mex.2020.1630
45. M. Li, Z. Wu, Preparation and performance of highly conductive phase change materials prepared with paraffin, expanded graphite, and diatomite. Int. J. Green Energy **8**(1), 121–129 (2011). https://doi.org/10.1080/15435075.2011.546759
46. Y. Yuan, Y. Yuan, N. Zhang, Y. Du, X. Cao, Preparation and thermal characterization of capric–myristic–palmitic acid/expanded graphite composite as phase change material for energy storage. Mater. Lett. **125**, 154–157 (2014). https://doi.org/10.1016/j.matlet.2014.04.002
47. Y. Liu, Y. Yang, Form-stable phase change material based on $Na_2CO_3{\cdot}10H_2O$-Na2HPO$_4{\cdot}12H_2O$ eutectic hydrated salt/expanded graphite oxide composite: the influence of chemical structures of expanded graphite oxide. Renew. Energy **115**, 734–740 (2018). https://doi.org/10.1016/j.renene.2017.08.097
48. Y. Li, X. Zhang, J.M. Munyalo, Z. Tian, J. Ji, Preparation and thermophysical properties of low temperature composite phase change material octanoic-lauric acid/expanded graphite. J. Mol. Liq. **277**, 577–583 (2019). https://doi.org/10.1016/j.molliq.2018.12.111
49. W. Fu, Y. Lu, R. Zhang, J. Liu, T. Zhang, Developing NaAc•$3H_2O$-based composite phase change material using glycine as temperature regulator and expanded graphite as supporting material for use in floor radiant heating. J. Mol. Liq. **317**, 113932 (2020). https://doi.org/10.1016/j.molliq.2020.113932
50. Y. Ren, C. Xu, M. Yuan, F. Ye, X. Ju, X. Du, Ca $(NO_3)_2$-$NaNO_3$/expanded graphite composite as a novel shape-stable phase change material for mid-to high-temperature thermal energy storage. Energy Convers. Manag. **163**, 50–58 (2018). https://doi.org/10.1016/j.enconman.2018.02.057
51. K. Dong, N. Sheng, D. Zou, C. Wang, K. Shimono, T. Akiyama, T. Nomura, A high-thermal-conductivity, high-durability phase-change composite using a carbon fibre sheet as a supporting matrix. Appl. Energy. **264**, 114685 (2020). https://doi.org/10.1016/j.apenergy.2020.114685
52. S.Y. Lee, H.K. Shin, M. Park, K.Y. Rhee, S.J. Park, Thermal characterization of erythritol/expanded graphite composites for high thermal storage capacity. Carbon **68**, 67–72 (2014). https://doi.org/10.1016/j.carbon.2013.09.053
53. H. Zhang, X. Gao, C. Chen, T. Xu, Y. Fang, Z. Zhang, A capric–palmitic–stearic acid ternary eutectic mixture/expanded graphite composite phase change material for thermal energy storage. Compos. Part A Appl. Sci. Manuf. **87**, 138–145 (2016). https://doi.org/10.1016/j.compositesa.2016.04.024
54. Q. Zhang, H. Wang, Z. Ling, X. Fang, Z. Zhang, RT100/expand graphite composite phase change material with excellent structure stability, photo-thermal performance and good thermal reliability. Sol. Energy Mater. Sol. Cells **140**, 158–166 (2015). https://doi.org/10.1016/j.solmat.2015.04.008
55. C. Li, B. Zhang, Q. Liu, N-eicosane/expanded graphite as composite phase change materials for electro-driven thermal energy storage. J. Energy Storage. **29**, 101339 (2020). https://doi.org/10.1016/j.est.2020.101339
56. L. Xu, X. Liu, R. Yang, Flame retardant paraffin-based shape-stabilized phase change material via expandable graphite-based flame-retardant coating. Molecules **25**, 2408 (2020). https://doi.org/10.3390/molecules25102408
57. Y. Deng, J. Li, T. Qian, W. Guan, X. Wang, Preparation and characterization of KNO_3/diatomite shape-stabilized composite phase change material for high temperature thermal energy storage. J. Mater. Sci. Technol. **33**, 198–203 (2017). https://doi.org/10.1016/j.jmst.2016.02.011
58. T. Qian, J. Li, X. Min, W. Guan, Y. Deng, L. Ning, Enhanced thermal conductivity of PEG/diatomite shape-stabilized phase change materials with Ag nanoparticles for thermal energy storage. J. Mater. Chem. A **3**, 8526–8536 (2015). https://doi.org/10.1039/C5TA00309A

59. C. Li, M. Wang, B. Xie, H. Ma, J. Chen, Enhanced properties of diatomite-based composite phase change materials for thermal energy storage. Renew. Energy **147**, 265–274 (2020). https://doi.org/10.1016/j.renene.2019.09.001
60. J. Han, S. Liu, Myristic acid-hybridized diatomite composite as a shape-stabilized phase change material for thermal energy storage. RSC Adv. **7**(36), 22170–22177 (2017). https://doi.org/10.1039/C7RA02385E
61. F. Tang, D. Su, Y. Tang, G. Fang, Synthesis and thermal properties of fatty acid eutectics and diatomite composites as shape-stabilized phase change materials with enhanced thermal conductivity. Sol. Energy Mater. Sol. Cells **141**, 218–224 (2015). https://doi.org/10.1016/j.solmat.2015.05.045
62. A. Sarı, A. Karaipekli, Fatty acid esters-based composite phase change materials for thermal energy storage in buildings. Appl. Therm. Eng. **37**, 208–216 (2012). https://doi.org/10.1016/j.applthermaleng.2011.11.017
63. P. Liu, X. Gu, L. Bian, X. Cheng, L. Peng, H. He, Thermal properties and enhanced thermal conductivity of capric acid/diatomite/carbon nanotube composites as form-stable phase change materials for thermal energy storage. ACS Omega **4**, 2964–2972 (2019). https://doi.org/10.1021/acsomega.8b03130
64. X. Fu, Z. Liu, Y. Xiao, J. Wang, J. Lei, Preparation and properties of lauric acid/diatomite composites as novel form-stable phase change materials for thermal energy storage. Energy Build. **104**, 244–249 (2015). https://doi.org/10.1016/j.enbuild.2015.06.059
65. T. Qian, J. Li, X. Min, Y. Deng, W. Guan, L. Ning, Diatomite: a promising natural candidate as carrier material for low, middle and high temperature phase change material. Energy Convers. Manag. **98**, 34–45 (2015). https://doi.org/10.1016/j.enconman.2015.03.071
66. Z. Rao, G. Zhang, T. Xu, K. Hong, Experimental study on a novel form-stable phase change materials based on diatomite for solar energy storage. Sol. Energy Mater. Sol. Cells **182**, 52–60 (2018). https://doi.org/10.1016/j.solmat.2018.03.016
67. X. Li, J.G. Sanjayan, J.L. Wilson, Fabrication and stability of form-stable diatomite/paraffin phase change material composites. Energy Build. **76**, 284–294 (2014). https://doi.org/10.1016/j.enbuild.2014.02.082
68. B. Xu, Z. Li, Paraffin/diatomite/multi-wall carbon nanotubes composite phase change material tailor-made for thermal energy storage cement-based composites. Energy **72**, 371–380 (2014). https://doi.org/10.1016/j.energy.2014.05.049
69. M. Li, B. Mu, Fabrication and characterization of capric acid/reduced graphene oxide decorated diatomite composite phase change materials for solar energy storage. Royal Soc. Open Sci. **6**, 181664 (2019). https://doi.org/10.1098/rsos.181664
70. A. Sarı, A. Bicer, F.A. Al-Sulaiman, A. Karaipekli, V.V. Tyagi, Diatomite/CNTs/PEG composite PCMs with shape-stabilized and improved thermal conductivity: preparation and thermal energy storage properties. Energy Build. **164**, 166–175 (2018). https://doi.org/10.1016/j.enbuild.2018.01.009
71. Z. Chen, D. Su, M. Qin, G. Fang, Preparation and characteristics of composite phase change material (CPCM) with SiO_2 and diatomite as endothermal-hydroscopic material. Energy Build. **86**, 1–6 (2015). https://doi.org/10.1016/j.enbuild.2014.10.013
72. S. Ramakrishnan, X. Wang, J. Sanjayan, J. Wilson, Assessing the feasibility of integrating form-stable phase change material composites with cementitious composites and prevention of PCM leakage. Mater. Lett. **192**, 88–91 (2017). https://doi.org/10.1016/j.matlet.2016.12.052
73. M. Rottmann, T. Beikircher, Pressure dependent effective thermal conductivity of pure and SiC-opacified expanded perlite between 293 K and 1073 K. Int. J. Therm. Sci. **179**, 107652 (2022). https://doi.org/10.1016/j.enbuild.2014.02.082
74. W. Pichór, A. Janiec, Thermal stability of expanded perlite modified by mullite. Ceram. Int. **35**, 527–530 (2009). https://doi.org/10.1016/j.ceramint.2007.10.008
75. F. Cheng, X. Zhang, R. Wen, Z. Huang, M. Fang, Thermal conductivity enhancement of form-stable tetradecanol/expanded perlite composite phase change materials by adding Cu powder and carbon fiber for thermal energy storage. Appl. Therm. Eng. **156**, 653–659 (2019). https://doi.org/10.1016/j.applthermaleng.2019.03.140

76. J. Zhang, X. Guan, X. Song, H. Hou, Z. Yang, J. Zhu, Preparation and properties of gypsum based energy storage materials with capric acid–palmitic acid/expanded perlite composite PCM. Energy Build. **92**, 155–160 (2015). https://doi.org/10.1016/j.enbuild.2015.01.063
77. X. Zhang, R. Wen, Z. Huang, C. Tang, Y. Huang, Enhancement of thermal conductivity by the introduction of carbon nanotubes as a filler in paraffin/expanded perlite form-stable phase-change materials. Energy Build. **149**, 463–470 (2017). https://doi.org/10.1016/j.enbuild.2017.05.037
78. X. Kong, C. Yao, P. Jie, Y. Liu, C. Qi, X. Rong, Development and thermal performance of an expanded perlite-based phase change material wallboard for passive cooling in building. Energy Build. **152**, 547–557 (2017). https://doi.org/10.1016/j.enbuild.2017.06.067
79. A. Karaipekli, A. Biçer, A. Sarı, V.V. Tyagi, Thermal characteristics of expanded perlite/paraffin composite phase change material with enhanced thermal conductivity using carbon nanotubes. Energy Convers. Manag. **134**, 373–381 (2017). https://doi.org/10.1016/j.enconman.2016.12.053
80. N. Zhang, Y. Yuan, Y. Yuan, T. Li, X. Cao, Lauric–palmitic–stearic acid/expanded perlite composite as form-stable phase change material: preparation and thermal properties. Energy build. **82**, 505–511 (2014). https://doi.org/10.1016/j.enbuild.2014.07.049
81. Z. Lu, B. Xu, J. Zhang, Y. Zhu, G. Sun, Z. Li, Preparation and characterization of expanded perlite/paraffin composite as form-stable phase change material. Sol. Energy **108**, 460–466 (2014). https://doi.org/10.1016/j.solener.2014.08.008
82. P. Lv, M. Ding, C. Liu, Z. Rao, Experimental investigation on thermal properties and thermal performance enhancement of octadecanol/expanded perlite form stable phase change materials for efficient thermal energy storage. Renew. Energy **131**, 911–922 (2019). https://doi.org/10.1016/j.renene.2018.07.102
83. X. Zuo, X. Zhao, J. Li, Y. Hu, H. Yang, D. Chen, Enhanced thermal conductivity of form-stable composite phase-change materials with graphite hybridizing expanded perlite/paraffin. Sol. Energy **209**, 85–95 (2020). https://doi.org/10.1016/j.solener.2020.08.082
84. W.M. Guan, J.H. Li, T.T. Qian, X. Wang, Y. Deng, Preparation of paraffin/expanded vermiculite with enhanced thermal conductivity by implanting network carbon in vermiculite layers. Chem. Eng. J. **277**, 56–63 (2015). https://doi.org/10.1016/j.cej.2015.04.077
85. P. Lv, C. Liu, Z. Rao, Experiment study on the thermal properties of paraffin/kaolin thermal energy storage form-stable phase change materials. Appl. Energy. **182**, 475–487 (2016). https://doi.org/10.1016/j.apenergy.2016.08.147
86. A. Sarı, R.K. Sharma, G. Hekimoğlu, V.V. Tyagi, Preparation, characterization, thermal energy storage properties and temperature control performance of form-stabilized sepiolite based composite phase change materials. Energy Build. **188**, 111–119 (2019). https://doi.org/10.1016/j.enbuild.2019.02.008
87. S. Song, F. Qiu, W. Zhu, Y. Guo, Y. Zhang, Polyethylene glycol/halloysite@Ag nanocomposite PCM for thermal energy storage: simultaneously high latent heat and enhanced thermal conductivity. Sol. Energy Mater. Sol. Cells **193**, 237–245 (2019). https://doi.org/10.1016/j.solmat.2019.01.023
88. J. Shi, M. Li, Surface modification effects in phase change material-infiltrated attapulgite. Mater. Chem. Phys. **254**, 123521 (2020). https://doi.org/10.1016/j.matchemphys.2020.123521
89. M. Li, Q. Guo, S. Nutt, Carbon nanotube/paraffin/montmorillonite composite phase change material for thermal energy storage. Sol. Energy **146**, 1–7 (2017). https://doi.org/10.1016/j.solener.2017.02.003
90. H. Zhang, L. Zhang, Q. Li, C. Huang, H. Guo, L. Xiong, X. Chen, Preparation and characterization of methyl palmitate/palygorskite composite phase change material for thermal energy storage in buildings. Constr. Build. Mater. **226**, 212–219 (2019). https://doi.org/10.1016/j.conbuildmat.2019.07.152
91. W. Liang, Y. Wu, H. Sun, Z. Zhu, P. Chen, B. Yang, A. Li, Halloysite clay nanotubes based phase change material composites with excellent thermal stability for energy saving and storage. RSC Adv. **6**, 19669–19675 (2016). https://doi.org/10.1039/C5RA27964J

92. Y. Deng, J. Li, Y. Deng, H. Nian, H. Jiang, Supercooling suppression and thermal conductivity enhancement of $Na_2HPO_4 \cdot 12H_2O$/expanded vermiculite form-stable composite phase change materials with alumina for heat storage. ACS Sustain. Chem. Eng. **6**, 6792–6801 (2018). https://doi.org/10.1021/acssuschemeng.8b00631
93. X. Li, H. Wei, X. Lin, X. Xie, Preparation of stearic acid/modified expanded vermiculite composite phase change material with simultaneously enhanced thermal conductivity and latent heat. Sol. Energy Mater. Sol. Cells **155**, 9–13 (2016). https://doi.org/10.1016/j.solmat.2016.04.057
94. J. Zhang, X. Zhang, Y. Wan, D. Mei, B. Zhang, Preparation and thermal energy properties of paraffin/halloysite nanotube composite as form-stable phase change material. Sol. Energy **86**, 1142–1148 (2012). https://doi.org/10.1016/j.solener.2012.01.002
95. J. Wang, M. Yang, Y. Lu, Z. Jin, L. Tan, Surface functionalization engineering driven crystallization behavior of polyethylene glycol confined in mesoporous silica for shape-stabilized phase change materials. Nano Energy **19**, 78–87 (2016). https://doi.org/10.1016/j.nanoen.2015.11.001
96. Y. Chen, H. Ding, B. Wang, Q. Shi, J. Gao, Z. Cui, Y. Wan, Dopamine functionalization for improving crystallization behaviour of polyethylene glycol in shape-stable phase change material with silica fume as the matrix. J. Clean. Prod. **208**, 951–959 (2019). https://doi.org/10.1016/j.jclepro.2018.10.207
97. S. Thanakkasaranee, J. Seo, Effect of halloysite nanotubes on shape stabilities of polyethylene glycol-based composite phase change materials. Int. J. Heat Mass Transf. **132**, 154–161 (2019). https://doi.org/10.1016/j.ijheatmasstransfer.2018.11.160
98. Q. Shen, J. Ouyang, Y. Zhang, H. Yang, Lauric acid/modified sepiolite composite as a form-stable phase change material for thermal energy storage. Appl. Clay Sci. **146**, 14–22 (2017). https://doi.org/10.1016/j.clay.2017.05.035
99. A. Hassan, A.H. Mourad, Y. Rashid, N. Ismail, M.S. Laghari, Thermal and structural performance of geopolymer concrete containing phase change material encapsulated in expanded clay. Energy Build. **191**, 72–81 (2019). https://doi.org/10.1016/j.enbuild.2019.03.005
100. R. Huang, J. Feng, Z. Ling, X. Fang, Z. Zhang, A sodium acetate trihydrate-formamide/expanded perlite composite with high latent heat and suitable phase change temperatures for use in building roof. Constr. Build. Mater. **226**, 859–867 (2019). https://doi.org/10.1016/j.conbuildmat.2019.07.331
101. C. Yao, X. Kong, Y. Li, Y. Du, C. Qi, Numerical and experimental research of cold storage for a novel expanded perlite-based shape-stabilized phase change material wallboard used in building. Energy Convers. Manag. **155**, 20–31 (2018). https://doi.org/10.1016/j.enconman.2017.10.052
102. X. Kong, L. Wang, H. Li, G. Yuan, C. Yao, Experimental study on a novel hybrid system of active composite PCM wall and solar thermal system for clean heating supply in winter. Sol. Energy **195**, 259–270 (2020). https://doi.org/10.1016/j.solener.2019.11.081
103. E. Meng, J. Wang, H. Yu, R. Cai, Y. Chen, B. Zhou, Experimental study of the thermal protection performance of the high reflectivity-phase change material (PCM) roof in summer. Build. Environ. **164**, 106381 (2019). https://doi.org/10.1016/j.buildenv.2019.106381
104. C. Liu, J. Bian, G. Zhang, D. Li, X. Liu, Influence of optical parameters on thermal and optical performance of multi-layer glazed roof filled with PCM. Appl. Therm. Eng. **134**, 615–625 (2018). https://doi.org/10.1016/j.applthermaleng.2018.01.117
105. W. Sun, Y. Zhang, Z. Ling, X. Fang, Z. Zhang, Experimental investigation on the thermal performance of double-layer PCM radiant floor system containing two types of inorganic composite PCMs. Energy Build. **211**, 109806 (2020). https://doi.org/10.1016/j.enbuild.2020.109806
106. Y. Fang, Y. Ding, Y. Tang, X. Liang, C. Jin, S. Wang, X. Gao, Z. Zhang, Thermal properties enhancement and application of a novel sodium acetate trihydrate-formamide/expanded graphite shape-stabilized composite phase change material for electric radiant floor heating. Appl. Therm. Eng. **150**, 1177–1185 (2019). https://doi.org/10.1016/j.applthermaleng.2019.01.069

107. M.J. Abden, Z. Tao, Z. Pan, L. George, R. Wuhrer, Inclusion of methyl stearate/diatomite composite in gypsum board ceiling for building energy conservation. Appl. Energy **259**, 114113 (2020). https://doi.org/10.1016/j.apenergy.2019.114113
108. T. Ma, H. Yang, Y. Zhang, L. Lu, X. Wang, Using phase change materials in photovoltaic systems for thermal regulation and electrical efficiency improvement: a review and outlook. Renew. Sust. Energ. Rev. **43**, 1273–1284 (2015). https://doi.org/10.1016/j.rser.2014.12.003
109. Y. Du, J. Chen, Z. Han, W. Liu, A review on solutions for improving rutting resistance of asphalt pavement and test methods. Constr. Build. Mater. **168**, 893–905 (2018). https://doi.org/10.1016/j.conbuildmat.2018.02.151
110. D. Zhang, M. Chen, S. Wu, Q. Liu, J. Wan, Preparation of expanded graphite/polyethylene glycol composite phase change material for thermoregulation of asphalt binder. Constr. Build. Mater. **169**, 513–521 (2018). https://doi.org/10.1016/j.conbuildmat.2018.02.167
111. W.L. Cheng, Y.K. Huo, Y.L. Nian, Performance of solar still using shape-stabilized PCM: experimental and theoretical investigation. Desalination **455**, 89–99 (2019). https://doi.org/10.1016/j.desal.2019.01.007
112. S.W. Sharshir, G. Peng, L. Wu, F.A. Essa, A.E. Kabeel, N. Yang, The effects of flake graphite nanoparticles, phase change material, and film cooling on the solar still performance. Appl Energy **191**, 358–366 (2017). https://doi.org/10.1016/j.apenergy.2017.01.067
113. W.L. Cheng, B.J. Mei, Y.N. Liu, Y.H. Huang, X.D. Yuan, A novel household refrigerator with shape-stabilized PCM (Phase Change Material) heat storage condensers: an experimental investigation. Energy **36**, 5797–5804 (2011). https://doi.org/10.1016/j.energy.2011.08.050
114. W.L. Cheng, X.D. Yuan, Numerical analysis of a novel household refrigerator with shape-stabilized PCM (phase change material) heat storage condensers. Energy **59**, 265–276 (2013). https://doi.org/10.1016/j.energy.2013.06.045
115. K. Merlin, J. Soto, D. Delaunay, L. Traonvouez, Industrial waste heat recovery using an enhanced conductivity latent heat thermal energy storage. Appl. Energy **183**, 491–503 (2016). https://doi.org/10.1016/j.apenergy.2016.09.007
116. H. Mehling, L.F. Cabeza, S. Hippeli, S. Hiebler, PCM-module to improve hot water heat stores with stratification. Renew. Energy **28**, 699–711 (2003). https://doi.org/10.1016/S0960-1481(02)00108-8
117. X. Gu, K. Dong, L. Peng, L. Bian, Q. Sun, W. Luo, B. Zhang, Round-the-clock interfacial solar vapor generator enabled by form-stable phase change materials with enhanced photothermal conversion capacity. Energy Conver. Manag. **277**, 116634 (2023). https://doi.org/10.1016/j.enconman.2022.116634

Index

S. J. Ikhmayies (ed.), *Advances in Minerals Research*, Advances in Material Research and Technology, https://doi.org/10.1007/978-3-031-49175-7

B

Q

T

U

The manufacturer's authorised representative in the EU is Springer Nature Customer Service Centre GmbH, Europaplatz 3, 69115 Heidelberg, Germany. If you have any concerns regarding our products, please contact ProductSafety@springernature.com

Printed and bound by CPI Group (UK) Ltd, Croydon, CR0 4YY
01/10/2026
02225855-0009